Student Solution Manual for
Essential Mathematical Methods for the Physical Sciences

This *Student Solution Manual* provides complete solutions to all the odd-numbered problems in *Essential Mathematical Methods for the Physical Sciences*. It takes students through each problem step by step, so they can clearly see how the solution is reached, and understand any mistakes in their own working. Students will learn by example how to select an appropriate method, improving their problem-solving skills.

K. F. RILEY read mathematics at the University of Cambridge and proceeded to a Ph.D. there in theoretical and experimental nuclear physics. He became a Research Associate in elementary particle physics at Brookhaven, and then, having taken up a lectureship at the Cavendish Laboratory, Cambridge, continued this research at the Rutherford Laboratory and Stanford; in particular he was involved in the experimental discovery of a number of the early baryonic resonances. As well as having been Senior Tutor at Clare College, where he has taught physics and mathematics for over 40 years, he has served on many committees concerned with the teaching and examining of these subjects at all levels of tertiary and undergraduate education. He is also one of the authors of *200 Puzzling Physics Problems* (Cambridge University Press, 2001).

M. P. HOBSON read natural sciences at the University of Cambridge, specializing in theoretical physics, and remained at the Cavendish Laboratory to complete a Ph.D. in the physics of star formation. As a Research Fellow at Trinity Hall, Cambridge, and subsequently an Advanced Fellow of the Particle Physics and Astronomy Research Council, he developed an interest in cosmology, and in particular in the study of fluctuations in the cosmic microwave background. He was involved in the first detection of these fluctuations using a ground-based interferometer. Currently a University Reader at the Cavendish Laboratory, his research interests include both theoretical and observational aspects of cosmology, and he is the principal author of *General Relativity: An Introduction for Physicists* (Cambridge University Press, 2006). He is also a Director of Studies in Natural Sciences at Trinity Hall and enjoys an active role in the teaching of undergraduate physics and mathematics.

Essential Mathematical Methods for the Physical Sciences

Student Solution Manual

K. F. RILEY
University of Cambridge

M. P. HOBSON
University of Cambridge

CAMBRIDGE
UNIVERSITY PRESS

CAMBRIDGE UNIVERSITY PRESS
Cambridge, New York, Melbourne, Madrid, Cape Town, Singapore,
São Paulo, Delhi, Dubai, Tokyo, Mexico City

Cambridge University Press
The Edinburgh Building, Cambridge CB2 8RU, UK

Published in the United States of America by Cambridge University Press, New York

www.cambridge.org
Information on this title: www.cambridge.org/9780521141024

First published 2011

Printed in the United Kingdom at the University Press, Cambridge

A catalogue record for this publication is available from the British Library

ISBN 978-0-521-14102-4 Paperback

Contents

Preface

For reasons that are explained in the preface to *Essential Mathematical Methods for the Physical Sciences* the text of the third edition of *Mathematical Methods for Physics and Engineering (MMPE)* (Cambridge: Cambridge University Press, 2006) by Riley, Hobson and Bence, after a number of additions and omissions, has been republished as two slightly overlapping texts. *Essential Mathematical Methods for the Physical Sciences (EMMPS)* contains most of the more advanced material, and specifically develops mathematical *methods* that can be applied throughout the physical sciences; an augmented version of the more introductory material, principally concerned with mathematical *tools* rather than methods, is available as *Foundation Mathematics for the Physical Sciences*. The full text of *MMPE*, including all of the more specialized and advanced topics, is still available under its original title.

As in the third edition of *MMPE*, the penultimate subsection of each chapter of *EMMPS* consists of a significant number of problems, nearly all of which are based on topics drawn from several sections of that chapter. Also as in the third edition, hints and outline answers are given in the final subsection, but only to the odd-numbered problems, leaving all even-numbered problems free to be set as unaided homework.

This book is the solutions manual for the problems in *EMMPS*. For the 230 plus *odd-numbered* problems it contains, complete solutions are available, to both students and their teachers, in the form of this manual; these are in addition to the hints and outline answers given in the main text. For each problem, the original question is reproduced and then followed by a fully worked solution. For those original problems that make internal reference to the main text or to other (even-numbered) problems not included in this solutions manual, the questions have been reworded, usually by including additional information, so that the questions can stand alone. Some further minor rewording has been included to improve the page layout.

In many cases the solution given is even fuller than one that might be expected of a good student who has understood the material. This is because we have aimed to make the solutions instructional as well as utilitarian. To this end, we have included comments that are intended to show how the plan for the solution is formulated and have provided the justifications for particular intermediate steps (something not always done, even by the best of students). We have also tried to write each individual substituted formula in the form that best indicates how it was obtained, before simplifying it at the next or a subsequent stage. Where several lines of algebraic manipulation or calculus are needed to obtain a final result, they are normally included in full; this should enable the student to determine whether an incorrect answer is due to a misunderstanding of principles or to a technical error.

As noted above, the original questions are reproduced in full, or in a suitably modified stand-alone form, at the start of each problem. Reference to the main text is not needed provided that standard formulae are known (and a set of tables is available for a few of the statistical and numerical problems). This means that, although it is not its prime purpose, this manual could be used as a test or quiz book by a student who has learned, or thinks that he or she has learned, the material covered in the main text.

1

Matrices and vector spaces

1.1 Which of the following statements about linear vector spaces are true? Where a statement is false, give a counter-example to demonstrate this.

(a) Non-singular $N \times N$ matrices form a vector space of dimension N^2.
(b) Singular $N \times N$ matrices form a vector space of dimension N^2.
(c) Complex numbers form a vector space of dimension 2.
(d) Polynomial functions of x form an infinite-dimensional vector space.
(e) Series $\{a_0, a_1, a_2, \ldots, a_N\}$ for which $\sum_{n=0}^{N} |a_n|^2 = 1$ form an N-dimensional vector space.
(f) Absolutely convergent series form an infinite-dimensional vector space.
(g) Convergent series with terms of alternating sign form an infinite-dimensional vector space.

We first remind ourselves that for a set of entities to form a vector space, they must pass five tests: (i) closure under commutative and associative addition; (ii) closure under multiplication by a scalar; (iii) the existence of a null vector in the set; (iv) multiplication by unity leaves any vector unchanged; (v) each vector has a corresponding negative vector.

(a) False. The matrix $\mathbf{0}_N$, the $N \times N$ null matrix, required by (iii) is *not* non-singular and is therefore not in the set.

(b) Consider the sum of $\begin{pmatrix} 1 & 0 \\ 0 & 0 \end{pmatrix}$ and $\begin{pmatrix} 0 & 0 \\ 0 & 1 \end{pmatrix}$. The sum is the unit matrix which is not singular and so the set is not closed; this violates requirement (i). The statement is false.

(c) The space is closed under addition and multiplication by a scalar; multiplication by unity leaves a complex number unchanged; there is a null vector ($= 0 + i0$) and a negative complex number for each vector. All the necessary conditions are satisfied and the statement is true.

(d) As in the previous case, all the conditions are satisfied and the statement is true.

(e) This statement is false. To see why, consider $b_n = a_n + a_n$ for which $\sum_{n=0}^{N} |b_n|^2 = 4 \neq 1$, i.e. the set is not closed (violating (i)), or note that there is no zero vector with unit norm (violating (iii)).

(f) True. Note that an absolutely convergent series remains absolutely convergent when the signs of all of its terms are reversed.

(g) False. Consider the two series defined by

$$a_0 = \tfrac{1}{2}, \quad a_n = 2\left(-\tfrac{1}{2}\right)^n \text{ for } n \geq 1; \quad b_n = -\left(-\tfrac{1}{2}\right)^n \text{ for } n \geq 0.$$

The series that is the sum of $\{a_n\}$ and $\{b_n\}$ does not have alternating signs and so closure (required by (i)) does not hold.

1.3 By considering the matrices

$$A = \begin{pmatrix} 1 & 0 \\ 0 & 0 \end{pmatrix}, \qquad B = \begin{pmatrix} 0 & 0 \\ 3 & 4 \end{pmatrix},$$

show that $AB = 0$ does *not* imply that either A or B is the zero matrix but that it does imply that at least one of them is singular.

We have

$$AB = \begin{pmatrix} 1 & 0 \\ 0 & 0 \end{pmatrix} \begin{pmatrix} 0 & 0 \\ 3 & 4 \end{pmatrix} = \begin{pmatrix} 0 & 0 \\ 0 & 0 \end{pmatrix}.$$

Thus AB is the zero matrix 0 without either $A = 0$ or $B = 0$.

However, $AB = 0 \Rightarrow |A||B| = |0| = 0$ and therefore either $|A| = 0$ or $|B| = 0$ (or both).

1.5 Using the properties of determinants, solve with a minimum of calculation the following equations for x:

(a) $$\begin{vmatrix} x & a & a & 1 \\ a & x & b & 1 \\ a & b & x & 1 \\ a & b & c & 1 \end{vmatrix} = 0,$$ (b) $$\begin{vmatrix} x+2 & x+4 & x-3 \\ x+3 & x & x+5 \\ x-2 & x-1 & x+1 \end{vmatrix} = 0.$$

(a) In view of the similarities between some rows and some columns, the property most likely to be useful here is that if a determinant has two rows/columns equal (or multiples of each other) then its value is zero.

(i) We note that setting $x = a$ makes the first and fourth columns multiples of each other and hence makes the value of the determinant 0; thus $x = a$ is one solution to the equation.

(ii) Setting $x = b$ makes the second and third rows equal, and again the determinant vanishes; thus b is another root of the equation.

(iii) Setting $x = c$ makes the third and fourth rows equal, and yet again the determinant vanishes; thus c is also a root of the equation.

Since the determinant contains no x in its final column, it is a cubic polynomial in x and there will be exactly three roots to the equation. We have already found all three!

(b) Here, the presence of x multiplied by unity in every entry means that subtracting rows/columns will lead to a simplification. After (i) subtracting the first column from each of the others, and then (ii) subtracting the first row from each of the others, the determinant becomes

$$\begin{vmatrix} x+2 & 2 & -5 \\ x+3 & -3 & 2 \\ x-2 & 1 & 3 \end{vmatrix} = \begin{vmatrix} x+2 & 2 & -5 \\ 1 & -5 & 7 \\ -4 & -1 & 8 \end{vmatrix}$$

$$= (x+2)(-40+7) + 2(-28-8) - 5(-1-20)$$

$$= -33(x+2) - 72 + 105$$

$$= -33x - 33.$$

Thus $x = -1$ is the only solution to the original (linear!) equation.

1.7 Prove the following results involving Hermitian matrices.

(a) If A is Hermitian and U is unitary then $U^{-1}AU$ is Hermitian.
(b) If A is anti-Hermitian then iA is Hermitian.
(c) The product of two Hermitian matrices A and B is Hermitian if and only if A and B commute.
(d) If S is a real antisymmetric matrix then $A = (I - S)(I + S)^{-1}$ is orthogonal. If A is given by

$$A = \begin{pmatrix} \cos\theta & \sin\theta \\ -\sin\theta & \cos\theta \end{pmatrix}$$

then find the matrix S that is needed to express A in the above form.
(e) If K is skew-Hermitian, i.e. $K^\dagger = -K$, then $V = (I + K)(I - K)^{-1}$ is unitary.

The general properties of matrices that we will need are $(A^\dagger)^{-1} = (A^{-1})^\dagger$ and

$$(AB \cdots C)^T = C^T \cdots B^T A^T, \qquad (AB \cdots C)^\dagger = C^\dagger \cdots B^\dagger A^\dagger.$$

(a) Given that $A = A^\dagger$ and $U^\dagger U = I$, consider

$$(U^{-1}AU)^\dagger = U^\dagger A^\dagger (U^{-1})^\dagger = U^{-1}A(U^\dagger)^{-1} = U^{-1}A(U^{-1})^{-1} = U^{-1}AU,$$

i.e. $U^{-1}AU$ is Hermitian.

(b) Given $A^\dagger = -A$, consider

$$(iA)^\dagger = -iA^\dagger = -i(-A) = iA,$$

i.e. iA is Hermitian.

(c) Given $A = A^\dagger$ and $B = B^\dagger$.

(i) Suppose $AB = BA$, then

$$(AB)^\dagger = B^\dagger A^\dagger = BA = AB,$$

i.e. AB is Hermitian.

(ii) Now suppose that $(AB)^\dagger = AB$. Then

$$BA = B^\dagger A^\dagger = (AB)^\dagger = AB,$$

i.e. A and B commute.

Thus, AB is Hermitian $\Longleftrightarrow$ A and B commute.

(d) Given that S is real and $S^T = -S$ with $A = (I - S)(I + S)^{-1}$, consider

$$\begin{aligned}
A^T A &= [(I - S)(I + S)^{-1}]^T [(I - S)(I + S)^{-1}] \\
&= [(I + S)^{-1}]^T (I + S)(I - S)(I + S)^{-1} \\
&= (I - S)^{-1}(I + S - S - S^2)(I + S)^{-1} \\
&= (I - S)^{-1}(I - S)(I + S)(I + S)^{-1} \\
&= I I = I,
\end{aligned}$$

i.e. A is orthogonal.

If $A = (I - S)(I + S)^{-1}$, then $A + AS = I - S$ and $(A + I)S = I - A$, giving

$$S = (A + I)^{-1}(I - A)$$

$$= \begin{pmatrix} 1 + \cos\theta & \sin\theta \\ -\sin\theta & 1 + \cos\theta \end{pmatrix}^{-1} \begin{pmatrix} 1 - \cos\theta & -\sin\theta \\ \sin\theta & 1 - \cos\theta \end{pmatrix}$$

$$= \frac{1}{2 + 2\cos\theta} \begin{pmatrix} 1 + \cos\theta & -\sin\theta \\ \sin\theta & 1 + \cos\theta \end{pmatrix} \begin{pmatrix} 1 - \cos\theta & -\sin\theta \\ \sin\theta & 1 - \cos\theta \end{pmatrix}$$

$$= \frac{1}{4\cos^2(\theta/2)} \begin{pmatrix} 0 & -2\sin\theta \\ 2\sin\theta & 0 \end{pmatrix}$$

$$= \begin{pmatrix} 0 & -\tan(\theta/2) \\ \tan(\theta/2) & 0 \end{pmatrix}.$$

(e) This proof is almost identical to the first section of part (d) but with S replaced by $-K$ and transposed matrices replaced by Hermitian conjugate matrices.

1.9 The *commutator* $[X, Y]$ of two matrices is defined by the equation

$$[X, Y] = XY - YX.$$

Two anticommuting matrices A and B satisfy

$$A^2 = I, \qquad B^2 = I, \qquad [A, B] = 2iC.$$

(a) Prove that $C^2 = I$ and that $[B, C] = 2iA$.
(b) Evaluate $[[[A, B], [B, C]], [A, B]]$.

(a) From $AB - BA = 2iC$ and $AB = -BA$ it follows that $AB = iC$. Thus,

$$-C^2 = iCiC = ABAB = A(-AB)B = -(AA)(BB) = -II = -I,$$

i.e. $C^2 = I$. In deriving the above result we have used the associativity of matrix multiplication.

For the commutator of B and C,

$$[B, C] = BC - CB$$
$$= B(-iAB) - (-i)ABB$$
$$= -i(BA)B + iAI$$
$$= -i(-AB)B + iA$$
$$= iA + iA = 2iA.$$

(b) To evaluate this multiple-commutator expression we must work outwards from the innermost "explicit" commutators. There are three such commutators at the first stage. We also need the result that $[C, A] = 2iB$; this can be proved in the same way as that for $[B, C]$ in part (a), or by making the cyclic replacements $A \rightarrow B \rightarrow C \rightarrow A$ in the assumptions and their consequences, as proved in part (a). Then we have

$$[[[\mathsf{A}, \mathsf{B}], [\mathsf{B}, \mathsf{C}]], [\mathsf{A}, \mathsf{B}]] = [[2i\mathsf{C}, 2i\mathsf{A}], 2i\mathsf{C}]$$
$$= -4[[\mathsf{C}, \mathsf{A}], 2i\mathsf{C}]$$
$$= -4[2i\mathsf{B}, 2i\mathsf{C}]$$
$$= (-4)(-4)[\mathsf{B}, \mathsf{C}] = 32i\mathsf{A}.$$

1.11 A general triangle has angles α, β and γ and corresponding opposite sides a, b and c. Express the length of each side in terms of the lengths of the other two sides and the relevant cosines, writing the relationships in matrix and vector form, using the vectors having components a, b, c and $\cos\alpha$, $\cos\beta$, $\cos\gamma$. Invert the matrix and hence deduce the cosine-law expressions involving α, β and γ.

By considering each side of the triangle as the sum of the projections onto it of the other two sides, we have the three simultaneous equations:

$$a = b\cos\gamma + c\cos\beta,$$
$$b = c\cos\alpha + a\cos\gamma,$$
$$c = b\cos\alpha + a\cos\beta.$$

Written in matrix and vector form, $\mathsf{A}\mathbf{x} = \mathbf{y}$, they become

$$\begin{pmatrix} 0 & c & b \\ c & 0 & a \\ b & a & 0 \end{pmatrix} \begin{pmatrix} \cos\alpha \\ \cos\beta \\ \cos\gamma \end{pmatrix} = \begin{pmatrix} a \\ b \\ c \end{pmatrix}.$$

The matrix A is non-singular, since $|\mathsf{A}| = 2abc \neq 0$, and therefore has an inverse given by

$$\mathsf{A}^{-1} = \frac{1}{2abc} \begin{pmatrix} -a^2 & ab & ac \\ ab & -b^2 & bc \\ ac & bc & -c^2 \end{pmatrix}.$$

And so, writing $\mathbf{x} = \mathsf{A}^{-1}\mathbf{y}$, we have

$$\begin{pmatrix} \cos\alpha \\ \cos\beta \\ \cos\gamma \end{pmatrix} = \frac{1}{2abc} \begin{pmatrix} -a^2 & ab & ac \\ ab & -b^2 & bc \\ ac & bc & -c^2 \end{pmatrix} \begin{pmatrix} a \\ b \\ c \end{pmatrix}.$$

From this we can read off the cosine-law equation

$$\cos\alpha = \frac{1}{2abc}(-a^3 + ab^2 + ac^2) = \frac{b^2 + c^2 - a^2}{2bc},$$

and the corresponding expressions for $\cos\beta$ and $\cos\gamma$.

1.13 Determine which of the matrices below are mutually commuting, and, for those that are, demonstrate that they have a complete set of eigenvectors in common:

$$A = \begin{pmatrix} 6 & -2 \\ -2 & 9 \end{pmatrix}, \quad B = \begin{pmatrix} 1 & 8 \\ 8 & -11 \end{pmatrix},$$

$$C = \begin{pmatrix} -9 & -10 \\ -10 & 5 \end{pmatrix}, \quad D = \begin{pmatrix} 14 & 2 \\ 2 & 11 \end{pmatrix}.$$

To establish the result we need to examine all pairs of products.

$$AB = \begin{pmatrix} 6 & -2 \\ -2 & 9 \end{pmatrix} \begin{pmatrix} 1 & 8 \\ 8 & -11 \end{pmatrix}$$

$$= \begin{pmatrix} -10 & 70 \\ 70 & -115 \end{pmatrix}$$

$$= \begin{pmatrix} 1 & 8 \\ 8 & -11 \end{pmatrix} \begin{pmatrix} 6 & -2 \\ -2 & 9 \end{pmatrix} = BA.$$

$$AC = \begin{pmatrix} 6 & -2 \\ -2 & 9 \end{pmatrix} \begin{pmatrix} -9 & -10 \\ -10 & 5 \end{pmatrix}$$

$$= \begin{pmatrix} -34 & -70 \\ -72 & 65 \end{pmatrix} \neq \begin{pmatrix} -34 & -72 \\ -70 & 65 \end{pmatrix}$$

$$= \begin{pmatrix} -9 & -10 \\ -10 & 5 \end{pmatrix} \begin{pmatrix} 6 & -2 \\ -2 & 9 \end{pmatrix} = CA.$$

Continuing in this way, we find:

$$AD = \begin{pmatrix} 80 & -10 \\ -10 & 95 \end{pmatrix} = DA.$$

$$BC = \begin{pmatrix} -89 & 30 \\ 38 & -135 \end{pmatrix} \neq \begin{pmatrix} -89 & 38 \\ 30 & -135 \end{pmatrix} = CB.$$

$$BD = \begin{pmatrix} 30 & 90 \\ 90 & -105 \end{pmatrix} = DB.$$

$$CD = \begin{pmatrix} -146 & -128 \\ -130 & 35 \end{pmatrix} \neq \begin{pmatrix} -146 & -130 \\ -128 & 35 \end{pmatrix} = DC.$$

These results show that whilst A, B and D are mutually commuting, none of them commutes with C.

We could use any of the three mutually commuting matrices to find the common set (actually a pair, as they are 2×2 matrices) of eigenvectors. We arbitrarily choose A. The eigenvalues of A satisfy

$$\begin{vmatrix} 6 - \lambda & -2 \\ -2 & 9 - \lambda \end{vmatrix} = 0,$$

$$\lambda^2 - 15\lambda + 50 = 0,$$

$$(\lambda - 5)(\lambda - 10) = 0.$$

For $\lambda = 5$, an eigenvector $(x \quad y)^{\mathrm{T}}$ must satisfy $x - 2y = 0$, whilst, for $\lambda = 10$, $4x + 2y = 0$. Thus a pair of independent eigenvectors of A are $(2 \quad 1)^{\mathrm{T}}$ and $(1 \quad -2)^{\mathrm{T}}$. Direct substitution verifies that they are also eigenvectors of B and D with pairs of eigenvalues $5, -15$ and $15, 10$, respectively.

1.15 Solve the simultaneous equations

$$2x + 3y + z = 11,$$
$$x + y + z = 6,$$
$$5x - y + 10z = 34.$$

To eliminate z, (i) subtract the second equation from the first and (ii) subtract 10 times the second equation from the third.

$$x + 2y = 5,$$
$$-5x - 11y = -26.$$

To eliminate x add 5 times the first equation to the second

$$-y = -1.$$

Thus $y = 1$ and, by resubstitution, $x = 3$ and $z = 2$.

1.17 Show that the following equations have solutions only if $\eta = 1$ or 2, and find them in these cases:

$$x + y + z = 1, \qquad \text{(i)}$$
$$x + 2y + 4z = \eta, \qquad \text{(ii)}$$
$$x + 4y + 10z = \eta^2. \qquad \text{(iii)}$$

Expressing the equations in the form $\mathsf{A}\mathbf{x} = \mathbf{b}$, we first need to evaluate $|\mathsf{A}|$ as a preliminary to determining A^{-1}. However, we find that $|\mathsf{A}| = 1(20 - 16) + 1(4 - 10) + 1(4 - 2) = 0$. This result implies both that A is singular and has no inverse, and that the equations must be linearly dependent.

Either by observation or by solving for the combination coefficients, we see that for the LHS this linear dependence is expressed by

$$2 \times \text{(i)} + 1 \times \text{(iii)} - 3 \times \text{(ii)} = 0.$$

For a consistent solution, this must also be true for the RHSs, i.e.

$$2 + \eta^2 - 3\eta = 0.$$

This quadratic equation has solutions $\eta = 1$ and $\eta = 2$, which are therefore the only values of η for which the original equations have a solution. As the equations are linearly dependent, we may use any two to find these allowed solutions; for simplicity we use the first two in each case.

For $\eta = 1$,
$$x + y + z = 1, \quad x + 2y + 4z = 1 \Rightarrow \mathbf{x}^1 = (1 + 2\alpha \quad -3\alpha \quad \alpha)^{\mathrm{T}}.$$

For $\eta = 2$,
$$x + y + z = 1, \quad x + 2y + 4z = 2 \Rightarrow \mathbf{x}^2 = (2\alpha \quad 1 - 3\alpha \quad \alpha)^{\mathrm{T}}.$$

In both cases there is an infinity of solutions as α may take any finite value.

1.19 Make an LU decomposition of the matrix
$$\mathsf{A} = \begin{pmatrix} 3 & 6 & 9 \\ 1 & 0 & 5 \\ 2 & -2 & 16 \end{pmatrix}$$

and hence solve $\mathsf{A}\mathbf{x} = \mathbf{b}$, where (i) $\mathbf{b} = (21 \quad 9 \quad 28)^{\mathrm{T}}$, (ii) $\mathbf{b} = (21 \quad 7 \quad 22)^{\mathrm{T}}$.

Using the notation
$$\mathsf{A} = \begin{pmatrix} 1 & 0 & 0 \\ L_{21} & 1 & 0 \\ L_{31} & L_{32} & 1 \end{pmatrix} \begin{pmatrix} U_{11} & U_{12} & U_{13} \\ 0 & U_{22} & U_{23} \\ 0 & 0 & U_{33} \end{pmatrix},$$

and considering rows and columns alternately in the usual way for an LU decomposition, we require the following to be satisfied.

1st row: $U_{11} = 3$, $U_{12} = 6$, $U_{13} = 9$.
1st col: $L_{21}U_{11} = 1$, $L_{31}U_{11} = 2 \Rightarrow L_{21} = \frac{1}{3}$, $L_{31} = \frac{2}{3}$.
2nd row: $L_{21}U_{12} + U_{22} = 0$, $L_{21}U_{13} + U_{23} = 5 \Rightarrow U_{22} = -2$, $U_{23} = 2$.
2nd col: $L_{31}U_{12} + L_{32}U_{22} = -2 \Rightarrow L_{32} = 3$.
3rd row: $L_{31}U_{13} + L_{32}U_{23} + U_{33} = 16 \Rightarrow U_{33} = 4$.

Thus
$$\mathsf{L} = \begin{pmatrix} 1 & 0 & 0 \\ \frac{1}{3} & 1 & 0 \\ \frac{2}{3} & 3 & 1 \end{pmatrix} \quad \text{and} \quad \mathsf{U} = \begin{pmatrix} 3 & 6 & 9 \\ 0 & -2 & 2 \\ 0 & 0 & 4 \end{pmatrix}.$$

To solve $\mathsf{A}\mathbf{x} = \mathbf{b}$ with $\mathsf{A} = \mathsf{LU}$, we first determine $\mathbf{y}$ from $\mathsf{L}\mathbf{y} = \mathbf{b}$ and then solve $\mathsf{U}\mathbf{x} = \mathbf{y}$ for $\mathbf{x}$.

(i) For $\mathsf{A}\mathbf{x} = (21 \quad 9 \quad 28)^{\mathrm{T}}$, we first solve
$$\begin{pmatrix} 1 & 0 & 0 \\ \frac{1}{3} & 1 & 0 \\ \frac{2}{3} & 3 & 1 \end{pmatrix} \begin{pmatrix} y_1 \\ y_2 \\ y_3 \end{pmatrix} = \begin{pmatrix} 21 \\ 9 \\ 28 \end{pmatrix}.$$

This can be done, almost by inspection, to give $\mathbf{y} = (21 \quad 2 \quad 8)^{\mathrm{T}}$.
We can now write $\mathsf{U}\mathbf{x} = \mathbf{y}$ explicitly as
$$\begin{pmatrix} 3 & 6 & 9 \\ 0 & -2 & 2 \\ 0 & 0 & 4 \end{pmatrix} \begin{pmatrix} x_1 \\ x_2 \\ x_3 \end{pmatrix} = \begin{pmatrix} 21 \\ 2 \\ 8 \end{pmatrix}$$

to give, equally easily, that the solution to the original matrix equation is $\mathbf{x} = (-1 \quad 1 \quad 2)^{\mathrm{T}}$.

(ii) To solve $Ax = (21\ \ 7\ \ 22)^T$ we use exactly the same forms for L and U, but the new values for the components of b, to obtain $y = (21\ \ 0\ \ 8)^T$ leading to the solution $x = (-3\ \ 2\ \ 2)^T$.

1.21 Use the Cholesky decomposition method to determine whether the following matrices are positive definite. For each that is, determine the corresponding lower diagonal matrix L :

$$A = \begin{pmatrix} 2 & 1 & 3 \\ 1 & 3 & -1 \\ 3 & -1 & 1 \end{pmatrix}, \qquad B = \begin{pmatrix} 5 & 0 & \sqrt{3} \\ 0 & 3 & 0 \\ \sqrt{3} & 0 & 3 \end{pmatrix}.$$

The matrix A is real and so we seek a real lower-diagonal matrix L such that $LL^T = A$. In order to avoid a lot of subscripts, we use lower-case letters as the non-zero elements of L:

$$\begin{pmatrix} a & 0 & 0 \\ b & c & 0 \\ d & e & f \end{pmatrix} \begin{pmatrix} a & b & d \\ 0 & c & e \\ 0 & 0 & f \end{pmatrix} = \begin{pmatrix} 2 & 1 & 3 \\ 1 & 3 & -1 \\ 3 & -1 & 1 \end{pmatrix}.$$

Firstly, from A_{11}, $a^2 = 2$. Since an overall negative sign multiplying the elements of L is irrelevant, we may choose $a = +\sqrt{2}$. Next, $ba = A_{12} = 1$, implying that $b = 1/\sqrt{2}$. Similarly, $d = 3/\sqrt{2}$.

From the second row of A we have

$$b^2 + c^2 = 3 \quad \Rightarrow \quad c = \sqrt{\tfrac{5}{2}},$$

$$bd + ce = -1 \quad \Rightarrow \quad e = \sqrt{\tfrac{2}{5}}\left(-1-\tfrac{3}{2}\right) = -\sqrt{\tfrac{5}{2}}.$$

And, from the final row,

$$d^2 + e^2 + f^2 = 1 \quad \Rightarrow \quad f = \left(1-\tfrac{9}{2}-\tfrac{5}{2}\right)^{1/2} = \sqrt{-6}.$$

That f is imaginary shows that A is not a positive definite matrix.

The corresponding argument (keeping the same symbols but with different numerical values) for the matrix B is as follows.

Firstly, from A_{11}, $a^2 = 5$. Since an overall negative sign multiplying the elements of L is irrelevant, we may choose $a = +\sqrt{5}$. Next, $ba = B_{12} = 0$, implying that $b = 0$. Similarly, $d = \sqrt{3}/\sqrt{5}$.

From the second row of B we have

$$b^2 + c^2 = 3 \quad \Rightarrow \quad c = \sqrt{3},$$

$$bd + ce = 0 \quad \Rightarrow \quad e = \sqrt{\tfrac{1}{3}(0-0)} = 0.$$

And, from the final row,

$$d^2 + e^2 + f^2 = 3 \quad \Rightarrow \quad f = (3 - \tfrac{3}{5} - 0)^{1/2} = \sqrt{\tfrac{12}{5}}.$$

Thus all the elements of L have been calculated and found to be real and, in summary,

$$L = \begin{pmatrix} \sqrt{5} & 0 & 0 \\ 0 & \sqrt{3} & 0 \\ \sqrt{\frac{3}{5}} & 0 & \sqrt{\frac{12}{5}} \end{pmatrix}.$$

That $LL^T = B$ can be confirmed by substitution.

1.23 Find three real orthogonal column matrices, each of which is a simultaneous eigenvector of

$$A = \begin{pmatrix} 0 & 0 & 1 \\ 0 & 1 & 0 \\ 1 & 0 & 0 \end{pmatrix} \quad \text{and} \quad B = \begin{pmatrix} 0 & 1 & 1 \\ 1 & 0 & 1 \\ 1 & 1 & 0 \end{pmatrix}.$$

We first note that

$$AB = \begin{pmatrix} 1 & 1 & 0 \\ 1 & 0 & 1 \\ 0 & 1 & 1 \end{pmatrix} = BA.$$

The two matrices commute and so they *will* have a common set of eigenvectors.
The eigenvalues of A are given by

$$\begin{vmatrix} -\lambda & 0 & 1 \\ 0 & 1-\lambda & 0 \\ 1 & 0 & -\lambda \end{vmatrix} = (1-\lambda)(\lambda^2 - 1) = 0,$$

i.e. $\lambda = 1$, $\lambda = 1$ and $\lambda = -1$, with corresponding eigenvectors $e^1 = (1 \quad y_1 \quad 1)^T$, $e^2 = (1 \quad y_2 \quad 1)^T$ and $e^3 = (1 \quad 0 \quad -1)^T$. For these to be mutually orthogonal requires that $y_1 y_2 = -2$.

The third vector, e^3, is clearly an eigenvector of B with eigenvalue $\mu_3 = -1$. For e^1 or e^2 to be an eigenvector of B with eigenvalue μ requires

$$\begin{pmatrix} 0-\mu & 1 & 1 \\ 1 & 0-\mu & 1 \\ 1 & 1 & 0-\mu \end{pmatrix} \begin{pmatrix} 1 \\ y \\ 1 \end{pmatrix} = \begin{pmatrix} 0 \\ 0 \\ 0 \end{pmatrix};$$

i.e. $-\mu + y + 1 = 0,$
and $1 - \mu y + 1 = 0,$
giving $-\dfrac{2}{y} + y + 1 = 0,$
$\Rightarrow \quad y^2 + y - 2 = 0,$
$\Rightarrow \quad y = 1 \quad \text{or} \quad -2.$

Thus, $y_1 = 1$ with $\mu_1 = 2$, whilst $y_2 = -2$ with $\mu_2 = -1$.
The common eigenvectors are thus

$$e^1 = (1 \quad 1 \quad 1)^T, \quad e^2 = (1 \quad -2 \quad 1)^T, \quad e^3 = (1 \quad 0 \quad -1)^T.$$

We note, as a check, that $\sum_i \mu_i = 2 + (-1) + (-1) = 0 = \text{Tr } B$.

1.25 Given that A is a real symmetric matrix with normalized eigenvectors e^i, obtain the coefficients α_i involved when column matrix x, which is the solution of

$$\text{Ax} - \mu\text{x} = \text{v}, \qquad\qquad (*)$$

is expanded as $\text{x} = \sum_i \alpha_i e^i$. Here μ is a given constant and v is a given column matrix.

(a) Solve $(*)$ when

$$A = \begin{pmatrix} 2 & 1 & 0 \\ 1 & 2 & 0 \\ 0 & 0 & 3 \end{pmatrix},$$

$\mu = 2$ and $\text{v} = (1 \quad 2 \quad 3)^{\text{T}}$.

(b) Would $(*)$ have a solution if (i) $\mu = 1$ and $\text{v} = (1 \quad 2 \quad 3)^{\text{T}}$, (ii) $\text{v} = (2 \quad 2 \quad 3)^{\text{T}}$? Where it does, find it.

Let $\text{x} = \sum_i \alpha_i e^i$, where $Ae^i = \lambda_i e^i$. Then

$$\text{Ax} - \mu\text{x} = \text{v},$$

$$\sum_i A\alpha_i e^i - \sum_i \mu\alpha_i e^i = \text{v},$$

$$\sum_i \left(\lambda_i \alpha_i e^i - \mu\alpha_i e^i \right) = \text{v},$$

$$\alpha_j = \frac{(e^j)^\dagger \text{v}}{\lambda_j - \mu}.$$

To obtain the last line we have used the mutual orthogonality of the eigenvectors. We note, in passing, that if $\mu = \lambda_j$ for any j there is no solution unless $(e^j)^\dagger \text{v} = 0$.

(a) To obtain the eigenvalues of the given matrix A, consider

$$0 = |A - \lambda I| = (3 - \lambda)(4 - 4\lambda + \lambda^2 - 1) = (3 - \lambda)(3 - \lambda)(1 - \lambda).$$

The eigenvalues, and a possible set of corresponding normalized eigenvectors, are therefore,

$$\begin{aligned}
\text{for} \quad \lambda = 3, \quad & e^1 = (0 \quad 0 \quad 1)^{\text{T}}; \\
\text{for} \quad \lambda = 3, \quad & e^2 = 2^{-1/2}(1 \quad 1 \quad 0)^{\text{T}}; \\
\text{for} \quad \lambda = 1, \quad & e^3 = 2^{-1/2}(1 \quad -1 \quad 0)^{\text{T}}.
\end{aligned}$$

Since $\lambda = 3$ is a degenerate eigenvalue, there are infinitely many acceptable pairs of orthogonal eigenvectors corresponding to it; any pair of vectors of the form (a_i, a_i, b_i) with $2a_1 a_2 + b_1 b_2 = 0$ will suffice. The pair given is just about the simplest choice possible.

With $\mu = 2$ and $\text{v} = (1 \quad 2 \quad 3)^{\text{T}}$,

$$\alpha_1 = \frac{3}{3 - 2}, \qquad \alpha_2 = \frac{3/\sqrt{2}}{3 - 2}, \qquad \alpha_3 = \frac{-1/\sqrt{2}}{1 - 2}.$$

Thus the solution vector is

$$\mathbf{x} = 3 \begin{pmatrix} 0 \\ 0 \\ 1 \end{pmatrix} + \frac{3}{\sqrt{2}} \frac{1}{\sqrt{2}} \begin{pmatrix} 1 \\ 1 \\ 0 \end{pmatrix} + \frac{1}{\sqrt{2}} \frac{1}{\sqrt{2}} \begin{pmatrix} 1 \\ -1 \\ 0 \end{pmatrix} = \begin{pmatrix} 2 \\ 1 \\ 3 \end{pmatrix}.$$

(b) If $\mu = 1$ then it is equal to the third eigenvalue and a solution is only possible if $(\mathbf{e}^3)^\dagger \mathbf{v} = 0$.

For (i) $\mathbf{v} = (1 \quad 2 \quad 3)^\mathrm{T}$, $(\mathbf{e}^3)^\dagger \mathbf{v} = -1/\sqrt{2}$ and so no solution is possible.

For (ii) $\mathbf{v} = (2 \quad 2 \quad 3)^\mathrm{T}$, $(\mathbf{e}^3)^\dagger \mathbf{v} = 0$, and so a solution is possible. The other scalar products needed are $(\mathbf{e}^1)^\dagger \mathbf{v} = 3$ and $(\mathbf{e}^2)^\dagger \mathbf{v} = 2\sqrt{2}$. For this vector $\mathbf{v}$ the solution to the equation is

$$\mathbf{x} = \frac{3}{3-1} \begin{pmatrix} 0 \\ 0 \\ 1 \end{pmatrix} + \frac{2\sqrt{2}}{3-1} \frac{1}{\sqrt{2}} \begin{pmatrix} 1 \\ 1 \\ 0 \end{pmatrix} = \begin{pmatrix} 1 \\ 1 \\ \frac{3}{2} \end{pmatrix}.$$

[The solutions to both parts can be checked by resubstitution.]

1.27 By finding the eigenvectors of the Hermitian matrix

$$\mathbf{H} = \begin{pmatrix} 10 & 3i \\ -3i & 2 \end{pmatrix},$$

construct a unitary matrix $\mathbf{U}$ such that $\mathbf{U}^\dagger \mathbf{H} \mathbf{U} = \Lambda$, where Λ is a real diagonal matrix.

We start by finding the eigenvalues of $\mathbf{H}$ using

$$\begin{vmatrix} 10 - \lambda & 3i \\ -3i & 2 - \lambda \end{vmatrix} = 0,$$

$$20 - 12\lambda + \lambda^2 - 3 = 0,$$

$$\lambda = 1 \quad \text{or} \quad 11.$$

As expected for an Hermitian matrix, the eigenvalues are real.

For $\lambda = 1$ and normalized eigenvector $(x \quad y)^\mathrm{T}$,

$$9x + 3iy = 0 \quad \Rightarrow \quad \mathbf{x}^1 = (10)^{-1/2} (1 \quad 3i)^\mathrm{T}.$$

For $\lambda = 11$ and normalized eigenvector $(x \quad y)^\mathrm{T}$,

$$-x + 3iy = 0 \quad \Rightarrow \quad \mathbf{x}^2 = (10)^{-1/2} (3i \quad 1)^\mathrm{T}.$$

Again as expected, $(\mathbf{x}^1)^\dagger \mathbf{x}^2 = 0$, thus verifying the mutual orthogonality of the eigenvectors. It should be noted that the normalization factor is determined by $(\mathbf{x}^i)^\dagger \mathbf{x}^i = 1$ (and *not* by $(\mathbf{x}^i)^\mathrm{T} \mathbf{x}^i = 1$).

We now use these normalized eigenvectors of H as the columns of the matrix U and check that it is unitary:

$$U = \frac{1}{\sqrt{10}} \begin{pmatrix} 1 & 3i \\ 3i & 1 \end{pmatrix}, \quad U^\dagger = \frac{1}{\sqrt{10}} \begin{pmatrix} 1 & -3i \\ -3i & 1 \end{pmatrix},$$

$$UU^\dagger = \frac{1}{10} \begin{pmatrix} 1 & 3i \\ 3i & 1 \end{pmatrix} \begin{pmatrix} 1 & -3i \\ -3i & 1 \end{pmatrix} = \frac{1}{10} \begin{pmatrix} 10 & 0 \\ 0 & 10 \end{pmatrix} = I.$$

U has the further property that

$$U^\dagger HU = \frac{1}{\sqrt{10}} \begin{pmatrix} 1 & -3i \\ -3i & 1 \end{pmatrix} \begin{pmatrix} 10 & 3i \\ -3i & 2 \end{pmatrix} \frac{1}{\sqrt{10}} \begin{pmatrix} 1 & 3i \\ 3i & 1 \end{pmatrix}$$

$$= \frac{1}{10} \begin{pmatrix} 1 & -3i \\ -3i & 1 \end{pmatrix} \begin{pmatrix} 1 & 33i \\ 3i & 11 \end{pmatrix}$$

$$= \frac{1}{10} \begin{pmatrix} 10 & 0 \\ 0 & 110 \end{pmatrix} = \begin{pmatrix} 1 & 0 \\ 0 & 11 \end{pmatrix} = \Lambda.$$

That the diagonal entries of Λ are the eigenvalues of H is in accord with the general theory of normal matrices.

1.29 Given that the matrix

$$A = \begin{pmatrix} 2 & -1 & 0 \\ -1 & 2 & -1 \\ 0 & -1 & 2 \end{pmatrix}$$

has two eigenvectors of the form $(1 \quad y \quad 1)^T$, use the stationary property of the expression $J(x) = x^T Ax/(x^T x)$ to obtain the corresponding eigenvalues. Deduce the third eigenvalue.

Since A is real and symmetric, each eigenvalue λ is real. Further, from the first component of $Ax = \lambda x$, we have that $2 - y = \lambda$, showing that y is also real. Considered as a function of a general vector of the form $(1 \quad y \quad 1)^T$, the quadratic form $x^T Ax$ can be written explicitly as

$$x^T Ax = (1\ y\ 1) \begin{pmatrix} 2 & -1 & 0 \\ -1 & 2 & -1 \\ 0 & -1 & 2 \end{pmatrix} \begin{pmatrix} 1 \\ y \\ 1 \end{pmatrix}$$

$$= (1\ y\ 1) \begin{pmatrix} 2 - y \\ 2y - 2 \\ 2 - y \end{pmatrix}$$

$$= 2y^2 - 4y + 4.$$

The scalar product $x^T x$ has the value $2 + y^2$, and so we need to find the stationary values of

$$I = \frac{2y^2 - 4y + 4}{2 + y^2}.$$

These are given by

$$0 = \frac{dI}{dy} = \frac{(2 + y^2)(4y - 4) - (2y^2 - 4y + 4)2y}{(2 + y^2)^2}$$

$$0 = 4y^2 - 8,$$

$$y = \pm\sqrt{2}.$$

The corresponding eigenvalues are the values of I at the stationary points:

$$\text{for} \quad y = \sqrt{2}, \quad \lambda_1 = \frac{2(2) - 4\sqrt{2} + 4}{2 + 2} = 2 - \sqrt{2};$$

$$\text{for} \quad y = -\sqrt{2}, \quad \lambda_2 = \frac{2(2) + 4\sqrt{2} + 4}{2 + 2} = 2 + \sqrt{2}.$$

The final eigenvalue can be found using the fact that the sum of the eigenvalues is equal to the trace of the matrix; so

$$\lambda_3 = (2 + 2 + 2) - (2 - \sqrt{2}) - (2 + \sqrt{2}) = 2.$$

1.31 The equation of a particular conic section is

$$Q \equiv 8x_1^2 + 8x_2^2 - 6x_1x_2 = 110.$$

Determine the type of conic section this represents, the orientation of its principal axes, and relevant lengths in the directions of these axes.

The eigenvalues of the matrix $\begin{pmatrix} 8 & -3 \\ -3 & 8 \end{pmatrix}$ associated with the quadratic form on the LHS (without any prior scaling) are given by

$$0 = \begin{vmatrix} 8 - \lambda & -3 \\ -3 & 8 - \lambda \end{vmatrix}$$

$$= \lambda^2 - 16\lambda + 55$$

$$= (\lambda - 5)(\lambda - 11).$$

Referred to the corresponding eigenvectors as axes, the conic section (an ellipse since both eigenvalues are positive) will take the form

$$5y_1^2 + 11y_2^2 = 110 \quad \text{or, in standard form,} \quad \frac{y_1^2}{22} + \frac{y_2^2}{10} = 1.$$

Thus the semi-axes are of lengths $\sqrt{22}$ and $\sqrt{10}$; the former is in the direction of the vector $(x_1 \quad x_2)^T$ given by $(8 - 5)x_1 - 3x_2 = 0$, i.e. it is the line $x_1 = x_2$. The other principal axis will be the line at right angles to this, namely the line $x_1 = -x_2$.

1.33 Find the direction of the axis of symmetry of the quadratic surface

$$7x^2 + 7y^2 + 7z^2 - 20yz - 20xz + 20xy = 3.$$

The straightforward, but longer, solution to this problem is as follows.

Consider the characteristic polynomial of the matrix associated with the quadratic surface, namely,

$$f(\lambda) = \begin{vmatrix} 7-\lambda & 10 & -10 \\ 10 & 7-\lambda & -10 \\ -10 & -10 & 7-\lambda \end{vmatrix}$$

$$= (7-\lambda)(-51 - 14\lambda + \lambda^2) + 10(30 + 10\lambda) - 10(-30 - 10\lambda)$$

$$= -\lambda^3 + 21\lambda^2 + 153\lambda + 243.$$

If the quadratic surface has an axis of symmetry, it must have two equal major axes (perpendicular to it), and hence the characteristic equation must have a repeated root. This same root will therefore also be a root of $df/d\lambda = 0$, i.e. of

$$-3\lambda^2 + 42\lambda + 153 = 0,$$

$$\lambda^2 - 14\lambda - 51 = 0,$$

$$\lambda = 17 \quad \text{or} \quad -3.$$

Substitution shows that -3 is a root (and therefore a double root) of $f(\lambda) = 0$, but that 17 is not. The non-repeated root can be calculated as the trace of the matrix minus the repeated roots, i.e. $21 - (-3) - (-3) = 27$. It is the eigenvector that corresponds to this eigenvalue that gives the direction $(x \quad y \quad z)^T$ of the axis of symmetry. Its components must satisfy

$$(7 - 27)x + 10y - 10z = 0,$$

$$10x + (7 - 27)y - 10z = 0.$$

The axis of symmetry is therefore in the direction $(1 \quad 1 \quad -1)^T$.

A more subtle solution is obtained by noting that setting $\lambda = -3$ makes *all three* of the rows (or columns) of the determinant multiples of each other, i.e. it reduces the determinant to rank one. Thus -3 is a repeated root of the characteristic equation and the third root is $21 - 2(-3) = 27$. The rest of the analysis is as above.

We note in passing that, as two eigenvalues are negative and equal, the surface is the hyperboloid of revolution obtained by rotating a (two-branched) hyperbola about its axis of symmetry. Referred to this axis and two others forming a mutually orthogonal set, the equation of the quadratic surface takes the form $-3\chi^2 - 3\eta^2 + 27\zeta^2 = 3$ and so the tips of the two "nose cones" ($\chi = \eta = 0$) are separated by $\frac{2}{3}$ of a unit.

1.35 This problem demonstrates the reverse of the usual procedure of diagonalizing a matrix.

(a) Rearrange the result $A' = S^{-1}AS$ (which shows how to make a change of basis that diagonalizes A) so as to express the original matrix A in terms of the unitary matrix S and the diagonal matrix A'. Hence show how to construct a matrix A that has given eigenvalues and given (orthogonal) column matrices as its eigenvectors.

(b) Find the matrix that has as eigenvectors $(1 \quad 2 \quad 1)^T$, $(1 \quad -1 \quad 1)^T$ and $(1 \quad 0 \quad -1)^T$ and corresponding eigenvalues λ, μ and ν.

(c) Try a particular case, say $\lambda = 3$, $\mu = -2$ and $\nu = 1$, and verify by explicit solution that the matrix so found does have these eigenvalues.

(a) Since S is unitary, we can multiply the given result on the left by S and on the right by $S^\dagger$ to obtain

$$SA'S^\dagger = SS^{-1}ASS^\dagger = (I)\,A\,(I) = A.$$

More explicitly, in terms of the eigenvalues and normalized eigenvectors x^i of A,

$$A = (x^1 \quad x^2 \quad \cdots \quad x^n)\Lambda(x^1 \quad x^2 \quad \cdots \quad x^n)^\dagger.$$

Here Λ is the diagonal matrix that has the eigenvalues of A as its diagonal elements.

Now, given normalized orthogonal column matrices and n specified values, we can use this result to construct a matrix that has the column matrices as eigenvectors and the values as eigenvalues.

(b) The normalized versions of the given column vectors are

$$\frac{1}{\sqrt{6}}(1 \quad 2 \quad 1)^T, \quad \frac{1}{\sqrt{3}}(1 \quad -1 \quad 1)^T, \quad \frac{1}{\sqrt{2}}(1 \quad 0 \quad -1)^T,$$

and the orthogonal matrix S can be constructed using these as its columns:

$$S = \frac{1}{\sqrt{6}}\begin{pmatrix} 1 & \sqrt{2} & \sqrt{3} \\ 2 & -\sqrt{2} & 0 \\ 1 & \sqrt{2} & -\sqrt{3} \end{pmatrix}.$$

The required matrix A can now be formed as $S\Lambda S^\dagger$:

$$A = \frac{1}{6}\begin{pmatrix} 1 & \sqrt{2} & \sqrt{3} \\ 2 & -\sqrt{2} & 0 \\ 1 & \sqrt{2} & -\sqrt{3} \end{pmatrix}\begin{pmatrix} \lambda & 0 & 0 \\ 0 & \mu & 0 \\ 0 & 0 & \nu \end{pmatrix}\begin{pmatrix} 1 & 2 & 1 \\ \sqrt{2} & -\sqrt{2} & \sqrt{2} \\ \sqrt{3} & 0 & -\sqrt{3} \end{pmatrix}$$

$$= \frac{1}{6}\begin{pmatrix} 1 & \sqrt{2} & \sqrt{3} \\ 2 & -\sqrt{2} & 0 \\ 1 & \sqrt{2} & -\sqrt{3} \end{pmatrix}\begin{pmatrix} \lambda & 2\lambda & \lambda \\ \sqrt{2}\mu & -\sqrt{2}\mu & \sqrt{2}\mu \\ \sqrt{3}\nu & 0 & -\sqrt{3}\nu \end{pmatrix}$$

$$= \frac{1}{6}\begin{pmatrix} \lambda + 2\mu + 3\nu & 2\lambda - 2\mu & \lambda + 2\mu - 3\nu \\ 2\lambda - 2\mu & 4\lambda + 2\mu & 2\lambda - 2\mu \\ \lambda + 2\mu - 3\nu & 2\lambda - 2\mu & \lambda + 2\mu + 3\nu \end{pmatrix}.$$

(c) Setting $\lambda = 3$, $\mu = -2$ and $\nu = 1$, as a particular case, gives A as

$$A = \frac{1}{6}\begin{pmatrix} 2 & 10 & -4 \\ 10 & 8 & 10 \\ -4 & 10 & 2 \end{pmatrix}.$$

We complete the problem by solving for the eigenvalues of A in the usual way. To avoid working with fractions, and any confusion with the value $\lambda = 3$ used when constructing

A, we will find the eigenvalues of 6A and denote them by η.

$$0 = |6A - \eta I|$$

$$= \begin{vmatrix} 2 - \eta & 10 & -4 \\ 10 & 8 - \eta & 10 \\ -4 & 10 & 2 - \eta \end{vmatrix}$$

$$= (2 - \eta)(\eta^2 - 10\eta - 84) + 10(10\eta - 60) - 4(132 - 4\eta)$$

$$= -\eta^3 + 12\eta^2 + 180\eta - 1296$$

$$= -(\eta - 6)(\eta^2 - 6\eta - 216)$$

$$= -(\eta - 6)(\eta + 12)(\eta - 18).$$

Thus 6A has eigenvalues 6, -12 and 18; the values for A itself are 1, -2 and 3, as expected.

1.37 A more general form of expression for the determinant of a 3×3 matrix A than (1.45) is given by

$$|A|\epsilon_{lmn} = A_{li} A_{mj} A_{nk} \epsilon_{ijk}. \tag{1.1}$$

The former could, as stated earlier in this chapter, have been written as

$$|A| = \epsilon_{ijk} A_{i1} A_{j2} A_{k3}.$$

The more general form removes the explicit mention of $1, 2, 3$ at the expense of an additional Levi–Civita symbol; the form of (1.1) can be readily extended to cover a general $N \times N$ matrix.

   Use this more general form to prove properties (i), (iii), (v), (vi) and (vii) of determinants stated in Subsection 1.9.1. Property (iv) is obvious by inspection. For definiteness take $N = 3$, but convince yourself that your methods of proof would be valid for any positive integer N.

A full account of the answer to this problem is given in the *Hints and answers* section at the end of the chapter, almost as if it were part of the main text. The reader is referred there for the details.

1.39 Three coupled pendulums swing perpendicularly to the horizontal line containing their points of suspension, and the following equations of motion are satisfied:

$$-m\ddot{x}_1 = cmx_1 + d(x_1 - x_2),$$
$$-M\ddot{x}_2 = cMx_2 + d(x_2 - x_1) + d(x_2 - x_3),$$
$$-m\ddot{x}_3 = cmx_3 + d(x_3 - x_2),$$

where x_1, x_2 and x_3 are measured from the equilibrium points; m, M and m are the masses of the pendulum bobs; and c and d are positive constants. Find the normal frequencies of the system and sketch the corresponding patterns of oscillation. What happens as $d \to 0$ or $d \to \infty$?

In a normal mode all three coordinates x_i oscillate with the same frequency and with fixed relative phases. When this is represented by solutions of the form $x_i = X_i \cos \omega t$, where the X_i are fixed constants, the equations become, in matrix and vector form,

$$\begin{pmatrix} cm + d - m\omega^2 & -d & 0 \\ -d & cM + 2d - M\omega^2 & -d \\ 0 & -d & cm + d - m\omega^2 \end{pmatrix} \begin{pmatrix} X_1 \\ X_2 \\ X_3 \end{pmatrix} = \mathbf{0}.$$

For there to be a non-trivial solution to these simultaneous homogeneous equations, we need

$$0 = \begin{vmatrix} (c - \omega^2)m + d & -d & 0 \\ -d & (c - \omega^2)M + 2d & -d \\ 0 & -d & (c - \omega^2)m + d \end{vmatrix}$$

$$= \begin{vmatrix} (c - \omega^2)m + d & 0 & -(c - \omega^2)m - d \\ -d & (c - \omega^2)M + 2d & -d \\ 0 & -d & (c - \omega^2)m + d \end{vmatrix}$$

$$= [(c - \omega^2)m + d] \left\{ [(c - \omega^2)M + 2d] [(c - \omega^2)m + d] - d^2 - d^2 \right\}$$

$$= (cm - m\omega^2 + d)(c - \omega^2)[Mm(c - \omega^2) + 2dm + dM].$$

Thus, the normal (angular) frequencies are given by

$$\omega^2 = c, \quad \omega^2 = c + \frac{d}{m} \quad \text{and} \quad \omega^2 = c + \frac{2d}{M} + \frac{d}{m}.$$

If the solution column matrix is $\mathbf{X} = (X_1 \quad X_2 \quad X_3)^{\mathrm{T}}$, then
(i) for $\omega^2 = c$, the components of $\mathbf{X}$ must satisfy

$$dX_1 - dX_2 = 0,$$
$$-dX_1 + 2dX_2 - dX_3 = 0, \quad \Rightarrow \quad \mathbf{X}^1 = (1 \quad 1 \quad 1)^{\mathrm{T}};$$

(ii) for $\omega^2 = c + \dfrac{d}{m}$, we have

$$-dX_2 = 0,$$
$$-dX_1 + \left(-\frac{dM}{m} + 2d \right) X_2 - dX_3 = 0, \quad \Rightarrow \quad \mathbf{X}^2 = (1 \quad 0 \quad -1)^{\mathrm{T}};$$

(iii) for $\omega^2 = c + \dfrac{2d}{M} + \dfrac{d}{m}$, the components must satisfy

$$\left[\left(-\frac{2d}{M} - \frac{d}{m} \right) m + d \right] X_1 - dX_2 = 0,$$

$$-dX_2 + \left[\left(-\frac{2d}{M} - \frac{d}{m} \right) m + d \right] X_3 = 0, \quad \Rightarrow \quad \mathbf{X}^3 = \left(1 \quad -\frac{2m}{M} \quad 1 \right)^{\mathrm{T}}.$$

The corresponding patterns are shown in Figure 1.1.

If $d \to 0$, the three oscillations decouple and each pendulum swings independently with angular frequency $\sqrt{c}$.

If $d \to \infty$, the three pendulums become rigidly coupled. The second and third modes have (theoretically) infinite frequency and therefore zero amplitude. The only sustainable

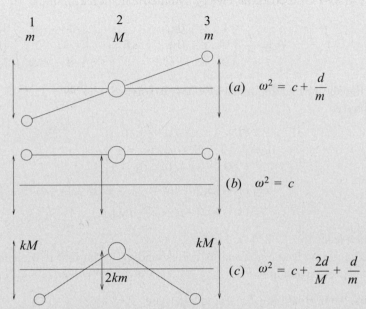

Figure 1.1 The normal modes, as viewed from above, of the coupled pendulums in Problem 1.39.

mode is the one shown as case (b) in the figure; one in which all the pendulums swing as a single entity with angular frequency $\sqrt{c}$.

1.41 Find the normal frequencies of a system consisting of three particles of masses $m_1 = m$, $m_2 = \mu\, m$, $m_3 = m$ connected in that order in a straight line by two equal light springs of force constant k. Describe the corresponding modes of oscillation.

Now consider the particular case in which $\mu = 2$.

(a) Show that the eigenvectors derived above have the expected orthogonality properties with respect to both the kinetic energy matrix A and the potential energy matrix B.

(b) For the situation in which the masses are released from rest with initial displacements (relative to their equilibrium positions) of $x_1 = 2\epsilon$, $x_2 = -\epsilon$ and $x_3 = 0$, determine their subsequent motions and maximum displacements.

Let the coordinates of the particles, x_1, x_2, x_3, be measured from their equilibrium positions, at which the springs are neither extended nor compressed.

The kinetic energy of the system is simply

$$T = \tfrac{1}{2}m \left(\dot{x}_1^2 + \mu\, \dot{x}_2^2 + \dot{x}_3^2 \right),$$

whilst the potential energy stored in the springs takes the form

$$V = \tfrac{1}{2}k \left[(x_2 - x_1)^2 + (x_3 - x_2)^2 \right].$$

The kinetic- and potential-energy symmetric matrices are thus

$$A = \frac{m}{2} \begin{pmatrix} 1 & 0 & 0 \\ 0 & \mu & 0 \\ 0 & 0 & 1 \end{pmatrix}, \qquad B = \frac{k}{2} \begin{pmatrix} 1 & -1 & 0 \\ -1 & 2 & -1 \\ 0 & -1 & 1 \end{pmatrix}.$$

To find the normal frequencies we have to solve $|B - \omega^2 A| = 0$. Thus, writing $m\omega^2/k = \lambda$, we have

$$\begin{aligned} 0 &= \begin{vmatrix} 1 - \lambda & -1 & 0 \\ -1 & 2 - \mu\lambda & -1 \\ 0 & -1 & 1 - \lambda \end{vmatrix} \\ &= (1 - \lambda)(2 - \mu\lambda - 2\lambda + \mu\lambda^2 - 1) + (-1 + \lambda) \\ &= (1 - \lambda)\lambda(-\mu - 2 + \mu\lambda), \end{aligned}$$

which leads to $\lambda = 0$, 1 or $1 + 2/\mu$.

The normalized eigenvectors corresponding to the first two eigenvalues can be found by inspection and are

$$x^1 = \frac{1}{\sqrt{3}} \begin{pmatrix} 1 \\ 1 \\ 1 \end{pmatrix}, \qquad x^2 = \frac{1}{\sqrt{2}} \begin{pmatrix} 1 \\ 0 \\ -1 \end{pmatrix}.$$

The components of the third eigenvector must satisfy

$$-\frac{2}{\mu} x_1 - x_2 = 0 \quad \text{and} \quad x_2 - \frac{2}{\mu} x_3 = 0.$$

The normalized third eigenvector is therefore

$$x^3 = \frac{1}{\sqrt{2 + (4/\mu^2)}} \begin{pmatrix} 1 & -\dfrac{2}{\mu} & 1 \end{pmatrix}^T.$$

The physical motions associated with these normal modes are as follows.

The first, with $\lambda = \omega = 0$ and all the x_i equal, merely describes bodily translation of the whole system, with no (i.e. zero-frequency) internal oscillations.

In the second solution, the central particle remains stationary, $x_2 = 0$, whilst the other two oscillate with equal amplitudes in antiphase with each other. This motion has frequency $\omega = (k/m)^{1/2}$, the same as that for the oscillations of a single mass m suspended from a single spring of force constant k.

The final and most complicated of the three normal modes has angular frequency $\omega = \{[(\mu + 2)/\mu](k/m)\}^{1/2}$, and involves a motion of the central particle which is in antiphase with that of the two outer ones and which has an amplitude $2/\mu$ times as great. In this motion the two springs are compressed and extended in turn. We also note that in the second and third normal modes the center of mass of the system remains stationary.

Now setting $\mu = 2$, we have as the three normal (angular) frequencies 0, Ω and $\sqrt{2}\Omega$, where $\Omega^2 = k/m$. The corresponding (unnormalized) eigenvectors are

$$x^1 = (1 \quad 1 \quad 1)^T, \qquad x^2 = (1 \quad 0 \quad -1)^T, \qquad x^3 = (1 \quad -1 \quad 1)^T.$$

(a) The matrices A and B have the forms

$$A = \begin{pmatrix} 1 & 0 & 0 \\ 0 & 2 & 0 \\ 0 & 0 & 1 \end{pmatrix}, \quad B = \begin{pmatrix} 1 & -1 & 0 \\ -1 & 2 & -1 \\ 0 & -1 & 1 \end{pmatrix}.$$

To verify the standard orthogonality relations we need to show that the quadratic forms $(\mathbf{x}^i)^\dagger A \mathbf{x}^j$ and $(\mathbf{x}^i)^\dagger B \mathbf{x}^j$ have zero value for $i \neq j$. Direct evaluation of all the separate cases is as follows:

$$(\mathbf{x}^1)^\dagger A \mathbf{x}^2 = 1 + 0 - 1 = 0,$$
$$(\mathbf{x}^1)^\dagger A \mathbf{x}^3 = 1 - 2 + 1 = 0,$$
$$(\mathbf{x}^2)^\dagger A \mathbf{x}^3 = 1 + 0 - 1 = 0,$$
$$(\mathbf{x}^1)^\dagger B \mathbf{x}^2 = (\mathbf{x}^1)^\dagger \, (1 \quad 0 \quad -1)^T = 1 + 0 - 1 = 0,$$
$$(\mathbf{x}^1)^\dagger B \mathbf{x}^3 = (\mathbf{x}^1)^\dagger \, (2 \quad -4 \quad 2)^T = 2 - 4 + 2 = 0,$$
$$(\mathbf{x}^2)^\dagger B \mathbf{x}^3 = (\mathbf{x}^2)^\dagger \, (2 \quad -4 \quad 2)^T = 2 + 0 - 2 = 0.$$

If $(\mathbf{x}^i)^\dagger A \mathbf{x}^j$ has zero value then so does $(\mathbf{x}^j)^\dagger A \mathbf{x}^i$ (and similarly for B). So there is no need to investigate the other six possibilities and the verification is complete.

(b) In order to determine the behavior of the system we need to know which modes are present in the initial configuration. Each contributory mode will subsequently oscillate with its own frequency. In order to carry out this initial decomposition we write

$$(2\epsilon \quad -\epsilon \quad 0)^T = a \, (1 \quad 1 \quad 1)^T + b \, (1 \quad 0 \quad -1)^T + c \, (1 \quad -1 \quad 1)^T,$$

from which it is clear that $a = 0$, $b = \epsilon$ and $c = \epsilon$. As each mode vibrates with its own frequency, the subsequent displacements are given by

$$x_1 = \epsilon(\cos \Omega t + \cos \sqrt{2}\Omega t),$$
$$x_2 = -\epsilon \cos \sqrt{2}\Omega t,$$
$$x_3 = \epsilon(-\cos \Omega t + \cos \sqrt{2}\Omega t).$$

Since Ω and $\sqrt{2}\Omega$ are not rationally related, at some times the two modes will, for all practical purposes (but not mathematically), be in phase and, at other times, be out of phase. Thus the maximum displacements will be $x_1(\max) = 2\epsilon$, $x_2(\max) = \epsilon$ and $x_3(\max) = 2\epsilon$.

1.43 It is shown in physics and engineering textbooks that circuits containing capacitors and inductors can be analyzed by replacing a capacitor of capacitance C by a "complex impedance" $1/(i\omega C)$ and an inductor of inductance L by an impedance $i\omega L$, where ω is the angular frequency of the currents flowing and $i^2 = -1$.

Use this approach and Kirchhoff's circuit laws to analyze the circuit shown in Figure 1.2 and obtain three linear equations governing the currents I_1, I_2 and I_3. Show that the only possible frequencies of self-sustaining currents satisfy either (a) $\omega^2 LC = 1$ or (b) $3\omega^2 LC = 1$. Find the corresponding current patterns and, in each case, by identifying parts of the circuit in which no current flows, draw an equivalent circuit that contains only one capacitor and one inductor.

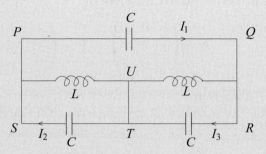

Figure 1.2 The circuit and notation for Problem 1.43.

We apply Kirchhoff's laws to the three closed loops $PQUP$, $SUTS$ and $TURT$ and obtain, respectively,

$$\frac{1}{i\omega C}I_1 + i\omega L(I_1 - I_3) + i\omega L(I_1 - I_2) = 0,$$

$$i\omega L(I_2 - I_1) + \frac{1}{i\omega C}I_2 = 0,$$

$$i\omega L(I_3 - I_1) + \frac{1}{i\omega C}I_3 = 0.$$

For these simultaneous homogeneous linear equations to be consistent, it is necessary that

$$0 = \begin{vmatrix} \dfrac{1}{i\omega C} + 2i\omega L & -i\omega L & -i\omega L \\ -i\omega L & \dfrac{1}{i\omega C} + i\omega L & 0 \\ -i\omega L & 0 & \dfrac{1}{i\omega C} + i\omega L \end{vmatrix} = \begin{vmatrix} \lambda - 2 & 1 & 1 \\ 1 & \lambda - 1 & 0 \\ 1 & 0 & \lambda - 1 \end{vmatrix},$$

where, after dividing all entries by $-i\omega L$, we have written the combination $(LC\omega^2)^{-1}$ as λ to save space. Expanding the determinant gives

$$0 = (\lambda - 2)(\lambda - 1)^2 - (\lambda - 1) - (\lambda - 1)$$
$$= (\lambda - 1)(\lambda^2 - 3\lambda + 2 - 2)$$
$$= \lambda(\lambda - 1)(\lambda - 3).$$

Only the non-zero roots are of practical physical interest, and these are $\lambda = 1$ and $\lambda = 3$.

(a) The first of these eigenvalues has an eigenvector $\mathsf{l}^1 = (I_1 \quad I_2 \quad I_3)^{\mathrm{T}}$ that satisfies

$$-I_1 + I_2 + I_3 = 0,$$
$$I_1 = 0 \quad \Rightarrow \quad \mathsf{l}^1 = (0 \quad 1 \quad -1)^{\mathrm{T}}.$$

Thus there is no current in PQ and the capacitor in that link can be ignored. Equal currents circulate, in opposite directions, in the other two loops and, although the link TU carries both, there is no transfer between the two loops. Each loop is therefore equivalent to a capacitor of capacitance C in parallel with an inductor of inductance L.

(b) The second eigenvalue has an eigenvector $l^2 = (l_1 \quad l_2 \quad l_3)^T$ that satisfies

$$I_1 + I_2 + I_3 = 0,$$
$$I_1 + 2I_2 = 0 \quad \Rightarrow \quad l^2 = (-2 \quad 1 \quad 1)^T.$$

In this mode there is no current in TU and the circuit is equivalent to an inductor of inductance $L + L$ in parallel with a capacitor of capacitance $3C/2$; this latter capacitance is made up of C in parallel with the capacitance equivalent to two capacitors C in series, i.e. in parallel with $\frac{1}{2}C$. Thus, the equivalent single components are an inductance of $2L$ and a capacitance of $3C/2$.

1.45 A double pendulum consists of two identical uniform rods, each of length ℓ and mass M, smoothly jointed together and suspended by attaching the free end of one rod to a fixed point. The system makes small oscillations in a vertical plane, with the angles made with the vertical by the upper and lower rods denoted by θ_1 and θ_2, respectively. The expressions for the kinetic energy T and the potential energy V of the system are (to second order in the θ_i)

$$T \approx Ml^2 \left(\tfrac{8}{3}\dot{\theta}_1^2 + 2\dot{\theta}_1\dot{\theta}_2 + \tfrac{2}{3}\dot{\theta}_2^2 \right),$$
$$V \approx Mgl \left(\tfrac{3}{2}\theta_1^2 + \tfrac{1}{2}\theta_2^2 \right).$$

Determine the normal frequencies of the system and find new variables ξ and η that will reduce these two expressions to diagonal form, i.e. to

$$a_1\dot{\xi}^2 + a_2\dot{\eta}^2 \quad \text{and} \quad b_1\xi^2 + b_2\eta^2.$$

To find the new variables we will use the following result. If the reader is not familiar with it, a standard textbook should be consulted.

If $Q_1 = u^T A u$ and $Q_2 = u^T B u$ are two real symmetric quadratic forms and u^n are those column matrices that satisfy

$$B u^n = \lambda_n A u^n,$$

then the matrix P whose columns are the vectors u^n is such that the change of variables $u = Pv$ reduces both quadratic forms simultaneously to sums of squares, i.e. $Q_1 = v^T C v$ and $Q_2 = v^T D v$, with both C and D diagonal.

Further points to note are:

(i) that for the u^i as determined above, $(u^m)^T A u^n = 0$ if $m \neq n$ and similarly if A is replaced by B;

(ii) that P is not in general an orthogonal matrix, even if the vectors u^n are normalized.

(iii) In the special case that A is the identity matrix I: the above procedure is the same as diagonalizing B; P *is* an orthogonal matrix if normalized vectors are used; mutual orthogonality of the eigenvectors takes on its usual form.

This problem is a physical example to which the above mathematical result can be applied, the two real symmetric (actually positive-definite) matrices being the kinetic and potential energy matrices.

$$A = \begin{pmatrix} \tfrac{8}{3} & 1 \\ 1 & \tfrac{2}{3} \end{pmatrix}, \quad B = \begin{pmatrix} \tfrac{3}{2} & 0 \\ 0 & \tfrac{1}{2} \end{pmatrix} \quad \text{with} \quad \lambda_i = \frac{\omega_i^2 l}{g}.$$

We find the normal frequencies by solving

$$0 = |B - \lambda A|$$

$$= \begin{vmatrix} \frac{3}{2} - \frac{8}{3}\lambda & -\lambda \\ -\lambda & \frac{1}{2} - \frac{2}{3}\lambda \end{vmatrix}$$

$$= \frac{3}{4} - \frac{7}{3}\lambda + \frac{16}{9}\lambda^2 - \lambda^2$$

$$\Rightarrow \quad 0 = 28\lambda^2 - 84\lambda + 27.$$

Thus, $\lambda = 2.634$ or $\lambda = 0.3661$, and the normal frequencies are $(2.634g/l)^{1/2}$ and $(0.3661g/l)^{1/2}$.

The corresponding column vectors u^i have components that satisfy the following.
(i) For $\lambda = 0.3661$,

$$\left(\frac{3}{2} - \frac{8}{3}0.3661\right)\theta_1 - 0.3661\theta_2 = 0 \quad \Rightarrow \quad u^1 = (1 \quad 1.431)^T.$$

(ii) For $\lambda = 2.634$,

$$\left(\frac{3}{2} - \frac{8}{3}2.634\right)\theta_1 - 2.634\theta_2 = 0 \quad \Rightarrow \quad u^2 = (1 \quad -2.097)^T.$$

We can now construct P as

$$P = \begin{pmatrix} 1 & 1 \\ 1.431 & -2.097 \end{pmatrix}$$

and define new variables (ξ, η) by $(\theta_1 \quad \theta_2)^T = P(\xi \quad \eta)^T$. When the substitutions $\theta_1 = \xi + \eta$ and $\theta_2 = 1.431\xi - 2.097\eta \equiv \alpha\xi - \beta\eta$ are made into the expressions for T and V, they both take on diagonal forms. This can be checked by computing the coefficients of $\xi\eta$ in the two expressions. They are as follows.

$$\text{For } V: \quad 3 - \alpha\beta = 0, \quad \text{and} \quad \text{for } T: \quad \frac{16}{3} + 2(\alpha - \beta) - \frac{4}{3}\alpha\beta = 0.$$

As an example, the full expression for the potential energy becomes $V = Mg\ell(2.524\,\xi^2 + 3.699\,\eta^2)$.

1.47 Three particles each of mass m are attached to a light horizontal string having fixed ends, the string being thus divided into four equal portions, each of length a and under a tension T. Show that for small transverse vibrations the amplitudes x^i of the normal modes satisfy $Bx = (ma\omega^2/T)x$, where B is the matrix

$$\begin{pmatrix} 2 & -1 & 0 \\ -1 & 2 & -1 \\ 0 & -1 & 2 \end{pmatrix}.$$

Estimate the lowest and highest eigenfrequencies using trial vectors $(3 \quad 4 \quad 3)^T$ and $(3 \quad -4 \quad 3)^T$. Use also the exact vectors $\begin{pmatrix} 1 & \sqrt{2} & 1 \end{pmatrix}^T$ and $\begin{pmatrix} 1 & -\sqrt{2} & 1 \end{pmatrix}^T$ and compare the results.

For the ith mass, with displacement y_i, the force it experiences as a result of the tension in the string connecting it to the $(i + 1)$th mass is the resolved component of that tension perpendicular to the equilibrium line, i.e. $f = \dfrac{y_{i+1} - y_i}{a}T$. Similarly the force due to the tension in the string connecting it to the $(i - 1)$th mass is $f = \dfrac{y_{i-1} - y_i}{a} T$. Because the ends of the string are fixed the notional zeroth and fourth masses have $y_0 = y_4 = 0$.

The equations of motion are, therefore,

$$m\ddot{x}_1 = \frac{T}{a}[(0 - x_1) + (x_2 - x_1)],$$

$$m\ddot{x}_2 = \frac{T}{a}[(x_1 - x_2) + (x_3 - x_2)],$$

$$m\ddot{x}_3 = \frac{T}{a}[(x_2 - x_3) + (0 - x_3)].$$

If the displacements are written as $x_i = X_i \cos \omega t$ and $\mathbf{x} = (X_1 \quad X_2 \quad X_3)^{\mathrm{T}}$, then these equations become

$$-\frac{ma\omega^2}{T}X_1 = -2X_1 + X_2,$$

$$-\frac{ma\omega^2}{T}X_2 = X_1 - 2X_2 + X_3,$$

$$-\frac{ma\omega^2}{T}X_3 = X_2 - 2X_3.$$

This set of equations can be written as $\mathbf{Bx} = \dfrac{ma\omega^2}{T}\mathbf{x}$, with

$$\mathbf{B} = \begin{pmatrix} 2 & -1 & 0 \\ -1 & 2 & -1 \\ 0 & -1 & 2 \end{pmatrix}.$$

The Rayleigh–Ritz method shows that any estimate λ of $\dfrac{\mathbf{x}^{\mathrm{T}}\mathbf{Bx}}{\mathbf{x}^{\mathrm{T}}\mathbf{x}}$ always lies between the lowest and highest possible values of $ma\omega^2/T$.

Using the suggested *trial* vectors gives the following estimates for λ.

(i) For $\mathbf{x} = (3 \quad 4 \quad 3)^{\mathrm{T}}$

$$\lambda = [(3, 4, 3)\mathbf{B}(3 \quad 4 \quad 3)^{\mathrm{T}}]/34$$
$$= [(3, 4, 3)(2 \quad 2 \quad 2)^{\mathrm{T}}]/34$$
$$= 20/34 = 0.588.$$

(ii) For $\mathbf{x} = (3 \quad -4 \quad 3)^{\mathrm{T}}$

$$\lambda = [(3, -4, 3)\mathbf{B}(3 \quad -4 \quad 3)^{\mathrm{T}}]/34$$
$$= [(3, -4, 3)(10 \quad -14 \quad 10)^{\mathrm{T}}]/34$$
$$= 116/34 = 3.412.$$

Using, instead, the *exact* vectors yields the exact values of λ as follows.

(i) For the eigenvector corresponding to the lowest eigenvalue, $\mathbf{x} = (1,\ \sqrt{2},\ 1)^T$,

$$\lambda = \left[(1,\ \sqrt{2},\ 1)\mathsf{B}(1,\ \sqrt{2},\ 1)^T\right]/4$$
$$= \left[(1,\ \sqrt{2},\ 1)(2-\sqrt{2},\ 2\sqrt{2}-2,\ 2-\sqrt{2})^T\right]/4$$
$$= 2-\sqrt{2} = 0.586.$$

(ii) For the eigenvector corresponding to the highest eigenvalue, $\mathbf{x} = (1,\ -\sqrt{2},\ 1)^T$,

$$\lambda = \left[(1,\ -\sqrt{2},\ 1)\mathsf{B}(1,\ -\sqrt{2},\ 1)^T\right]/4$$
$$= \left[(1,\ -\sqrt{2},\ 1)(2+\sqrt{2},\ -2\sqrt{2}-2,\ 2+\sqrt{2})^T\right]/4$$
$$= 2+\sqrt{2} = 3.414.$$

As can be seen, the (crude) trial vectors give excellent approximations to the lowest and highest eigenfrequencies.

2

Vector calculus

2.1 Evaluate the integral

$$\int \left[\mathbf{a}(\dot{\mathbf{b}} \cdot \mathbf{a} + \mathbf{b} \cdot \dot{\mathbf{a}}) + \dot{\mathbf{a}}(\mathbf{b} \cdot \mathbf{a}) - 2(\dot{\mathbf{a}} \cdot \mathbf{a})\mathbf{b} - \dot{\mathbf{b}}|\mathbf{a}|^2 \right] dt$$

in which $\dot{\mathbf{a}}$ and $\dot{\mathbf{b}}$ are the derivatives of the real vectors $\mathbf{a}$ and $\mathbf{b}$ with respect to t.

In order to evaluate this integral, we need to group the terms in the integrand so that each is a part of the total derivative of a product of factors. Clearly, the first three terms are the derivative of $\mathbf{a}(\mathbf{b} \cdot \mathbf{a})$, i.e.

$$\frac{d}{dt}[\mathbf{a}(\mathbf{b} \cdot \mathbf{a})] = \dot{\mathbf{a}}(\mathbf{b} \cdot \mathbf{a}) + \mathbf{a}(\dot{\mathbf{b}} \cdot \mathbf{a}) + \mathbf{a}(\mathbf{b} \cdot \dot{\mathbf{a}}).$$

Remembering that the scalar product is commutative, and that $|\mathbf{a}|^2 = \mathbf{a} \cdot \mathbf{a}$, we also have

$$\frac{d}{dt}[\mathbf{b}(\mathbf{a} \cdot \mathbf{a})] = \dot{\mathbf{b}}(\mathbf{a} \cdot \mathbf{a}) + \mathbf{b}(\dot{\mathbf{a}} \cdot \mathbf{a}) + \mathbf{b}(\mathbf{a} \cdot \dot{\mathbf{a}})$$

$$= \dot{\mathbf{b}}(\mathbf{a} \cdot \mathbf{a}) + 2\mathbf{b}(\dot{\mathbf{a}} \cdot \mathbf{a}).$$

Hence,

$$I = \int \left\{ \frac{d}{dt}[\mathbf{a}(\mathbf{b} \cdot \mathbf{a})] - \frac{d}{dt}[\mathbf{b}(\mathbf{a} \cdot \mathbf{a})] \right\} dt$$

$$= \mathbf{a}(\mathbf{b} \cdot \mathbf{a}) - \mathbf{b}(\mathbf{a} \cdot \mathbf{a}) + \mathbf{h}$$

$$= \mathbf{a} \times (\mathbf{a} \times \mathbf{b}) + \mathbf{h},$$

where $\mathbf{h}$ is the (vector) constant of integration. To obtain the final line above, we used a special case of the expansion of a vector triple product.

2.3 The general equation of motion of a (non-relativistic) particle of mass m and charge q when it is placed in a region where there is a magnetic field $\mathbf{B}$ and an electric field $\mathbf{E}$ is

$$m\ddot{\mathbf{r}} = q(\mathbf{E} + \dot{\mathbf{r}} \times \mathbf{B});$$

here $\mathbf{r}$ is the position of the particle at time t and $\dot{\mathbf{r}} = d\mathbf{r}/dt$, etc. Write this as three separate equations in terms of the Cartesian components of the vectors involved.

For the simple case of crossed uniform fields $\mathbf{E} = E\mathbf{i}$, $\mathbf{B} = B\mathbf{j}$, in which the particle starts from the origin at $t = 0$ with $\dot{\mathbf{r}} = v_0\mathbf{k}$, find the equations of motion and show the following:

(a) if $v_0 = E/B$ then the particle continues its initial motion;
(b) if $v_0 = 0$ then the particle follows the space curve given in terms of the parameter ξ by

$$x = \frac{mE}{B^2 q}(1 - \cos \xi), \qquad y = 0, \qquad z = \frac{mE}{B^2 q}(\xi - \sin \xi).$$

Interpret this curve geometrically and relate ξ to t. Show that the total distance traveled by the particle after time t is given by

$$\frac{2E}{B} \int_0^t \left| \sin \frac{Bqt'}{2m} \right| dt'.$$

Expressed in Cartesian coordinates, the components of the vector equation read

$$m\ddot{x} = qE_x + q(\dot{y}B_z - \dot{z}B_y),$$
$$m\ddot{y} = qE_y + q(\dot{z}B_x - \dot{x}B_z),$$
$$m\ddot{z} = qE_z + q(\dot{x}B_y - \dot{y}B_x).$$

For $E_x = E$, $B_y = B$ and all other field components zero, the equations reduce to

$$m\ddot{x} = qE - qB\dot{z}, \qquad m\ddot{y} = 0, \qquad m\ddot{z} = qB\dot{x}.$$

The second of these, together with the initial conditions $y(0) = \dot{y}(0) = 0$, implies that $y(t) = 0$ for all t. The final equation can be integrated directly to give

$$m\dot{z} = qBx + mv_0, \qquad (*)$$

which can now be substituted into the first to give a differential equation for x:

$$m\ddot{x} = qE - qB\left(\frac{qB}{m}x + v_0\right),$$

$$\Rightarrow \quad \ddot{x} + \left(\frac{qB}{m}\right)^2 x = \frac{q}{m}(E - v_0 B).$$

(i) If $v_0 = E/B$ then the equation for x is that of simple harmonic motion and

$$x(t) = A \cos \omega t + B \sin \omega t,$$

where $\omega = qB/m$. However, in the present case, the initial conditions $x(0) = \dot{x}(0) = 0$ imply that $x(t) = 0$ for all t. Thus, there is no motion in either the x- or the y-direction and, as is then shown by $(*)$, the particle continues with its initial speed v_0 in the z-direction.
  (ii) If $v_0 = 0$, the equation of motion is

$$\ddot{x} + \omega^2 x = \frac{qE}{m},$$

which again has sinusoidal solutions but has a non-zero RHS. The full solution consists of the same complementary function as in part (i) together with the simplest possible particular integral, namely $x = qE/m\omega^2$. It is therefore

$$x(t) = A \cos \omega t + B \sin \omega t + \frac{qE}{m\omega^2}.$$

The initial condition $x(0) = 0$ implies that $A = -qE/(m\omega^2)$, whilst $\dot{x}(0) = 0$ requires that $B = 0$. Thus,

$$x = \frac{qE}{m\omega^2}(1 - \cos \omega t),$$

$$\Rightarrow \quad \dot{z} = \frac{qB}{m}x = \omega\frac{qE}{m\omega^2}(1 - \cos \omega t) = \frac{qE}{m\omega}(1 - \cos \omega t).$$

Since $z(0) = 0$, straightforward integration gives

$$z = \frac{qE}{m\omega}\left(t - \frac{\sin \omega t}{\omega}\right) = \frac{qE}{m\omega^2}(\omega t - \sin \omega t).$$

Thus, since $qE/m\omega^2 = mE/B^2q$, the path is of the given parametric form with $\xi = \omega t$. It is a cycloid in the plane $y = 0$; the x-coordinate varies in the restricted range $0 \le x \le 2qE/(m\omega^2)$, whilst the z-coordinate continually increases, though not at a uniform rate.

The element of path length is given by $ds^2 = dx^2 + dy^2 + dz^2$. In this case, writing $qE/(m\omega) = E/B$ as μ,

$$ds = \left[\left(\frac{dx}{dt}\right)^2 + \left(\frac{dz}{dt}\right)^2\right]^{1/2} dt$$

$$= \left[\mu^2 \sin^2 \omega t + \mu^2(1 - \cos \omega t)^2\right]^{1/2} dt$$

$$= \left[2\mu^2(1 - \cos \omega t)\right]^{1/2} dt = 2\mu|\sin \tfrac{1}{2}\omega t|\, dt.$$

Thus the total distance traveled after time t is given by

$$s = \int_0^t 2\mu|\sin \tfrac{1}{2}\omega t'|\, dt' = \frac{2E}{B}\int_0^t \left|\sin \frac{qBt'}{2m}\right| dt'.$$

2.5 If two systems of coordinates with a common origin O are rotating with respect to each other, the measured accelerations differ in the two systems. Denoting by $\mathbf{r}$ and $\mathbf{r}'$ position vectors in frames $OXYZ$ and $OX'Y'Z'$, respectively, the connection between the two is

$$\ddot{\mathbf{r}}' = \ddot{\mathbf{r}} + \dot{\boldsymbol{\omega}} \times \mathbf{r} + 2\boldsymbol{\omega} \times \dot{\mathbf{r}} + \boldsymbol{\omega} \times (\boldsymbol{\omega} \times \mathbf{r}),$$

where $\boldsymbol{\omega}$ is the angular velocity vector of the rotation of $OXYZ$ with respect to $OX'Y'Z'$ (taken as fixed). The third term on the RHS is known as the Coriolis acceleration, whilst the final term gives rise to a centrifugal force.

Consider the application of this result to the firing of a shell of mass m from a stationary ship on the steadily rotating earth, working to the first order in ω ($= 7.3 \times 10^{-5}$ rad s^{-1}). If the shell is fired with velocity $\mathbf{v}$ at time $t = 0$ and only reaches a height that is small compared with the radius of the earth, show that its acceleration, as recorded on the ship, is given approximately by

$$\ddot{\mathbf{r}} = \mathbf{g} - 2\boldsymbol{\omega} \times (\mathbf{v} + \mathbf{g}t),$$

where $m\mathbf{g}$ is the weight of the shell measured on the ship's deck.

The shell is fired at another stationary ship (a distance $\mathbf{s}$ away) and $\mathbf{v}$ is such that the shell would have hit its target had there been no Coriolis effect.

(a) Show that without the Coriolis effect the time of flight of the shell would have been $\tau = -2\mathbf{g} \cdot \mathbf{v}/g^2$.

(b) Show further that when the shell actually hits the sea it is off-target by approximately

$$\frac{2\tau}{g^2}[(\mathbf{g} \times \boldsymbol{\omega}) \cdot \mathbf{v}](\mathbf{g}\tau + \mathbf{v}) - (\boldsymbol{\omega} \times \mathbf{v})\tau^2 - \frac{1}{3}(\boldsymbol{\omega} \times \mathbf{g})\tau^3.$$

(c) Estimate the order of magnitude Δ of this miss for a shell for which the initial speed v is $300\,\mathrm{m\,s^{-1}}$, firing close to its maximum range ($\mathbf{v}$ makes an angle of $\pi/4$ with the vertical) in a northerly direction, whilst the ship is stationed at latitude $45°$ North.

As the earth is rotating steadily $\dot{\boldsymbol{\omega}} = \mathbf{0}$, and for the mass at rest on the deck,

$$m\ddot{\mathbf{r}}' = m\mathbf{g} + \mathbf{0} + 2\boldsymbol{\omega} \times \dot{\mathbf{0}} + m\boldsymbol{\omega} \times (\boldsymbol{\omega} \times \mathbf{r}).$$

This, including the centrifugal effect, defines $\mathbf{g}$ which is assumed constant throughout the trajectory.

For the moving mass ($\ddot{\mathbf{r}}'$ is unchanged),

$$m\mathbf{g} + \boldsymbol{\omega} \times (\boldsymbol{\omega} \times \mathbf{r}) = m\ddot{\mathbf{r}} + 2m\boldsymbol{\omega} \times \dot{\mathbf{r}} + m\boldsymbol{\omega} \times (\boldsymbol{\omega} \times \mathbf{r}),$$

i.e.
$$\ddot{\mathbf{r}} = \mathbf{g} - 2\boldsymbol{\omega} \times \dot{\mathbf{r}}.$$

Now, $\omega\dot{r} \ll g$ and so to zeroth order in ω

$$\ddot{\mathbf{r}} = \mathbf{g} \quad \Rightarrow \quad \dot{\mathbf{r}} = \mathbf{g}t + \mathbf{v}.$$

Resubstituting this into the Coriolis term gives, to first order in ω,

$$\ddot{\mathbf{r}} = \mathbf{g} - 2\boldsymbol{\omega} \times (\mathbf{v} + \mathbf{g}t).$$

(a) With no Coriolis force,

$$\dot{\mathbf{r}} = \mathbf{g}t + \mathbf{v} \quad \text{and} \quad \mathbf{r} = \tfrac{1}{2}\mathbf{g}t^2 + \mathbf{v}t.$$

Let $\mathbf{s} = \tfrac{1}{2}\mathbf{g}\tau^2 + \mathbf{v}\tau$ and use the observation that $\mathbf{s} \cdot \mathbf{g} = 0$, giving

$$\tfrac{1}{2}g^2\tau^2 + \mathbf{v} \cdot \mathbf{g}\tau = 0 \quad \Rightarrow \quad \tau = -\frac{2\mathbf{v} \cdot \mathbf{g}}{g^2}.$$

(b) With Coriolis force,

$$\ddot{\mathbf{r}} = \mathbf{g} - 2(\boldsymbol{\omega} \times \mathbf{g})t - 2(\boldsymbol{\omega} \times \mathbf{v}),$$
$$\dot{\mathbf{r}} = \mathbf{g}t - (\boldsymbol{\omega} \times \mathbf{g})t^2 - 2(\boldsymbol{\omega} \times \mathbf{v})t + \mathbf{v},$$
$$\mathbf{r} = \tfrac{1}{2}\mathbf{g}t^2 - \tfrac{1}{3}(\boldsymbol{\omega} \times \mathbf{g})t^3 - (\boldsymbol{\omega} \times \mathbf{v})t^2 + \mathbf{v}t. \qquad (*)$$

If the shell hits the sea at time T in the position $\mathbf{r} = \mathbf{s} + \boldsymbol{\Delta}$, then $(\mathbf{s} + \boldsymbol{\Delta}) \cdot \mathbf{g} = 0$, i.e.

$$0 = (\mathbf{s} + \boldsymbol{\Delta}) \cdot \mathbf{g} = \tfrac{1}{2}g^2 T^2 - 0 - (\boldsymbol{\omega} \times \mathbf{v}) \cdot \mathbf{g} T^2 + \mathbf{v} \cdot \mathbf{g} T,$$
$$\Rightarrow \quad -\mathbf{v} \cdot \mathbf{g} = T(\tfrac{1}{2}g^2 - (\boldsymbol{\omega} \times \mathbf{v}) \cdot \mathbf{g}),$$

$$\Rightarrow \quad T = -\frac{\mathbf{v} \cdot \mathbf{g}}{\tfrac{1}{2}g^2}\left[1 - \frac{(\boldsymbol{\omega} \times \mathbf{v}) \cdot \mathbf{g}}{\tfrac{1}{2}g^2}\right]^{-1}$$

$$\approx \tau\left(1 + \frac{2(\boldsymbol{\omega} \times \mathbf{v}) \cdot \mathbf{g}}{g^2} + \cdots\right).$$

Working to first order in ω, we may put $T = \tau$ in those terms in (∗) that involve another factor ω, namely $\omega \times \mathbf{v}$ and $\omega \times \mathbf{g}$. We then find, to this order, that

$$
\mathbf{s} + \Delta = \frac{1}{2}\mathbf{g}\left(\tau^2 + \frac{4(\omega \times \mathbf{v}) \cdot \mathbf{g}}{g^2}\tau^2 + \cdots\right) - \frac{1}{3}(\omega \times \mathbf{g})\tau^3
$$
$$
- (\omega \times \mathbf{v})\tau^2 + \mathbf{v}\tau + 2\frac{(\omega \times \mathbf{v}) \cdot \mathbf{g}}{g^2}\mathbf{v}\tau
$$
$$
= \mathbf{s} + \frac{(\omega \times \mathbf{v}) \cdot \mathbf{g}}{g^2}(2g\tau^2 + 2\mathbf{v}\tau) - \frac{1}{3}(\omega \times \mathbf{g})\tau^3 - (\omega \times \mathbf{v})\tau^2.
$$

Hence, as stated in the question,

$$
\Delta = \frac{2\tau}{g^2}[(\mathbf{g} \times \omega) \cdot \mathbf{v}](\mathbf{g}\tau + \mathbf{v}) - (\omega \times \mathbf{v})\tau^2 - \frac{1}{3}(\omega \times \mathbf{g})\tau^3.
$$

(c) With the ship at latitude 45° and firing the shell at close to 45° to the local horizontal, $\mathbf{v}$ and ω are almost parallel and the $\omega \times \mathbf{v}$ term can be set to zero. Further, with $\mathbf{v}$ in a northerly direction, $(\mathbf{g} \times \omega) \cdot \mathbf{v} = 0$.

Thus we are left with only the cubic term in τ. In this,

$$
\tau = \frac{2 \times 300 \cos(\pi/4)}{9.8} = 43.3 \text{ s},
$$

and $\omega \times \mathbf{g}$ is in a westerly direction (recall that ω is directed northwards and $\mathbf{g}$ is directed downwards, towards the origin) and of magnitude $7\ 10^{-5} \times 9.8 \times \sin(\pi/4) = 4.85\ 10^{-4}$ m s^{-3}. Thus the miss is by approximately

$$
-\tfrac{1}{3} \times 4.85\ 10^{-4} \times (43.3)^3 = -13 \text{ m},
$$

i.e. some 10–15 m to the East of its intended target.

2.7 Parameterizing the hyperboloid

$$
\frac{x^2}{a^2} + \frac{y^2}{b^2} - \frac{z^2}{c^2} = 1
$$

by $x = a \cos\theta \sec\phi$, $y = b \sin\theta \sec\phi$, $z = c \tan\phi$, show that an area element on its surface is

$$
dS = \sec^2\phi \left[c^2 \sec^2\phi \left(b^2 \cos^2\theta + a^2 \sin^2\theta\right) + a^2b^2 \tan^2\phi\right]^{1/2} d\theta\, d\phi.
$$

Use this formula to show that the area of the curved surface $x^2 + y^2 - z^2 = a^2$ between the planes $z = 0$ and $z = 2a$ is

$$
\pi a^2 \left(6 + \frac{1}{\sqrt{2}}\sinh^{-1} 2\sqrt{2}\right).
$$

With $x = a \cos\theta \sec\phi$, $y = b \sin\theta \sec\phi$ and $z = c \tan\phi$, the tangent vectors to the surface are given in Cartesian coordinates by

$$
\frac{d\mathbf{r}}{d\theta} = (-a \sin\theta \sec\phi,\ b \cos\theta \sec\phi,\ 0),
$$
$$
\frac{d\mathbf{r}}{d\phi} = (a \cos\theta \sec\phi \tan\phi,\ b \sin\theta \sec\phi \tan\phi,\ c \sec^2\phi),
$$

and the element of area by

$$dS = \left| \frac{d\mathbf{r}}{d\theta} \times \frac{d\mathbf{r}}{d\phi} \right| d\theta\, d\phi$$
$$= \left| (bc \cos\theta \sec^3\phi,\ ac \sin\theta \sec^3\phi,\ -ab \sec^2\phi \tan\phi) \right| d\theta\, d\phi$$
$$= \sec^2\phi \left[c^2 \sec^2\phi \left(b^2 \cos^2\theta + a^2 \sin^2\theta \right) + a^2 b^2 \tan^2\phi \right]^{1/2} d\theta\, d\phi.$$

We set $b = c = a$ and note that the plane $z = 2a$ corresponds to $\phi = \tan^{-1} 2$. The ranges of integration are therefore $0 \le \theta < 2\pi$ and $0 \le \phi \le \tan^{-1} 2$, whilst

$$dS = \sec^2\phi (a^4 \sec^2\phi + a^4 \tan^2\phi)^{1/2}\, d\theta\, d\phi,$$

i.e. it is independent of θ.

To evaluate the integral of dS, we set $\tan\phi = \sinh\psi/\sqrt{2}$, with

$$\sec^2\phi\, d\phi = \frac{1}{\sqrt{2}} \cosh\psi\, d\psi \quad \text{and} \quad \sec^2\phi = 1 + \tfrac{1}{2} \sinh^2\psi.$$

The upper limit for ψ will be given by $\Psi = \sinh^{-1} 2\sqrt{2}$; we note that $\cosh\Psi = 3$. Integrating over θ and making the above substitutions yields

$$S = 2\pi \int_0^\Psi \frac{1}{\sqrt{2}} \cosh\psi\, d\psi\, a^2 \left(1 + \frac{1}{2} \sinh^2\psi + \frac{1}{2} \sinh^2\psi \right)^{1/2}$$
$$= \sqrt{2}\pi a^2 \int_0^\Psi \cosh^2\psi\, d\psi$$
$$= \frac{\sqrt{2}\pi a^2}{2} \int_0^\Psi (\cosh 2\psi + 1)\, d\psi$$
$$= \frac{\sqrt{2}\pi a^2}{2} \left[\frac{\sinh 2\psi}{2} + \psi \right]_0^\Psi$$
$$= \frac{\pi a^2}{\sqrt{2}} [\sinh\psi \cosh\psi + \psi]_0^\Psi$$
$$= \frac{\pi a^2}{\sqrt{2}} [(2\sqrt{2})(3) + \sinh^{-1} 2\sqrt{2}] = \pi a^2 \left(6 + \frac{1}{\sqrt{2}} \sinh^{-1} 2\sqrt{2} \right).$$

2.9 Verify by direct calculation that

$$\nabla \cdot (\mathbf{a} \times \mathbf{b}) = \mathbf{b} \cdot (\nabla \times \mathbf{a}) - \mathbf{a} \cdot (\nabla \times \mathbf{b}).$$

The proof of this standard result for the divergence of a vector product is most easily carried out in Cartesian coordinates though, of course, the result is valid in any three-dimensional coordinate system.

$$\text{LHS} = \nabla \cdot (\mathbf{a} \times \mathbf{b})$$

$$= \frac{\partial}{\partial x}(a_y b_z - a_z b_y) + \frac{\partial}{\partial y}(a_z b_x - a_x b_z) + \frac{\partial}{\partial z}(a_x b_y - a_y b_x)$$

$$= a_x \left(-\frac{\partial b_z}{\partial y} + \frac{\partial b_y}{\partial z} \right) + a_y \left(\frac{\partial b_z}{\partial x} - \frac{\partial b_x}{\partial z} \right) + a_z \left(-\frac{\partial b_y}{\partial x} + \frac{\partial b_x}{\partial y} \right)$$

$$+ b_x \left(\frac{\partial a_z}{\partial y} - \frac{\partial a_y}{\partial z} \right) + b_y \left(-\frac{\partial a_z}{\partial x} + \frac{\partial a_x}{\partial z} \right) + b_z \left(\frac{\partial a_y}{\partial x} - \frac{\partial a_x}{\partial y} \right)$$

$$= -\mathbf{a} \cdot (\nabla \times \mathbf{b}) + \mathbf{b} \cdot (\nabla \times \mathbf{a}) = \text{RHS}.$$

2.11 Evaluate the Laplacian of the function

$$\psi(x, y, z) = \frac{zx^2}{x^2 + y^2 + z^2}$$

(a) directly in Cartesian coordinates, and (b) after changing to a spherical polar coordinate system. Verify that, as they must, the two methods give the same result.

(a) In Cartesian coordinates we need to evaluate

$$\nabla^2 \psi = \frac{\partial^2 \psi}{\partial x^2} + \frac{\partial^2 \psi}{\partial y^2} + \frac{\partial^2 \psi}{\partial z^2}.$$

The required derivatives are

$$\frac{\partial \psi}{\partial x} = \frac{2xz(y^2 + z^2)}{(x^2 + y^2 + z^2)^2}, \qquad \frac{\partial^2 \psi}{\partial x^2} = \frac{(y^2 + z^2)(2zy^2 + 2z^3 - 6x^2 z)}{(x^2 + y^2 + z^2)^3},$$

$$\frac{\partial \psi}{\partial y} = \frac{-2x^2 yz}{(x^2 + y^2 + z^2)^2}, \qquad \frac{\partial^2 \psi}{\partial y^2} = -\frac{2zx^2(x^2 + z^2 - 3y^2)}{(x^2 + y^2 + z^2)^3},$$

$$\frac{\partial \psi}{\partial z} = \frac{x^2(x^2 + y^2 - z^2)}{(x^2 + y^2 + z^2)^2}, \qquad \frac{\partial^2 \psi}{\partial z^2} = -\frac{2zx^2(3x^2 + 3y^2 - z^2)}{(x^2 + y^2 + z^2)^3}.$$

Thus, writing $r^2 = x^2 + y^2 + z^2$,

$$\nabla^2 \psi = \frac{2z[(y^2 + z^2)(y^2 + z^2 - 3x^2) - 4x^4]}{(x^2 + y^2 + z^2)^3}$$

$$= \frac{2z[(r^2 - x^2)(r^2 - 4x^2) - 4x^4]}{r^6}$$

$$= \frac{2z(r^2 - 5x^2)}{r^4}.$$

(b) In spherical polar coordinates,

$$\psi(r, \theta, \phi) = \frac{r \cos\theta\, r^2 \sin^2\theta \cos^2\phi}{r^2} = r \cos\theta \sin^2\theta \cos^2\phi.$$

The three contributions to $\nabla^2 \psi$ in spherical polars are

$$(\nabla^2 \psi)_r = \frac{1}{r^2} \frac{\partial}{\partial r} \left(r^2 \frac{\partial \psi}{\partial r} \right)$$

$$= \frac{2}{r} \cos\theta \sin^2\theta \cos^2\phi,$$

$$(\nabla^2 \psi)_\theta = \frac{1}{r^2 \sin\theta} \frac{\partial}{\partial\theta} \left(\sin\theta \frac{\partial \psi}{\partial\theta} \right)$$

$$= \frac{1}{r} \frac{\cos^2\phi}{\sin\theta} \frac{\partial}{\partial\theta} \left[\sin\theta \frac{\partial}{\partial\theta} (\cos\theta \sin^2\theta) \right]$$

$$= \frac{\cos^2\phi}{r} (4\cos^3\theta - 8\sin^2\theta \cos\theta),$$

$$(\nabla^2 \psi)_\phi = \frac{1}{r^2 \sin^2\theta} \frac{\partial^2 \psi}{\partial\phi^2}$$

$$= \frac{\cos\theta}{r} (-2\cos^2\phi + 2\sin^2\phi).$$

Thus, the full Laplacian in spherical polar coordinates reads

$$\nabla^2 \psi = \frac{\cos\theta}{r} (2\sin^2\theta \cos^2\phi + 4\cos^2\theta \cos^2\phi$$

$$- 8\sin^2\theta \cos^2\phi - 2\cos^2\phi + 2\sin^2\phi)$$

$$= \frac{\cos\theta}{r} (4\cos^2\phi - 10\sin^2\theta \cos^2\phi - 2\cos^2\phi + 2\sin^2\phi)$$

$$= \frac{\cos\theta}{r} (2 - 10\sin^2\theta \cos^2\phi)$$

$$= \frac{2r\cos\theta(r^2 - 5r^2 \sin^2\theta \cos^2\phi)}{r^4}.$$

Rewriting this last expression in terms of Cartesian coordinates, one finally obtains

$$\nabla^2 \psi = \frac{2z(r^2 - 5x^2)}{r^4},$$

which establishes the equivalence of the two approaches.

2.13 The (Maxwell) relationship between a time-independent magnetic field $\mathbf{B}$ and the current density $\mathbf{J}$ (measured in SI units in $A\,m^{-2}$) producing it,

$$\nabla \times \mathbf{B} = \mu_0 \mathbf{J},$$

can be applied to a long cylinder of conducting ionized gas which, in cylindrical polar coordinates, occupies the region $\rho < a$.

(a) Show that a uniform current density $(0, C, 0)$ and a magnetic field $(0, 0, B)$, with B constant ($= B_0$) for $\rho > a$ and $B = B(\rho)$ for $\rho < a$, are consistent with this equation. Given that $B(0) = 0$ and that $\mathbf{B}$ is continuous at $\rho = a$, obtain expressions for C and $B(\rho)$ in terms of B_0 and a.

(b) The magnetic field can be expressed as $\mathbf{B} = \nabla \times \mathbf{A}$, where $\mathbf{A}$ is known as the vector potential. Show that a suitable $\mathbf{A}$ can be found which has only one non-vanishing component, $A_\phi(\rho)$, and obtain explicit expressions for $A_\phi(\rho)$ for both $\rho < a$ and $\rho > a$. Like $\mathbf{B}$, the vector potential is continuous at $\rho = a$.

(c) The gas pressure $p(\rho)$ satisfies the hydrostatic equation $\nabla p = \mathbf{J} \times \mathbf{B}$ and vanishes at the outer wall of the cylinder. Find a general expression for p.

(a) In cylindrical polars with $\mathbf{B} = (0, 0, B(\rho))$, for $\rho \leq a$ we have

$$\mu_0(0, \; C, \; 0) = \nabla \times \mathbf{B} = \left(\frac{1}{\rho}\frac{\partial B}{\partial \phi}, \; -\frac{\partial B}{\partial \rho}, \; 0 \right).$$

As expected, $\partial B / \partial \phi = 0$. The azimuthal component of the equation gives

$$-\frac{\partial B}{\partial \rho} = \mu_0 C \quad \text{for} \quad \rho \leq a \quad \Rightarrow \quad B(\rho) = B(0) - \mu_0 C \rho.$$

Since $\mathbf{B}$ has to be differentiable at the origin of ρ and have no ϕ-dependence, $B(0)$ must be zero. This, together with $B = B_0$ for $\rho > a$ requires that $C = -B_0/(a\mu_0)$ and $B(\rho) = B_0 \rho / a$ for $0 \leq \rho \leq a$.

(b) With $\mathbf{B} = \nabla \times \mathbf{A}$, consider $\mathbf{A}$ of the form $\mathbf{A} = (0, \; A(\rho), \; 0)$. Then

$$(0, \; 0, \; B(\rho)) = \frac{1}{\rho}\left(\frac{\partial}{\partial z}(\rho A), \; 0, \; \frac{\partial}{\partial \rho}(\rho A) \right)$$

$$= \left(0, \; 0, \; \frac{1}{\rho}\frac{\partial}{\partial \rho}(\rho A) \right).$$

We now equate the only non-vanishing component on each side of the above equation, treating inside and outside the cylinder separately.

For $0 < \rho \leq a$,

$$\frac{1}{\rho}\frac{\partial}{\partial \rho}(\rho A) = \frac{B_0 \rho}{a},$$

$$\rho A = \frac{B_0 \rho^3}{3a} + D,$$

$$A(\rho) = \frac{B_0 \rho^2}{3a} + \frac{D}{\rho}.$$

Since $A(0)$ must be finite (so that A is differentiable there), $D = 0$.

For $\rho > a$,

$$\frac{1}{\rho}\frac{\partial}{\partial \rho}(\rho A) = B_0,$$

$$\rho A = \frac{B_0 \rho^2}{2} + E,$$

$$A(\rho) = \frac{1}{2}B_0 \rho + \frac{E}{\rho}.$$

At $\rho = a$, the continuity of $\mathbf{A}$ requires

$$\frac{B_0 a^2}{3a} = \frac{1}{2} B_0 a + \frac{E}{a} \Rightarrow E = -\frac{B_0 a^2}{6}.$$

Thus, to summarize,

$$A(\rho) = \frac{B_0 \rho^2}{3a} \quad \text{for} \quad 0 \le \rho \le a,$$

$$\text{and} \quad A(\rho) = B_0 \left(\frac{\rho}{2} - \frac{a^2}{6\rho} \right) \quad \text{for} \quad \rho \ge a.$$

(c) For the gas pressure $p(\rho)$ in the region $0 < \rho \le a$, we have $\nabla p = \mathbf{J} \times \mathbf{B}$. In component form,

$$\left(\frac{dp}{d\rho}, 0, 0 \right) = \left(0, -\frac{B_0}{a\mu_0}, 0 \right) \times \left(0, 0, \frac{B_0 \rho}{a} \right),$$

with $p(a) = 0$.

$$\frac{dp}{d\rho} = -\frac{B_0^2 \rho}{\mu_0 a^2} \quad \Rightarrow \quad p(\rho) = \frac{B_0^2}{2\mu_0} \left[1 - \left(\frac{\rho}{a} \right)^2 \right].$$

2.15 Maxwell's equations for electromagnetism in free space (i.e. in the absence of charges, currents and dielectric or magnetic media) can be written

(i) $\nabla \cdot \mathbf{B} = 0$,   (ii) $\nabla \cdot \mathbf{E} = 0$,

(iii) $\nabla \times \mathbf{E} + \dfrac{\partial \mathbf{B}}{\partial t} = \mathbf{0}$,   (iv) $\nabla \times \mathbf{B} - \dfrac{1}{c^2} \dfrac{\partial \mathbf{E}}{\partial t} = \mathbf{0}$.

A vector $\mathbf{A}$ is defined by $\mathbf{B} = \nabla \times \mathbf{A}$, and a scalar ϕ by $\mathbf{E} = -\nabla \phi - \partial \mathbf{A}/\partial t$. Show that if the condition

$$(\text{v}) \quad \nabla \cdot \mathbf{A} + \frac{1}{c^2} \frac{\partial \phi}{\partial t} = 0$$

is imposed (this is known as choosing the Lorentz gauge), then $\mathbf{A}$ and ϕ satisfy wave equations as follows.

$$(\text{vi}) \quad \nabla^2 \phi - \frac{1}{c^2} \frac{\partial^2 \phi}{\partial t^2} = 0,$$

$$(\text{vii}) \quad \nabla^2 \mathbf{A} - \frac{1}{c^2} \frac{\partial^2 \mathbf{A}}{\partial t^2} = \mathbf{0}.$$

The reader is invited to proceed as follows.

(a) Verify that the expressions for $\mathbf{B}$ and $\mathbf{E}$ in terms of $\mathbf{A}$ and ϕ are consistent with (i) and (iii).
(b) Substitute for $\mathbf{E}$ in (ii) and use the derivative with respect to time of (v) to eliminate $\mathbf{A}$ from the resulting expression. Hence obtain (vi).
(c) Substitute for $\mathbf{B}$ and $\mathbf{E}$ in (iv) in terms of $\mathbf{A}$ and ϕ. Then use the gradient of (v) to simplify the resulting equation and so obtain (vii).

(a) Substituting for $\mathbf{B}$ in (i),

$$\nabla \cdot \mathbf{B} = \nabla \cdot (\nabla \times \mathbf{A}) = 0, \quad \text{as it is for } \textit{any} \text{ vector } \mathbf{A}.$$

Substituting for **E** and **B** in (iii),

$$\nabla \times \mathbf{E} + \frac{\partial \mathbf{B}}{\partial t} = -(\nabla \times \nabla \phi) - \nabla \times \frac{\partial \mathbf{A}}{\partial t} + \frac{\partial}{\partial t}(\nabla \times \mathbf{A}) = 0.$$

Here we have used the facts that $\nabla \times \nabla \phi = 0$ for any scalar, and that, since $\partial/\partial t$ and ∇ act on different variables, the order in which they are applied to **A** can be reversed. Thus (i) and (iii) are automatically satisfied if **E** and **B** are represented in terms of **A** and ϕ.

(b) Substituting for **E** in (ii) and taking the time derivative of (v),

$$0 = \nabla \cdot \mathbf{E} = -\nabla^2 \phi - \frac{\partial}{\partial t}(\nabla \cdot \mathbf{A}),$$

$$0 = \frac{\partial}{\partial t}(\nabla \cdot \mathbf{A}) + \frac{1}{c^2}\frac{\partial^2 \phi}{\partial t^2}.$$

Adding these equations gives

$$0 = -\nabla^2 \phi + \frac{1}{c^2}\frac{\partial^2 \phi}{\partial t^2}.$$

This is result (vi), the wave equation for ϕ.

(c) Substituting for **B** and **E** in (iv) and taking the gradient of (v),

$$\nabla \times (\nabla \times \mathbf{A}) - \frac{1}{c^2}\left(-\frac{\partial}{\partial t}\nabla\phi - \frac{\partial^2 \mathbf{A}}{\partial t^2}\right) = 0,$$

$$\nabla(\nabla \cdot \mathbf{A}) - \nabla^2 \mathbf{A} + \frac{1}{c^2}\frac{\partial}{\partial t}(\nabla\phi) + \frac{1}{c^2}\frac{\partial^2 \mathbf{A}}{\partial t^2} = 0.$$

From (v),
$$\nabla(\nabla \cdot \mathbf{A}) + \frac{1}{c^2}\frac{\partial}{\partial t}(\nabla\phi) = 0.$$

Subtracting these gives
$$-\nabla^2 \mathbf{A} + \frac{1}{c^2}\frac{\partial^2 \mathbf{A}}{\partial t^2} = 0.$$

In the second line we have used the vector identity

$$\nabla^2 \mathbf{F} = \nabla(\nabla \cdot \mathbf{F}) - \nabla \times (\nabla \times \mathbf{F})$$

to replace $\nabla \times (\nabla \times \mathbf{A})$. The final equation is result (vii).

2.17 Paraboloidal coordinates u, v, ϕ are defined in terms of Cartesian coordinates by

$$x = uv\cos\phi, \qquad y = uv\sin\phi, \qquad z = \tfrac{1}{2}(u^2 - v^2).$$

Identify the coordinate surfaces in the u, v, ϕ system. Verify that each coordinate surface ($u =$ constant, say) intersects every coordinate surface on which one of the other two coordinates (v, say) is constant. Show further that the system of coordinates is an orthogonal one and determine its scale factors. Prove that the u-component of $\nabla \times \mathbf{a}$ is given by

$$\frac{1}{(u^2 + v^2)^{1/2}}\left(\frac{a_\phi}{v} + \frac{\partial a_\phi}{\partial v}\right) - \frac{1}{uv}\frac{\partial a_v}{\partial \phi}.$$

To find a surface of constant u we eliminate v from the given relationships:

$$x^2 + y^2 = u^2 v^2 \quad \Rightarrow \quad 2z = u^2 - \frac{x^2 + y^2}{u^2}.$$

This is an inverted paraboloid of revolution about the z-axis. The range of z is $-\infty < z \leq \frac{1}{2}u^2$.

Similarly, the surface of constant v is given by

$$2z = \frac{x^2 + y^2}{v^2} - v^2.$$

This is also a paraboloid of revolution about the z-axis, but this time it is not inverted. The range of z is $-\frac{1}{2}v^2 \leq z < \infty$.

Since every constant-u paraboloid has some part of its surface in the region $z > 0$ and every constant-v paraboloid has some part of its surface in the region $z < 0$, it follows that every member of the first set intersects each member of the second, and vice versa.

The surfaces of constant ϕ, $y = x \tan \phi$, are clearly (half-) planes containing the z-axis; each cuts the members of the other two sets in parabolic lines.

We now determine (the Cartesian components of) the tangential vectors and test their orthogonality:

$$\mathbf{e}_1 = \frac{\partial \mathbf{r}}{\partial u} = (v \cos \phi, \ v \sin \phi, \ u),$$

$$\mathbf{e}_2 = \frac{\partial \mathbf{r}}{\partial v} = (u \cos \phi, \ u \sin \phi, \ -v),$$

$$\mathbf{e}_3 = \frac{\partial \mathbf{r}}{\partial \phi} = (-uv \sin \phi, \ uv \cos \phi, \ 0),$$

$$\mathbf{e}_1 \cdot \mathbf{e}_2 = uv(\cos \phi \cos \phi + \sin \phi \sin \phi) - uv = 0,$$

$$\mathbf{e}_2 \cdot \mathbf{e}_3 = u^2 v(-\cos \phi \sin \phi + \sin \phi \cos \phi) = 0,$$

$$\mathbf{e}_1 \cdot \mathbf{e}_3 = uv^2(-\cos \phi \sin \phi + \sin \phi \cos \phi) = 0.$$

This shows that all pairs of tangential vectors are orthogonal and therefore that the coordinate system is an orthogonal one. Its scale factors are given by the magnitudes of these tangential vectors:

$$h_u^2 = |\mathbf{e}_1|^2 = (v \cos \phi)^2 + (v \sin \phi)^2 + u^2 = u^2 + v^2,$$

$$h_v^2 = |\mathbf{e}_2|^2 = (u \cos \phi)^2 + (u \sin \phi)^2 + v^2 = u^2 + v^2,$$

$$h_\phi^2 = |\mathbf{e}_3|^2 = (uv \sin \phi)^2 + (uv \cos \phi)^2 = u^2 v^2.$$

Thus

$$h_u = h_v = \sqrt{u^2 + v^2}, \qquad h_\phi = uv.$$

The u-component of $\nabla \times \mathbf{a}$ is given by

$$[\nabla \times \mathbf{a}]_u = \frac{h_u}{h_u h_v h_\phi} \left[\frac{\partial}{\partial v} (h_\phi a_\phi) - \frac{\partial}{\partial \phi} (h_v a_v) \right]$$

$$= \frac{1}{uv \sqrt{u^2 + v^2}} \left[\frac{\partial}{\partial v} (uv a_\phi) - \frac{\partial}{\partial \phi} (\sqrt{u^2 + v^2} a_v) \right]$$

$$= \frac{1}{\sqrt{u^2 + v^2}} \left(\frac{a_\phi}{v} + \frac{\partial a_\phi}{\partial v} \right) - \frac{1}{uv} \frac{\partial a_v}{\partial \phi},$$

as stated in the question.

2.19 Hyperbolic coordinates u, v, ϕ are defined in terms of Cartesian coordinates by

$$x = \cosh u \cos v \cos \phi, \qquad y = \cosh u \cos v \sin \phi, \qquad z = \sinh u \sin v.$$

Sketch the coordinate curves in the $\phi = 0$ plane, showing that far from the origin they become concentric circles and radial lines. In particular, identify the curves $u = 0$, $v = 0$, $v = \pi/2$ and $v = \pi$. Calculate the tangent vectors at a general point, show that they are mutually orthogonal and deduce that the appropriate scale factors are

$$h_u = h_v = (\cosh^2 u - \cos^2 v)^{1/2}, \qquad h_\phi = \cosh u \cos v.$$

Find the most general function $\psi(u)$ of u only that satisfies Laplace's equation $\nabla^2 \psi = 0$.

In the plane $\phi = 0$, i.e. $y = 0$, the curves $u = $ constant have x and z connected by

$$\frac{x^2}{\cosh^2 u} + \frac{z^2}{\sinh^2 u} = 1.$$

This general form is that of an ellipse, with foci at $(\pm 1, 0)$. With $u = 0$, it is the line joining the two foci (covered twice). As $u \to \infty$, and $\cosh u \approx \sinh u$ the form becomes that of a circle of very large radius.

The curves $v = $ constant are expressed by

$$\frac{x^2}{\cos^2 v} - \frac{z^2}{\sin^2 v} = 1.$$

These curves are hyperbolae that, for large x and z and fixed v, approximate $z = \pm x \tan v$, i.e. radial lines. The curve $v = 0$ is the part of the x-axis $1 \le x \le \infty$ (covered twice), whilst the curve $v = \pi$ is its reflection in the z-axis. The curve $v = \pi/2$ is the z-axis.

In Cartesian coordinates a general point and its derivatives with respect to u, v and ϕ are given by

$$\mathbf{r} = \cosh u \cos v \cos \phi \, \mathbf{i} + \cosh u \cos v \sin \phi \, \mathbf{j} + \sinh u \sin v \, \mathbf{k},$$

$$\mathbf{e}_1 = \frac{\partial \mathbf{r}}{\partial u} = \sinh u \cos v \cos \phi \, \mathbf{i} + \sinh u \cos v \sin \phi \, \mathbf{j} + \cosh u \sin v \, \mathbf{k},$$

$$\mathbf{e}_2 = \frac{\partial \mathbf{r}}{\partial v} = -\cosh u \sin v \cos \phi \, \mathbf{i} - \cosh u \sin v \sin \phi \, \mathbf{j} + \sinh u \cos v \, \mathbf{k},$$

$$\mathbf{e}_3 = \frac{\partial \mathbf{r}}{\partial \phi} = \cosh u \cos v(-\sin \phi \, \mathbf{i} + \cos \phi \, \mathbf{j}).$$

Now consider the scalar products:

$$\mathbf{e}_1 \cdot \mathbf{e}_2 = \sinh u \cos v \cosh u \sin v(-\cos^2 \phi - \sin^2 \phi + 1) = 0,$$

$$\mathbf{e}_1 \cdot \mathbf{e}_3 = \sinh u \cos^2 v \cosh u(-\sin \phi \cos \phi + \sin \phi \cos \phi) = 0,$$

$$\mathbf{e}_2 \cdot \mathbf{e}_3 = \cosh^2 u \sin v \cos v(\sin \phi \cos \phi - \sin \phi \cos \phi) = 0.$$

As each is zero, the system is an orthogonal one.

The scale factors are given by $|\mathbf{e}_i|$ and are thus found from:

$$\begin{aligned}
|\mathbf{e}_1|^2 &= \sinh^2 u \cos^2 v(\cos^2 \phi + \sin^2 \phi) + \cosh^2 u \sin^2 v \\
&= (\cosh^2 u - 1)\cos^2 v + \cosh^2 u(1 - \cos^2 v) \\
&= \cosh^2 u - \cos^2 v; \\
|\mathbf{e}_2|^2 &= \cosh^2 u \sin^2 v(\cos^2 \phi + \sin^2 \phi) + \sinh^2 u \cos^2 v \\
&= \cosh^2 u(1 - \cos^2 v) + (\cosh^2 u - 1)\cos^2 v \\
&= \cosh^2 u - \cos^2 v; \\
|\mathbf{e}_3|^2 &= \cosh^2 u \cos^2 v(\sin^2 \phi + \cos^2 \phi) = \cosh^2 u \cos^2 v.
\end{aligned}$$

The immediate deduction is that

$$h_u = h_v = (\cosh^2 u - \cos^2 v)^{1/2}, \qquad h_\phi = \cosh u \cos v.$$

An alternative form for h_u and h_v is $(\sinh^2 u + \sin^2 v)^{1/2}$.

If a solution of Laplace's equation is to be a function, $\psi(u)$, of u only, then all differentiation with respect to v and ϕ can be ignored. The expression for $\nabla^2\psi$ reduces to

$$\begin{aligned}
\nabla^2\psi &= \frac{1}{h_u h_v h_\phi}\left[\frac{\partial}{\partial u}\left(\frac{h_v h_\phi}{h_u}\frac{\partial\psi}{\partial u}\right)\right] \\
&= \frac{1}{\cosh u \cos v(\cosh^2 u - \cos^2 v)}\left[\frac{\partial}{\partial u}\left(\cosh u \cos v \frac{\partial\psi}{\partial u}\right)\right].
\end{aligned}$$

Laplace's equation itself is even simpler and reduces to

$$\frac{\partial}{\partial u}\left(\cosh u \frac{\partial\psi}{\partial u}\right) = 0.$$

This can be rewritten as

$$\frac{\partial\psi}{\partial u} = \frac{k}{\cosh u} = \frac{2k}{e^u + e^{-u}} = \frac{2ke^u}{e^{2u} + 1},$$

$$d\psi = \frac{Ae^u\, du}{1 + (e^u)^2} \qquad \Rightarrow \qquad \psi = B\tan^{-1}e^u + c.$$

This is the most general function of u only that satisfies Laplace's equation.

3 Line, surface and volume integrals

3.1 The vector field $\mathbf{F}$ is defined by

$$\mathbf{F} = 2xz\mathbf{i} + 2yz^2\mathbf{j} + (x^2 + 2y^2z - 1)\mathbf{k}.$$

Calculate $\nabla \times \mathbf{F}$ and deduce that $\mathbf{F}$ can be written $\mathbf{F} = \nabla\phi$. Determine the form of ϕ.

With $\mathbf{F}$ as given, we calculate the curl of $\mathbf{F}$ to see whether or not it is the zero vector:

$$\nabla \times \mathbf{F} = (4yz - 4yz, \ 2x - 2x, \ 0 - 0) = \mathbf{0}.$$

The fact that it is implies that $\mathbf{F}$ can be written as $\nabla\phi$ for some scalar ϕ.

The form of $\phi(x, y, z)$ is found by integrating, in turn, the components of $\mathbf{F}$ until consistency is achieved, i.e. until a ϕ is found that has partial derivatives equal to the corresponding components of $\mathbf{F}$:

$$2xz = F_x = \frac{\partial\phi}{\partial x} \quad \Rightarrow \quad \phi(x, y, z) = x^2z + g(y, z),$$

$$2yz^2 = F_y = \frac{\partial}{\partial y}[x^2z + g(y, z)] \quad \Rightarrow \quad g(y, z) = y^2z^2 + h(z),$$

$$x^2 + 2y^2z - 1 = F_z \quad = \quad \frac{\partial}{\partial z}[x^2z + y^2z^2 + h(z)]$$

$$\Rightarrow \quad h(z) = -z + k.$$

Hence, to within an unimportant constant, the form of ϕ is

$$\phi(x, y, z) = x^2z + y^2z^2 - z.$$

3.3 A vector field $\mathbf{F}$ is given by $\mathbf{F} = xy^2\mathbf{i} + 2\mathbf{j} + x\mathbf{k}$ and L is a path parameterized by $x = ct$, $y = c/t$, $z = d$ for the range $1 \le t \le 2$. Evaluate the three integrals

(a) $\displaystyle\int_L \mathbf{F}\, dt$, (b) $\displaystyle\int_L \mathbf{F}\, dy$, (c) $\displaystyle\int_L \mathbf{F} \cdot d\mathbf{r}$.

Although all three integrals are along the same path L, they are not necessarily of the same type. The vector or scalar nature of the integral is determined by that of the integrand when it is expressed in a form containing the infinitesimal dt.

(a) This is a vector integral and contains three separate integrations. We express each of the integrands in terms of t, according to the parameterization of the integration path

L, before integrating:

$$\int_L \mathbf{F}\,dt = \int_1^2 \left(\frac{c^3}{t}\mathbf{i} + 2\mathbf{j} + ct\,\mathbf{k}\right)dt$$

$$= \left[c^3 \ln t\,\mathbf{i} + 2t\,\mathbf{j} + \frac{1}{2}ct^2\,\mathbf{k}\right]_1^2$$

$$= c^3 \ln 2\,\mathbf{i} + 2\mathbf{j} + \frac{3}{2}c\,\mathbf{k}.$$

(b) This is a similar vector integral but here we must also replace the infinitesimal dy by the infinitesimal $-c\,dt/t^2$ before integrating:

$$\int_L \mathbf{F}\,dy = \int_1^2 \left(\frac{c^3}{t}\mathbf{i} + 2\mathbf{j} + ct\,\mathbf{k}\right)\left(\frac{-c}{t^2}\right)dt$$

$$= \left[\frac{c^4}{2t^2}\mathbf{i} + \frac{2c}{t}\mathbf{j} - c^2 \ln t\,\mathbf{k}\right]_1^2$$

$$= -\frac{3c^4}{8}\mathbf{i} - c\mathbf{j} - c^2 \ln 2\,\mathbf{k}.$$

(c) This is a scalar integral and before integrating we must take the scalar product of $\mathbf{F}$ with $d\mathbf{r} = dx\,\mathbf{i} + dy\,\mathbf{j} + dz\,\mathbf{k}$ to give a single integrand:

$$\int_L \mathbf{F}\cdot d\mathbf{r} = \int_1^2 \left(\frac{c^3}{t}\mathbf{i} + 2\mathbf{j} + ct\,\mathbf{k}\right)\cdot\left(c\mathbf{i} - \frac{c}{t^2}\mathbf{j} + 0\mathbf{k}\right)dt$$

$$= \int_1^2 \left(\frac{c^4}{t} - \frac{2c}{t^2}\right)dt$$

$$= \left[c^4 \ln t + \frac{2c}{t}\right]_1^2$$

$$= c^4 \ln 2 - c.$$

3.5 Determine the point of intersection P, in the first quadrant, of the two ellipses

$$\frac{x^2}{a^2} + \frac{y^2}{b^2} = 1 \quad \text{and} \quad \frac{x^2}{b^2} + \frac{y^2}{a^2} = 1.$$

Taking $b < a$, consider the contour L that bounds the area in the first quadrant that is common to the two ellipses. Show that the parts of L that lie along the coordinate axes contribute nothing to the line integral around L of $x\,dy - y\,dx$. Using a parameterization of each ellipse of the general form $x = X\cos\phi$ and $y = Y\sin\phi$, evaluate the two remaining line integrals and hence find the total area common to the two ellipses.

Note: The line integral of $x\,dy - y\,dx$ around a general closed convex contour is equal to twice the area enclosed by that contour.

From the symmetry of the equations under the interchange of x and y, the point P must have $x = y$. Thus,

$$x^2 \left(\frac{1}{a^2} + \frac{1}{b^2} \right) = 1 \quad \Rightarrow \quad x = \frac{ab}{(a^2 + b^2)^{1/2}}.$$

Denoting as curve C_1 the part of

$$\frac{x^2}{a^2} + \frac{y^2}{b^2} = 1$$

that lies on the boundary of the common region, we parameterize it by $x = a \cos \theta_1$ and $y = b \sin \theta_1$. Curve C_1 starts from P and finishes on the y-axis. At P,

$$a \cos \theta_1 = x = \frac{ab}{(a^2 + b^2)^{1/2}} \quad \Rightarrow \quad \tan \theta_1 = \frac{a}{b}.$$

It follows that θ_1 lies in the range $\tan^{-1}(a/b) \le \theta_1 \le \pi/2$. Note that θ_1 is *not* the angle between the x-axis and the line joining the origin O to the corresponding point on the curve; for example, when the point is P itself then $\theta_1 = \tan^{-1} a/b$, whilst the line OP makes an angle of $\pi/4$ with the x-axis.

Similarly, referring to that part of

$$\frac{x^2}{b^2} + \frac{y^2}{a^2} = 1$$

that lies on the boundary of the common region as curve C_2, we parameterize it by $x = b \cos \theta_2$ and $y = a \sin \theta_2$ with $0 \le \theta_2 \le \tan^{-1}(b/a)$.

On the x-axis, both y and dy are zero and the integrand, $x \, dy - y \, dx$, vanishes. Similarly, the integrand vanishes at all points on the y-axis. Hence,

$$I = \oint_L (x \, dy - y \, dx)$$

$$= \int_{C_2} (x \, dy - y \, dx) + \int_{C_1} (x \, dy - y \, dx)$$

$$= \int_0^{\tan^{-1}(b/a)} [\, ab(\cos \theta_2 \cos \theta_2) - ab \sin \theta_2(- \sin \theta_2) \,] \, d\theta_2$$

$$+ \int_{\tan^{-1}(a/b)}^{\pi/2} [\, ab(\cos \theta_1 \cos \theta_1) - ab \sin \theta_1(- \sin \theta_1) \,] \, d\theta_1$$

$$= ab \tan^{-1} \frac{b}{a} + ab \left(\frac{\pi}{2} - \tan^{-1} \frac{a}{b} \right)$$

$$= 2ab \tan^{-1} \frac{b}{a}.$$

As noted in the question, the area enclosed by L is equal to $\frac{1}{2}$ of this value, i.e. the total common area in all four quadrants is

$$4 \times \frac{1}{2} \times 2ab \tan^{-1} \frac{b}{a} = 4ab \tan^{-1} \frac{b}{a}.$$

Note that if we let $b \to a$ then the two ellipses become identical circles and we recover the expected value of πa^2 for their common area.

3.7 Evaluate the line integral

$$I = \oint_C \left[y(4x^2 + y^2)\,dx + x(2x^2 + 3y^2)\,dy \right]$$

around the ellipse $x^2/a^2 + y^2/b^2 = 1$.

As it stands this integral is complicated and, in fact, it is the sum of two integrals. The form of the integrand, containing powers of x and y that can be differentiated easily, makes this problem one to which Green's theorem in a plane might usefully be applied. The theorem states that

$$\oint_C (P\,dx + Q\,dy) = \int\int_R \left(\frac{\partial Q}{\partial x} - \frac{\partial P}{\partial y} \right) dx\,dy,$$

where C is a closed contour enclosing the convex region R.

In the notation used above,

$$P(x, y) = y(4x^2 + y^2) \quad \text{and} \quad Q(x, y) = x(2x^2 + 3y^2).$$

It follows that

$$\frac{\partial P}{\partial y} = 4x^2 + 3y^2 \quad \text{and} \quad \frac{\partial Q}{\partial x} = 6x^2 + 3y^2,$$

leading to

$$\frac{\partial Q}{\partial x} - \frac{\partial P}{\partial y} = 2x^2.$$

This can now be substituted into Green's theorem and the y-integration carried out immediately as the integrand does not contain y. Hence,

$$
\begin{aligned}
I &= \int\int_R 2x^2\,dx\,dy \\
&= \int_{-a}^{a} 2x^2\, 2b \left(1 - \frac{x^2}{a^2} \right)^{1/2} dx \\
&= 4b \int_{\pi}^{0} a^2 \cos^2 \phi \, \sin \phi \, (-a \sin \phi \, d\phi), \quad \text{on setting } x = a \cos \phi, \\
&= -ba^3 \int_{\pi}^{0} \sin^2(2\phi)\,d\phi = \tfrac{1}{2}\pi ba^3.
\end{aligned}
$$

In the final line we have used the standard result for the integral of the square of a sinusoidal function.

3.9 A single-turn coil C of arbitrary shape is placed in a magnetic field $\mathbf{B}$ and carries a current I. Show that the couple acting upon the coil can be written as

$$\mathbf{M} = I \int_C (\mathbf{B} \cdot \mathbf{r}) \, d\mathbf{r} - I \int_C \mathbf{B}(\mathbf{r} \cdot d\mathbf{r}).$$

For a planar rectangular coil of sides $2a$ and $2b$ placed with its plane vertical and at an angle ϕ to a uniform horizontal field $\mathbf{B}$, show that $\mathbf{M}$ is, as expected, $4abBI \cos \phi \, \mathbf{k}$.

For an arbitrarily shaped coil the total couple acting can only be found by considering that on an infinitesimal element and then integrating this over the whole coil. The force on an element $d\mathbf{r}$ of the coil is $d\mathbf{F} = I \, d\mathbf{r} \times \mathbf{B}$, and the moment of this force about the origin is $d\mathbf{M} = \mathbf{r} \times \mathbf{F}$. Thus the total moment is given by

$$\mathbf{M} = \oint_C \mathbf{r} \times (I \, d\mathbf{r} \times \mathbf{B})$$

$$= I \oint_C (\mathbf{r} \cdot \mathbf{B}) \, d\mathbf{r} - I \oint_C \mathbf{B}(\mathbf{r} \cdot d\mathbf{r}).$$

To obtain this second form we have used the vector identity

$$\mathbf{a} \times (\mathbf{b} \times \mathbf{c}) = (\mathbf{a} \cdot \mathbf{c})\mathbf{b} - (\mathbf{a} \cdot \mathbf{b})\mathbf{c}.$$

To determine the couple acting on the rectangular coil we work in Cartesian coordinates with the z-axis vertical and choose the orientation of axes in the horizontal plane such that the edge of the rectangle of length $2a$ is in the x-direction. Then

$$\mathbf{B} = B \cos \phi \, \mathbf{i} + B \sin \phi \, \mathbf{j}.$$

Considering the first term in $\mathbf{M}$:
(i) for the horizontal sides

$$\mathbf{r} = x \mathbf{i} \pm b \mathbf{k}, \quad d\mathbf{r} = dx \, \mathbf{i}, \quad \mathbf{r} \cdot \mathbf{B} = xB \cos \phi,$$

$$\int (\mathbf{r} \cdot \mathbf{B}) \, d\mathbf{r} = B \cos \phi \, \mathbf{i} \left(\int_{-a}^{a} x \, dx + \int_{a}^{-a} x \, dx \right) = 0;$$

(ii) for the vertical sides

$$\mathbf{r} = \pm a \mathbf{i} + z \mathbf{k}, \quad d\mathbf{r} = dz \, \mathbf{k}, \quad \mathbf{r} \cdot \mathbf{B} = \pm aB \cos \phi,$$

$$\int (\mathbf{r} \cdot \mathbf{B}) \, d\mathbf{r} = B \cos \phi \, \mathbf{k} \left(\int_{-b}^{b} (+a) \, dz + \int_{b}^{-b} (-a) \, dz \right) = 4abB \cos \phi \, \mathbf{k}.$$

For the second term in $\mathbf{M}$, since the field is uniform it can be taken outside the integral as a (vector) constant. On the horizontal sides the remaining integral is

$$\int \mathbf{r} \cdot d\mathbf{r} = \pm \int_{-a}^{a} x \, dx = 0.$$

Similarly, the contribution from the vertical sides vanishes and the whole of the second term contributes nothing in this particular configuration.

The total moment is thus $4abB \cos \phi \, \mathbf{k}$, as expected.

3.11 An axially symmetric solid body with its axis AB vertical is immersed in an incompressible fluid of density ρ_0. Use the following method to show that, whatever the shape of the body, for $\rho = \rho(z)$ in cylindrical polars the Archimedean upthrust is, as expected, $\rho_0 g V$, where V is the volume of the body.

Express the vertical component of the resultant force $(- \int p \, d\mathbf{S}$, where p is the pressure) on the body in terms of an integral; note that $p = -\rho_0 g z$ and that for an annular surface element of width dl, $\mathbf{n} \cdot \mathbf{n}_z \, dl = -d\rho$. Integrate by parts and use the fact that $\rho(z_A) = \rho(z_B) = 0$.

We measure z negatively from the water's surface $z = 0$ so that the hydrostatic pressure is $p = -\rho_0 g z$. By symmetry, there is no net horizontal force acting on the body.

The upward force, F, is due to the net vertical component of the hydrostatic pressure acting upon the body's surface:

$$F = -\hat{\mathbf{n}}_z \cdot \int p \, d\mathbf{S}$$

$$= -\hat{\mathbf{n}}_z \cdot \int (-\rho_0 g z)(2\pi\rho \, \hat{\mathbf{n}} \, dl),$$

where $2\pi\rho \, dl$ is the area of the strip of surface lying between z and $z + dz$ and $\hat{\mathbf{n}}$ is the outward unit normal to that surface.

Now, from geometry, $\hat{\mathbf{n}}_z \cdot \hat{\mathbf{n}}$ is equal to minus the sine of the angle between dl and dz and so $\hat{\mathbf{n}}_z \cdot \hat{\mathbf{n}} \, dl$ is equal to $-d\rho$. Thus,

$$F = 2\pi\rho_0 g \int_{z_A}^{z_B} \rho z(-d\rho)$$

$$= -2\pi\rho_0 g \int_{z_A}^{z_B} \left(\rho \frac{\partial\rho}{\partial z}\right) z \, dz$$

$$= -2\pi\rho_0 g \left\{\left[z \frac{\rho^2}{2}\right]_{z_A}^{z_B} - \int_{z_A}^{z_B} \frac{\rho^2}{2} \, dz\right\}.$$

But $\rho(z_A) = \rho(z_B) = 0$, and so the first contribution vanishes, leaving

$$F = \rho_0 g \int_{z_A}^{z_B} \pi\rho^2 \, dz = \rho_0 g V,$$

where V is the volume of the solid. This is the mathematical form of Archimedes' principle. Of course, the result is also valid for a closed body of arbitrary shape, $\rho = \rho(z, \phi)$, but a different method would be needed to prove it.

3.13 A vector field $\mathbf{a}$ is given by $-zxr^{-3}\mathbf{i} - zyr^{-3}\mathbf{j} + (x^2 + y^2)r^{-3}\mathbf{k}$, where $r^2 = x^2 + y^2 + z^2$. Establish that the field is conservative (a) by showing that $\nabla \times \mathbf{a} = \mathbf{0}$, and (b) by constructing its potential function ϕ.

We are told that

$$\mathbf{a} = -\frac{zx}{r^3}\mathbf{i} - \frac{zy}{r^3}\mathbf{j} + \frac{x^2 + y^2}{r^3}\mathbf{k},$$

with $r^2 = x^2 + y^2 + z^2$. We will need to differentiate r^{-3} with respect to x, y and z, using the chain rule, and so note that $\partial r/\partial x = x/r$, etc.

(a) Consider $\nabla \times \mathbf{a}$, term by term:

$$[\nabla \times \mathbf{a}]_x = \frac{\partial}{\partial y}\left(\frac{x^2 + y^2}{r^3}\right) - \frac{\partial}{\partial z}\left(\frac{-zy}{r^3}\right)$$

$$= \frac{-3(x^2 + y^2)y}{r^4 r} + \frac{2y}{r^3} + \frac{y}{r^3} - \frac{3(zy)z}{r^4 r}$$

$$= \frac{3y}{r^5}(-x^2 - y^2 + x^2 + y^2 + z^2 - z^2) = 0;$$

$$[\nabla \times \mathbf{a}]_y = \frac{\partial}{\partial z}\left(\frac{-zx}{r^3}\right) - \frac{\partial}{\partial x}\left(\frac{x^2 + y^2}{r^3}\right)$$

$$= \frac{3(zx)z}{r^4 r} - \frac{x}{r^3} - \frac{2x}{r^3} + \frac{3(x^2 + y^2)x}{r^4 r}$$

$$= \frac{3x}{r^5}(z^2 - x^2 - y^2 - z^2 + x^2 + y^2) = 0;$$

$$[\nabla \times \mathbf{a}]_z = \frac{\partial}{\partial x}\left(\frac{-zy}{r^3}\right) - \frac{\partial}{\partial y}\left(\frac{-zx}{r^3}\right)$$

$$= \frac{3(zy)x}{r^4 r} - \frac{3(zx)y}{r^4 r} = 0.$$

Thus all three components of $\nabla \times \mathbf{a}$ are zero, showing that $\mathbf{a}$ is a conservative field.

(b) To construct its potential function we proceed as follows:

$$\frac{\partial \phi}{\partial x} = \frac{-zx}{(x^2 + y^2 + z^2)^{3/2}} \Rightarrow \phi = \frac{z}{(x^2 + y^2 + z^2)^{1/2}} + f(y, z),$$

$$\frac{\partial \phi}{\partial y} = \frac{-zy}{(x^2 + y^2 + z^2)^{3/2}} = \frac{-zy}{(x^2 + y^2 + z^2)^{3/2}} + \frac{\partial f}{\partial y} \Rightarrow f(y, z) = g(z),$$

$$\frac{\partial \phi}{\partial z} = \frac{x^2 + y^2}{(x^2 + y^2 + z^2)^{3/2}}$$

$$= \frac{1}{(x^2 + y^2 + z^2)^{1/2}} + \frac{-z z}{(x^2 + y^2 + z^2)^{3/2}} + \frac{\partial g}{\partial z}$$

$$\Rightarrow g(z) = c.$$

Thus,

$$\phi(x, y, z) = c + \frac{z}{(x^2 + y^2 + z^2)^{1/2}} = c + \frac{z}{r}.$$

The very fact that we can construct a potential function $\phi = \phi(x, y, z)$ whose derivatives are the components of the vector field shows that the field is conservative.

3.15 A force $\mathbf{F}(\mathbf{r})$ acts on a particle at $\mathbf{r}$. In which of the following cases can $\mathbf{F}$ be represented in terms of a potential? Where it can, find the potential.

(a) $\mathbf{F} = F_0 \left[\mathbf{i} - \mathbf{j} - \dfrac{2(x-y)}{a^2} \mathbf{r} \right] \exp\left(-\dfrac{r^2}{a^2}\right);$

(b) $\mathbf{F} = \dfrac{F_0}{a} \left[z\mathbf{k} + \dfrac{(x^2+y^2-a^2)}{a^2} \mathbf{r} \right] \exp\left(-\dfrac{r^2}{a^2}\right);$

(c) $\mathbf{F} = F_0 \left[\mathbf{k} + \dfrac{a(\mathbf{r} \times \mathbf{k})}{r^2} \right].$

(a) We first write the field entirely in terms of the Cartesian unit vectors using $\mathbf{r} = x\,\mathbf{i} + y\,\mathbf{j} + z\,\mathbf{k}$ and then attempt to construct a suitable potential function ϕ:

$$\mathbf{F} = F_0 \left[\mathbf{i} - \mathbf{j} - \frac{2(x-y)}{a^2}\mathbf{r} \right] \exp\left(-\frac{r^2}{a^2}\right)$$

$$= \frac{F_0}{a^2} \left[(a^2 - 2x^2 + 2xy)\mathbf{i} + (-a^2 - 2xy + 2y^2)\mathbf{j} \right.$$

$$\left. + (-2xz + 2yz)\mathbf{k} \right] \exp\left(-\frac{r^2}{a^2}\right).$$

Since the partial derivative of $\exp(-r^2/a^2)$ with respect to any Cartesian coordinate u is $\exp(-r^2/a^2)(-2r/a^2)(u/r)$, the z-component of $\mathbf{F}$ appears to be the most straightforward to tackle first:

$$\frac{\partial \phi}{\partial z} = \frac{F_0}{a^2}(-2xz + 2yz) \exp\left(-\frac{r^2}{a^2}\right)$$

$$\Rightarrow \phi(x, y, z) = F_0(x-y) \exp\left(-\frac{r^2}{a^2}\right) + f(x, y)$$

$$\equiv \phi_1(x, y, z) + f(x, y).$$

Next we examine the derivatives of $\phi = \phi_1 + f$ with respect to x and y to see how closely they generate F_x and F_y:

$$\frac{\partial \phi_1}{\partial x} = F_0 \left[\exp\left(-\frac{r^2}{a^2}\right) + (x-y)\exp\left(-\frac{r^2}{a^2}\right)\left(\frac{-2x}{a^2}\right) \right]$$

$$= \frac{F_0}{a^2}(a^2 - 2x^2 + 2xy)\exp(-r^2/a^2) = F_x \quad \text{(as given)},$$

and
$$\frac{\partial \phi_1}{\partial y} = F_0 \left[-\exp\left(-\frac{r^2}{a^2}\right) + (x-y)\exp\left(-\frac{r^2}{a^2}\right)\left(\frac{-2y}{a^2}\right) \right]$$

$$= \frac{F_0}{a^2}(-a^2 - 2xy + 2y^2)\exp(-r^2/a^2) = F_y \quad \text{(as given)}.$$

Thus, to within an arbitrary constant, $\phi_1(x, y, z) = F_0(x-y)\exp\left(-\dfrac{r^2}{a^2}\right)$ is a suitable potential function for the field, without the need for any additional function $f(x, y)$.

(b) We follow the same line of argument as in part (a). First, expressing $\mathbf{F}$ in terms of $\mathbf{i}$, $\mathbf{j}$ and $\mathbf{k}$,

$$\mathbf{F} = \frac{F_0}{a}\left[z\,\mathbf{k} + \frac{x^2 + y^2 - a^2}{a^2}\mathbf{r}\right]\exp\left(-\frac{r^2}{a^2}\right)$$

$$= \frac{F_0}{a^3}[x(x^2 + y^2 - a^2)\,\mathbf{i} + y(x^2 + y^2 - a^2)\,\mathbf{j}$$

$$+ z(x^2 + y^2)\,\mathbf{k}]\exp\left(-\frac{r^2}{a^2}\right),$$

and then constructing a possible potential function ϕ. Again starting with the z-component:

$$\frac{\partial \phi}{\partial z} = \frac{F_0 z}{a^3}(x^2 + y^2)\exp\left(-\frac{r^2}{a^2}\right),$$

$$\Rightarrow \quad \phi(x, y, z) = -\frac{F_0}{2a}(x^2 + y^2)\exp\left(-\frac{r^2}{a^2}\right) + f(x, y)$$

$$\equiv \phi_1(x, y, z) + f(x, y).$$

Then,
$$\frac{\partial \phi_1}{\partial x} = -\frac{F_0}{2a}\left[2x - \frac{2x(x^2 + y^2)}{a^2}\right]\exp\left(-\frac{r^2}{a^2}\right) = F_x \quad \text{(as given)},$$

and
$$\frac{\partial \phi_1}{\partial y} = -\frac{F_0}{2a}\left[2y - \frac{2y(x^2 + y^2)}{a^2}\right]\exp\left(-\frac{r^2}{a^2}\right) = F_y \quad \text{(as given)}.$$

Thus, $\phi_1(x, y, z) = \frac{F_0}{2a}(x^2 + y^2)\exp\left(-\frac{r^2}{a^2}\right)$, as it stands, is a suitable potential function for $\mathbf{F}(\mathbf{r})$ and establishes the conservative nature of the field.

(c) Again we express F in Cartesian components:

$$\mathbf{F} = F_0\left[\mathbf{k} + \frac{a(\mathbf{r} \times \mathbf{k})}{r^2}\right] = \frac{ay}{r^2}\mathbf{i} - \frac{ax}{r^2}\mathbf{j} + \mathbf{k}.$$

That the z-component of $\mathbf{F}$ has no dependence on y whilst its y-component does depend upon z suggests that the x-component of $\nabla \times \mathbf{F}$ may not be zero. To test this out we compute

$$(\nabla \times \mathbf{F})_x = \frac{\partial(1)}{\partial y} - \frac{\partial}{\partial z}\left(\frac{-ax}{r^2}\right) = 0 - \frac{2axz}{r^4} \neq 0,$$

and find that it is not. To have even one component of $\nabla \times \mathbf{F}$ non-zero is sufficient to show that $\mathbf{F}$ is not conservative and that no potential function can be found. There is no point in searching further!

The same conclusion can be reached by considering the implication of $F_z = \mathbf{k}$, namely that any possible potential function has to have the form $\phi(x, y, z) = z + f(x, y)$. However, $\partial\phi/\partial x$ is known to be $-ay/r^2 = -ay/(x^2 + y^2 + z^2)$. This yields a contradiction, as it requires $\partial f(x, y)/\partial x$ to depend on z, which is clearly impossible.

3.17 The vector field $\mathbf{f}$ has components $y\mathbf{i} - x\mathbf{j} + \mathbf{k}$ and γ is a curve given parametrically by

$$\mathbf{r} = (a - c + c\cos\theta)\mathbf{i} + (b + c\sin\theta)\mathbf{j} + c^2\theta\mathbf{k}, \quad 0 \le \theta \le 2\pi.$$

Describe the shape of the path γ and show that the line integral $\int_\gamma \mathbf{f} \cdot d\mathbf{r}$ vanishes. Does this result imply that $\mathbf{f}$ is a conservative field?

As θ increases from 0 to 2π, the x- and y-components of $\mathbf{r}$ vary sinusoidally and in quadrature about fixed values $a - c$ and b. Both variations have amplitude c and both return to their initial values when $\theta = 2\pi$. However, the z-component increases monotonically from 0 to a value of $2\pi c^2$. The curve γ is therefore one loop of a circular spiral of radius c and pitch $2\pi c^2$. Its axis is parallel to the z-axis and passes through the points $(a - c, \ b, \ z)$.

The line element $d\mathbf{r}$ has components $(-c\sin\theta\, d\theta, c\cos\theta\, d\theta, c^2\, d\theta)$ and so the line integral of f along γ is given by

$$\int_\gamma \mathbf{f} \cdot d\mathbf{r} = \int_0^{2\pi} \left[y(-c\sin\theta) - x(c\cos\theta) + c^2 \right] d\theta$$

$$= \int_0^{2\pi} \left[-c(b + c\sin\theta)\sin\theta - c(a - c + c\cos\theta)\cos\theta + c^2 \right] d\theta$$

$$= \int_0^{2\pi} \left(-bc\sin\theta - c^2\sin^2\theta - c(a - c)\cos\theta - c^2\cos^2\theta + c^2 \right) d\theta$$

$$= 0 - \pi c^2 - 0 - \pi c^2 + 2\pi c^2 = 0.$$

However, this does not imply that $\mathbf{f}$ is a conservative field since (i) γ is not a closed loop, and (ii) even if it were, the line integral has to vanish for *every* loop, not just for a particular one.

Further,

$$\nabla \times \mathbf{f} = (0 - 0, \ 0 - 0, \ -1 - 1) = (0, \ 0, \ -2) \ne \mathbf{0},$$

showing explicitly that $\mathbf{f}$ is not conservative.

3.19 Evaluate the surface integral $\int \mathbf{r} \cdot d\mathbf{S}$, where $\mathbf{r}$ is the position vector, over that part of the surface $z = a^2 - x^2 - y^2$ for which $z \ge 0$, by each of the following methods.

(a) Parameterize the surface as $x = a\sin\theta \cos\phi, y = a\sin\theta \sin\phi, z = a^2\cos^2\theta$, and show that

$$\mathbf{r} \cdot d\mathbf{S} = a^4(2\sin^3\theta\cos\theta + \cos^3\theta\sin\theta)\, d\theta\, d\phi.$$

(b) Apply the divergence theorem to the volume bounded by the surface and the plane $z = 0$.

(a) With $x = a\sin\theta \cos\phi, y = a\sin\theta \sin\phi, z = a^2\cos^2\theta$, we first check that this does parameterize the surface appropriately:

$$a^2 - x^2 - y^2 = a^2 - a^2\sin^2\theta(\cos^2\phi + \sin^2\phi) = a^2(1 - \sin^2\theta) = a^2\cos^2\theta = z.$$

We see that it does so for the relevant part of the surface, i.e. that which lies above the plane $z = 0$ with $0 \le \theta \le \pi/2$. It would not do so for the part with $z < 0$ for which $x^2 + y^2$ has to be greater than a^2; this is not catered for by the given parameterization.

Having carried out this check, we calculate expressions for $d\mathbf{S}$ and hence $\mathbf{r} \cdot d\mathbf{S}$ in terms of θ and ϕ as follows:

$$\mathbf{r} = a \sin\theta \cos\phi \, \mathbf{i} + a \sin\theta \sin\phi \, \mathbf{j} + a^2 \cos^2\theta \, \mathbf{k},$$

and the tangent vectors at the point (θ, ϕ) on the surface are given by

$$\frac{\partial \mathbf{r}}{\partial \theta} = a \cos\theta \cos\phi \, \mathbf{i} + a \cos\theta \sin\phi \, \mathbf{j} - 2a^2 \cos\theta \sin\theta \, \mathbf{k},$$

$$\frac{\partial \mathbf{r}}{\partial \phi} = -a \sin\theta \sin\phi \, \mathbf{i} + a \sin\theta \cos\phi \, \mathbf{j}.$$

The corresponding vector element of surface area is thus

$$d\mathbf{S} = \frac{\partial \mathbf{r}}{\partial \theta} \times \frac{\partial \mathbf{r}}{\partial \phi}$$
$$= 2a^3 \cos\theta \sin^2\theta \cos\phi \, \mathbf{i} + 2a^3 \cos\theta \sin^2\theta \sin\phi \, \mathbf{j} + a^2 \cos\theta \sin\theta \, \mathbf{k},$$

giving $\mathbf{r} \cdot d\mathbf{S}$ as

$$\mathbf{r} \cdot d\mathbf{S} = 2a^4 \cos\theta \sin^3\theta \cos^2\phi + 2a^4 \cos\theta \sin^3\theta \sin^2\phi + a^4 \cos^3\theta \sin\theta$$
$$= 2a^4 \cos\theta \sin^3\theta + a^4 \cos^3\theta \sin\theta.$$

This is to be integrated over the ranges $0 \le \phi < 2\pi$ and $0 \le \theta \le \pi/2$ as follows:

$$\int \mathbf{r} \cdot d\mathbf{S} = a^4 \int_0^{2\pi} d\phi \int_0^{\pi/2} (2\sin^3\theta \cos\theta + \cos^3\theta \sin\theta) \, d\theta$$
$$= 2\pi a^4 \left(2 \left[\frac{\sin^4\theta}{4} \right]_0^{\pi/2} + \left[\frac{-\cos^4\theta}{4} \right]_0^{\pi/2} \right)$$
$$= 2\pi a^4 \left(\frac{2}{4} + \frac{1}{4} \right) = \frac{3\pi a^4}{2}.$$

(b) The divergence of the vector field $\mathbf{r}$ is 3, a constant, and so the surface integral $\int \mathbf{r} \cdot d\mathbf{S}$ taken over the complete surface Σ (including the part that lies in the plane $z = 0$) is, by the divergence theorem, equal to three times the volume V of the region bounded by Σ. Now,

$$V = \int_0^{a^2} \pi\rho^2 \, dz = \int_0^{a^2} \pi(a^2 - z) \, dz = \pi(a^4 - \tfrac{1}{2}a^4) = \tfrac{1}{2}\pi a^4,$$

and so $\int_\Sigma \mathbf{r} \cdot d\mathbf{S} = 3\pi a^4/2$.

However, on the part of the surface lying in the plane $z = 0$, $\mathbf{r} = x\,\mathbf{i} + y\,\mathbf{j} + 0\,\mathbf{k}$, whilst $d\mathbf{S} = -dS\,\mathbf{k}$. Consequently the scalar product $\mathbf{r} \cdot d\mathbf{S} = 0$; in words, for any point on this face its position vector is orthogonal to the normal to the face. The surface integral over this face therefore contributes nothing to the total integral and the value obtained is that due to the curved surface alone, in agreement with the result in (a).

3.21 Use the result

$$\int_V \nabla\phi \, dV = \oint_S \phi \, d\mathbf{S},$$

together with an appropriately chosen scalar function ϕ, to prove that the position vector $\bar{\mathbf{r}}$ of the center of mass of an arbitrarily shaped body of volume V and uniform density can be written

$$\bar{\mathbf{r}} = \frac{1}{V} \oint_S \tfrac{1}{2} r^2 \, d\mathbf{S}.$$

The position vector of the center of mass is defined by

$$\bar{\mathbf{r}} \int_V \rho \, dV = \int_V \mathbf{r}\rho \, dV.$$

In order to make use of the given equation, we need to find a scalar function f that is such that $\nabla f = \mathbf{r}$; when this is substituted into the RHS of the above equation, the expression for $\bar{\mathbf{r}}$ can be transformed into a surface integral, rather than a volume integral.

A suitable function for this purpose is $f(r) = \tfrac{1}{2} r^2$. Writing $\mathbf{r}$ in this form and canceling the constant ρ, we have, using the given general result, that

$$\bar{\mathbf{r}} V = \int_V \nabla\left(\frac{1}{2} r^2\right) dV = \oint_S \tfrac{1}{2} r^2 \, d\mathbf{S}.$$

From this it follows immediately that

$$\bar{\mathbf{r}} = \frac{1}{V} \oint_S \tfrac{1}{2} r^2 \, d\mathbf{S}.$$

This result provides an alternative method of finding the center of mass $\bar{z}\mathbf{k}$ of the uniform hemisphere $r = a, 0 \leq \theta \leq \pi/2, 0 \leq \phi < 2\pi$. The curved surface contributes $3a/4$ to $\bar{z}$ and the plane surface contributes $-3a/8$, giving $\bar{z} = 3a/8$.

3.23 Demonstrate the validity of the divergence theorem:

(a) by calculating the flux of the vector

$$\mathbf{F} = \frac{\alpha\mathbf{r}}{(r^2 + a^2)^{3/2}}$$

through the spherical surface $|\mathbf{r}| = \sqrt{3}a$;

(b) by showing that

$$\nabla \cdot \mathbf{F} = \frac{3\alpha a^2}{(r^2 + a^2)^{5/2}}$$

and evaluating the volume integral of $\nabla \cdot \mathbf{F}$ over the interior of the sphere $|\mathbf{r}| = \sqrt{3}a$. The substitution $r = a \tan\theta$ will prove useful in carrying out the integration.

(a) The field is radial with

$$\mathbf{F} = \frac{\alpha\,\mathbf{r}}{(r^2 + a^2)^{3/2}} = \frac{\alpha\,r}{(r^2 + a^2)^{3/2}} \,\hat{\mathbf{e}}_r.$$

The total flux is therefore given by

$$\Phi = \frac{4\pi r^2 \, \alpha \, r}{(r^2 + a^2)^{3/2}}\bigg|_{r=a\sqrt{3}} = \frac{4\pi a^3 \, \alpha \, 3\sqrt{3}}{8a^3} = \frac{3\sqrt{3}\pi\alpha}{2}.$$

(b) From the divergence theorem, the total flux over the surface of the sphere is equal to the volume integral of its divergence within the sphere. The divergence is given by

$$\nabla \cdot \mathbf{F} = \frac{1}{r^2}\frac{\partial}{\partial r}(r^2 F_r) = \frac{1}{r^2}\frac{\partial}{\partial r}\left(\frac{r^2 \, \alpha \, r}{(r^2 + a^2)^{3/2}}\right)$$

$$= \frac{1}{r^2}\left[\frac{3\alpha r^2}{(r^2 + a^2)^{3/2}} - \frac{3\alpha r^4}{(r^2 + a^2)^{5/2}}\right]$$

$$= \frac{3\alpha a^2}{(r^2 + a^2)^{5/2}},$$

and on integrating over the sphere, we have

$$\int_V \nabla \cdot \mathbf{F} \, dV = \int_0^{\sqrt{3}a} \frac{3\alpha a^2}{(r^2 + a^2)^{5/2}} 4\pi r^2 \, dr, \ \text{set } r = a\tan\theta, 0 \le \theta \le \tfrac{\pi}{3},$$

$$= 12\pi\alpha a^2 \int_0^{\pi/3} \frac{a^2 \tan^2\theta \, a \sec^2\theta}{a^5 \sec^5\theta} \, d\theta$$

$$= 12\pi\alpha \int_0^{\pi/3} \sin^2\theta \cos\theta \, d\theta$$

$$= 12\pi\alpha \left[\frac{\sin^3\theta}{3}\right]_0^{\pi/3} = 12\pi\alpha\frac{\sqrt{3}}{8} = \frac{3\sqrt{3}\pi\alpha}{2}, \quad \text{as in (a).}$$

The equality of the results in parts (a) and (b) is in accordance with the divergence theorem.

3.25 In a uniform conducting medium with unit relative permittivity, charge density ρ, current density $\mathbf{J}$, electric field $\mathbf{E}$ and magnetic field $\mathbf{B}$, Maxwell's electromagnetic equations take the form (with $\mu_0\epsilon_0 = c^{-2}$)

$$\text{(i) } \nabla \cdot \mathbf{B} = 0, \qquad \text{(ii) } \nabla \cdot \mathbf{E} = \rho/\epsilon_0,$$

$$\text{(iii) } \nabla \times \mathbf{E} + \dot{\mathbf{B}} = \mathbf{0}, \quad \text{(iv) } \nabla \times \mathbf{B} - (\dot{\mathbf{E}}/c^2) = \mu_0\mathbf{J}.$$

The density of stored energy in the medium is given by $\frac{1}{2}(\epsilon_0 E^2 + \mu_0^{-1} B^2)$. Show that the rate of change of the total stored energy in a volume V is equal to

$$-\int_V \mathbf{J} \cdot \mathbf{E} \, dV - \frac{1}{\mu_0}\oint_S (\mathbf{E} \times \mathbf{B}) \cdot d\mathbf{S},$$

where S is the surface bounding V.

[The first integral gives the ohmic heating loss, whilst the second gives the electromagnetic energy flux out of the bounding surface. The vector $\mu_0^{-1}(\mathbf{E} \times \mathbf{B})$ is known as the Poynting vector.]

The total stored energy is equal to the volume integral of the energy density. Let R be its rate of change. Then, differentiating under the integral sign, we have

$$R = \frac{d}{dt} \int_V \left(\frac{\epsilon_0}{2} E^2 + \frac{1}{2\mu_0} B^2 \right) dV$$

$$= \int_V \left(\epsilon_0 \mathbf{E} \cdot \dot{\mathbf{E}} + \frac{1}{\mu_0} \mathbf{B} \cdot \dot{\mathbf{B}} \right) dV.$$

Now using (iv) and (iii), we have

$$R = \int_V \left[\epsilon_0 \mathbf{E} \cdot (-\mu_0 c^2 \mathbf{J} + c^2 \nabla \times \mathbf{B}) - \frac{1}{\mu_0} \mathbf{B} \cdot (\nabla \times \mathbf{E}) \right] dV$$

$$= -\int_V \mathbf{E} \cdot \mathbf{J} \, dV + \int_V \left[\epsilon_0 c^2 \mathbf{E} \cdot (\nabla \times \mathbf{B}) - \frac{1}{\mu_0} \mathbf{B} \cdot (\nabla \times \mathbf{E}) \right] dV$$

$$= -\int_V \mathbf{E} \cdot \mathbf{J} \, dV - \frac{1}{\mu_0} \int_V \nabla \cdot (\mathbf{E} \times \mathbf{B}) \, dV$$

$$= -\int_V \mathbf{E} \cdot \mathbf{J} \, dV - \frac{1}{\mu_0} \oint_S (\mathbf{E} \times \mathbf{B}) \cdot d\mathbf{S}, \quad \text{by the divergence theorem.}$$

To obtain the penultimate line we used the vector identity

$$\nabla \cdot (\mathbf{a} \times \mathbf{b}) = \mathbf{b} \cdot (\nabla \times \mathbf{a}) - \mathbf{a} \cdot (\nabla \times \mathbf{b}).$$

3.27 The vector field $\mathbf{F}$ is given by

$$\mathbf{F} = (3x^2 yz + y^3 z + xe^{-x})\mathbf{i} + (3xy^2 z + x^3 z + ye^x)\mathbf{j} + (x^3 y + y^3 x + xy^2 z^2)\mathbf{k}.$$

Calculate (a) directly, and (b) by using Stokes' theorem the value of the line integral $\int_L \mathbf{F} \cdot d\mathbf{r}$, where L is the (three-dimensional) closed contour $OABCDEO$ defined by the successive vertices $(0, 0, 0)$, $(1, 0, 0)$, $(1, 0, 1)$, $(1, 1, 1)$, $(1, 1, 0)$, $(0, 1, 0)$, $(0, 0, 0)$.

(a) This calculation is a piece-wise evaluation of the line integral, made up of a series of scalar products of the length of a straight piece of the contour and the component of $\mathbf{F}$ parallel to it (integrated if that component varies along the particular straight section).

On OA, $y = z = 0$ and $F_x = xe^{-x}$;

$$I_1 = \int_0^1 xe^{-x} \, dx = \left[-xe^{-x} \right]_0^1 + \int_0^1 e^{-x} \, dx = 1 - 2e^{-1}.$$

On AB, $x = 1$ and $y = 0$ and $F_z = 0$; the integral I_2 is zero.

On BC, $x = 1$ and $z = 1$ and $F_y = 3y^2 + 1 + ey$;

$$I_3 = \int_0^1 (3y^2 + 1 + ey) \, dy = 1 + 1 + \tfrac{1}{2}e.$$

On CD, $x = 1$ and $y = 1$ and $F_z = 1 + 1 + z^2$;

$$I_4 = \int_1^0 (1 + 1 + z^2) \, dz = -1 - 1 - \tfrac{1}{3}.$$

On DE, $y = 1$ and $z = 0$ and $F_x = xe^{-x}$;

$$I_5 = \int_1^0 xe^{-x}\, dx = -1 + 2e^{-1}.$$

On EO, $x = z = 0$ and $F_y = ye^0$;

$$I_6 = \int_1^0 ye^0\, dy = -\tfrac{1}{2}.$$

Adding up these six contributions shows that the complete line integral has the value $\dfrac{e}{2} - \dfrac{5}{6}$.

(b) As a simple sketch shows, the given contour is three-dimensional. However, it is equivalent to two plane square contours, one $OADEO$ (denoted by S_1) lying in the plane $z = 0$ and the other $ABCDA$ (S_2) lying in the plane $x = 1$; the latter is traversed in the negative sense. The common segment AD does not form part of the original contour but, as it is traversed in opposite senses in the two constituent contours, it (correctly) contributes nothing to the line integral.

To use Stokes' theorem we first need to calculate

$$(\nabla \times \mathbf{F})_x = x^3 + 3y^2x + 2yxz^2 - 3xy^2 - x^3 = 2yxz^2,$$
$$(\nabla \times \mathbf{F})_y = 3x^2y + y^3 - 3x^2y - y^3 - y^2z^2 = -y^2z^2,$$
$$(\nabla \times \mathbf{F})_z = 3y^2z + 3x^2z + ye^x - 3x^2z - 3y^2z = ye^x.$$

Now, S_1 has its normal in the positive z-direction and so only the z-component of $\nabla \times \mathbf{F}$ is needed in the first surface integral of Stokes' theorem. Likewise only the x-component of $\nabla \times \mathbf{F}$ is needed in the second integral, but its value must be subtracted because of the sense in which its contour is traversed:

$$\int_{OABCDEO} (\nabla \times \mathbf{F}) \cdot d\mathbf{r} = \int_{S_1} (\nabla \times \mathbf{F})_z\, dx\, dy - \int_{S_2} (\nabla \times \mathbf{F})_x\, dy\, dz$$

$$= \int_0^1 \int_0^1 ye^x\, dx\, dy - \int_0^1 \int_0^1 2y \times 1 \times z^2\, dy\, dz$$

$$= \frac{1}{2}(e - 1) - 2\frac{1}{2}\frac{1}{3} = \frac{e}{2} - \frac{5}{6}.$$

As they must, the two methods give the same value.

4 Fourier series

4.1 Prove the orthogonality relations that form the basis of the Fourier series representation of functions.

All of the results are based on the values of the integrals

$$S(n) = \int_{x_0}^{x_0+L} \sin\left(\frac{2\pi nx}{L}\right) dx \quad \text{and} \quad C(n) = \int_{x_0}^{x_0+L} \cos\left(\frac{2\pi nx}{L}\right) dx$$

for integer values of n. Since in all cases with $n \geq 1$ the integrand goes through a whole number of complete cycles, the "area under the curve" is zero. For the case $n = 0$, the integrand in $S(n)$ is zero and so therefore is $S(0)$; for $C(0)$ the integrand is unity and the value of $C(0)$ is L.

We now apply these observations to integrals whose integrands are the products of two sinusoidal functions with arguments that are multiples of a fundamental frequency. The integration interval is equal to the period of that fundamental frequency. To express the integrands in suitable forms, repeated use will be made of the expressions for the sums and differences of sinusoidal functions.

We consider first the product of a sine function and a cosine function:

$$I_1 = \int_{x_0}^{x_0+L} \sin\left(\frac{2\pi rx}{L}\right) \cos\left(\frac{2\pi px}{L}\right)$$

$$= \int_{x_0}^{x_0+L} \frac{1}{2}\left[\sin\left(\frac{2\pi(r+p)x}{L}\right) + \sin\left(\frac{2\pi(r-p)x}{L}\right)\right] dx$$

$$= \frac{1}{2}[S(r+p) + S(r-p)] = 0, \text{ for all } r \text{ and } p.$$

Next, we consider the product of two cosines:

$$I_2 = \int_{x_0}^{x_0+L} \cos\left(\frac{2\pi rx}{L}\right) \cos\left(\frac{2\pi px}{L}\right)$$

$$= \int_{x_0}^{x_0+L} \frac{1}{2}\left[\cos\left(\frac{2\pi(r+p)x}{L}\right) + \cos\left(\frac{2\pi(r-p)x}{L}\right)\right] dx$$

$$= \frac{1}{2}[C(r+p) + C(r-p)] = 0,$$

unless $r = p > 0$ when $I_2 = \frac{1}{2}L$. If r and p are both zero, then the integrand is unity and $I_2 = L$.

Finally, for the product of two sine functions:

$$
\begin{aligned}
I_3 &= \int_{x_0}^{x_0+L} \sin\left(\frac{2\pi r x}{L}\right) \sin\left(\frac{2\pi p x}{L}\right) \\
&= \int_{x_0}^{x_0+L} \frac{1}{2}\left[\cos\left(\frac{2\pi(r-p)x}{L}\right) - \cos\left(\frac{2\pi(r+p)x}{L}\right)\right] dx \\
&= \frac{1}{2}[C(r-p) - C(r+p)] = 0,
\end{aligned}
$$

unless $r = p > 0$ when $I_3 = \frac{1}{2}L$. If either of r and p is zero, then the integrand is zero and $I_3 = 0$.

In summary, all of the integrals have zero value except for those in which the integrand is the square of a single sinusoid. In these cases the integral has value $\frac{1}{2}L$ for all integers $r\ (=p)$ that are > 0. For $r\ (=p)$ equal to zero, the $\sin^2$ integral has value zero and the $\cos^2$ integral has value L.

4.3 Which of the following functions of x could be represented by a Fourier series over the range indicated?

(a) $\tanh^{-1}(x)$, $-\infty < x < \infty$;
(b) $\tan x$, $-\infty < x < \infty$;
(c) $|\sin x|^{-1/2}$, $-\infty < x < \infty$;
(d) $\cos^{-1}(\sin 2x)$, $-\infty < x < \infty$;
(e) $x \sin(1/x)$, $-\pi^{-1} < x \leq \pi^{-1}$, cyclically repeated.

The Dirichlet conditions that a function must satisfy before it can be represented by a Fourier series are:

(i) the function must be periodic;
(ii) it must be single-valued and continuous, except possibly at a finite number of finite discontinuities;
(iii) it must have only a finite number of maxima and minima within one period;
(iv) the integral over one period of $|f(x)|$ must converge.

We now test the given functions against these:

(a) $\tanh^{-1}(x)$ is not a periodic function, since it is only defined for $-1 \leq x \leq 1$ and changes (monotonically) from $-\infty$ to $+\infty$ as x varies over this restricted range. This function therefore fails condition (i) and *cannot* be represented as a Fourier series.

(b) $\tan x$ is a periodic function but its discontinuities are not finite, nor is its absolute modulus integrable. It therefore fails tests (ii) and (iv) and *cannot* be represented as a Fourier series.

(c) $|\sin x|^{-1/2}$ is a periodic function of period π and, although it becomes infinite at $x = n\pi$, there are no infinite discontinuities. Near $x = 0$, say, it behaves as $|x|^{-1/2}$ and its absolute modulus is therefore integrable. There is only one minimum in any one period. The function therefore satisfies all four Dirichlet conditions and *can* be represented as a Fourier series.

(d) $\cos^{-1}(\sin 2x)$ is clearly a multi-valued function and fails condition (ii); it *cannot* be represented as a Fourier series.

(e) $x \sin(1/x)$, for $-\pi^{-1} < x \le \pi^{-1}$ (cyclically repeated) is clearly cyclic (by definition), continuous, bounded, single-valued and integrable. However, since $\sin(1/x)$ oscillates with unlimited frequency near $x = 0$, there are an infinite number of maxima and minima in any region enclosing $x = 0$. Condition (iii) is therefore not satisfied and the function *cannot* be represented as a Fourier series.

4.5 Find the Fourier series of the function $f(x) = x$ in the range $-\pi < x \le \pi$. Hence show that

$$1 - \frac{1}{3} + \frac{1}{5} - \frac{1}{7} + \cdots = \frac{\pi}{4}.$$

This is an odd function in x and so a sine series with period 2π is appropriate. The coefficient of $\sin nx$ will be given by

$$b_n = \frac{2}{2\pi} \int_{-\pi}^{\pi} x \sin nx \, dx$$

$$= \frac{1}{\pi} \left\{ \left[-\frac{x \cos nx}{n} \right]_{-\pi}^{\pi} + \int_{-\pi}^{\pi} \frac{\cos nx}{n} \, dx \right\}$$

$$= \frac{1}{\pi} \left[-\frac{\pi(-1)^n - (-\pi)(-1)^n}{n} + 0 \right] = \frac{2(-1)^{n+1}}{n}.$$

Thus, $$x = f(x) = 2 \sum_{n=1}^{\infty} \frac{(-1)^{n+1}}{n} \sin nx.$$

We note in passing that although this series is convergent, as it must be, it has poor (i.e. n^{-1}) convergence; this can be put down to the periodic version of the function having a discontinuity (of 2π) at the end of each basic period.

To obtain the sum of a series from such a Fourier representation, we must make a judicious choice for the value of x – making such a choice is rather more of an art than a science! Here, setting $x = \pi/2$ gives

$$\frac{\pi}{2} = 2 \sum_{n=1}^{\infty} \frac{(-1)^{n+1} \sin(n\pi/2)}{n}$$

$$= 2 \sum_{n \text{ odd}} \frac{(-1)^{n+1}(-1)^{(n-1)/2}}{n},$$

$$\Rightarrow \quad \frac{\pi}{4} = \frac{1}{1} - \frac{1}{3} + \frac{1}{5} - \frac{1}{7} + \cdots.$$

4.7 For the function

$$f(x) = 1 - x, \qquad 0 \le x \le 1,$$

a Fourier sine series can be found by continuing it in the range $-1 < x \le 0$ as $f(x) = -1 - x$. The function thus has a discontinuity of 2 at $x = 0$. The series is

$$1 - x = f(x) = \frac{2}{\pi} \sum_{n=1}^{\infty} \frac{\sin n\pi x}{n}. \qquad (*)$$

In order to obtain a cosine series, the continuation has to be $f(x) = 1 + x$ in the range $-1 < x \le 0$. The function then has no discontinuity at $x = 0$ and the corresponding series is

$$1 - x = f(x) = \frac{1}{2} + \frac{4}{\pi^2} \sum_{n \text{ odd}} \frac{\cos n\pi x}{n^2}. \qquad (**)$$

For these continued functions and series, consider (i) their derivatives and (ii) their integrals. Do they give meaningful equations? You will probably find it helpful to sketch all the functions involved.

(i) *Derivatives*

  (a) The sine series. With the continuation given, the derivative df/dx has the value -1 everywhere, except at the origin where the function is not defined (though $f(0) = 0$ seems the only possible choice), continuous or differentiable. Differentiating the given series $(*)$ for $f(x)$ yields

$$\frac{df}{dx} = 2 \sum_{n=1}^{\infty} \cos n\pi x.$$

This series does not converge and the equation is not meaningful.

  (b) The cosine series. With the stated continuation for $f(x)$ the derivative is $+1$ for $-1 < x \le 0$ and is -1 for $0 \le x \le 1$. It is thus the negative of an odd (about $x = 0$) unit square-wave, whose Fourier series is

$$-\frac{4}{\pi} \sum_{n \text{ odd}} \frac{\sin n\pi x}{n}.$$

This is confirmed by differentiating $(**)$ term by term to obtain the same result:

$$\frac{df}{dx} = \frac{4}{\pi^2} \sum_{n \text{ odd}} \frac{-n\pi \sin n\pi x}{n^2} = -\frac{4}{\pi} \sum_{n \text{ odd}} \frac{\sin n\pi x}{n}.$$

(ii) *Integrals*

  Since integrals contain an arbitrary constant of integration, we will define $F(-1) = 0$, where $F(x)$ is the indefinite integral of $f(x)$.

  (a) The sine series. For $-1 \le x \le 0$,

$$F_a(x) = F(-1) + \int_{-1}^{x} (-1 - x)\, dx = -x - \tfrac{1}{2}x^2 - \tfrac{1}{2}.$$

For $0 \le x \le 1$,

$$F_a(x) = F(0) + \int_{0}^{x} (1 - x)\, dx = -\tfrac{1}{2} + \left[x - \tfrac{1}{2}x^2 \right]_0^x = x - \tfrac{1}{2}x^2 - \tfrac{1}{2}.$$

This is a continuous function and, like all indefinite integrals, is "smoother" than the function from which it is derived; this latter property will be reflected in the improved convergence of the derived series. Integrating term by term we find that its Fourier series is given by

$$F_a(x) = \frac{2}{\pi} \int_{-1}^{x} \sum_{n=1}^{\infty} \frac{\sin n\pi x'}{n} \, dx'$$

$$= \frac{2}{\pi} \sum_{n=1}^{\infty} \left[-\frac{\cos n\pi x'}{\pi n^2} \right]_{-1}^{x}$$

$$= \frac{2}{\pi^2} \sum_{n=1}^{\infty} \frac{(-1)^n - \cos n\pi x}{n^2}$$

$$= -\frac{1}{6} - \frac{2}{\pi^2} \sum_{n=1}^{\infty} \frac{\cos n\pi x}{n^2},$$

a series that has n^{-2} convergence. Here we have used the result that $\sum_{n=1}^{\infty}(-1)^n n^{-2} = -\pi^2/12$.

(b) The cosine series. The corresponding indefinite integral in this case is

$$F_b(x) = x + \tfrac{1}{2}x^2 + \tfrac{1}{2} \quad \text{for} \quad -1 \le x \le 0,$$
$$F_b(x) = x - \tfrac{1}{2}x^2 + \tfrac{1}{2} \quad \text{for} \quad 0 \le x \le 1,$$

and the corresponding integrated series, which has even better convergence (n^{-3}), is given by

$$\frac{1}{2}(x + 1) + \frac{4}{\pi^3} \sum_{n \text{ odd}} \frac{\sin n\pi x}{n^3}.$$

However, to have a true Fourier series expression, we must substitute a Fourier series for the $x/2$ term that arises from integrating the constant ($\tfrac{1}{2}$) in (**). This series must be that for $x/2$ across the complete range $-1 \le x \le 1$, and so neither (*) nor (**) can be rearranged for the purpose. A straightforward calculation (see Problem 4.25 part (b), if necessary) yields the poorly convergent sine series

$$x = 2 \sum_{n=1}^{\infty} \frac{(-1)^{n+1}}{n\pi} \sin n\pi x,$$

and makes the final expression for $F_b(x)$

$$\frac{1}{2} + \sum_{n=1}^{\infty} \frac{(-1)^{n+1}}{n\pi} \sin n\pi x + \frac{4}{\pi^3} \sum_{n \text{ odd}} \frac{\sin n\pi x}{n^3}.$$

As will be apparent from a simple sketch, the first series in the above expression dominates; all of its terms are present and it has only n^{-1} convergence. The second series has alternate terms missing and its convergence $\sim n^{-3}$.

4.9 Find the Fourier coefficients in the expansion of $f(x) = \exp x$ over the range $-1 < x < 1$. What value will the expansion have when $x = 2$?

Since the Fourier series will have period 2, we can say immediately that at $x = 2$ the series will converge to the value it has at $x = 0$, namely 1.

As the function $f(x) = \exp x$ is neither even nor odd, its Fourier series will contain both sine and cosine terms. The cosine coefficients are given by

$$a_n = \frac{2}{2} \int_{-1}^{1} e^x \cos(n\pi x)\, dx$$

$$= \left[\cos(n\pi x)\, e^x \right]_{-1}^{1} + \int_{-1}^{1} n\pi \sin(n\pi x)\, e^x\, dx$$

$$= (-1)^n (e^1 - e^{-1}) + \left[n\pi\, \sin(n\pi x)\, e^x \right]_{-1}^{1}$$

$$\qquad - \int_{-1}^{1} n^2\pi^2 \cos(n\pi x)\, e^x\, dx$$

$$= 2(-1)^n \sinh 1 - n^2\pi^2 a_n,$$

$$\Rightarrow \quad a_n = \frac{2(-1)^n \sinh 1}{1 + n^2\pi^2}.$$

Similarly, the sine coefficients are given by

$$b_n = \frac{2}{2} \int_{-1}^{1} e^x \sin(n\pi x)\, dx$$

$$= \left[\sin(n\pi x)\, e^x \right]_{-1}^{1} - \int_{-1}^{1} n\pi \cos(n\pi x)\, e^x\, dx$$

$$= 0 + \left[-n\pi \cos(n\pi x)\, e^x \right]_{-1}^{1} - \int_{-1}^{1} n^2\pi^2 \sin(n\pi x)\, e^x\, dx$$

$$= 2(-1)^{n+1} n\pi \sinh 1 - n^2\pi^2 b_n,$$

$$\Rightarrow \quad b_n = \frac{2(-1)^{n+1} n\pi \sinh 1}{1 + n^2\pi^2}.$$

4.11 Consider the function $f(x) = \exp(-x^2)$ in the range $0 \le x \le 1$. Show how it should be continued to give as its Fourier series a series (the actual form is not wanted) (a) with only cosine terms, (b) with only sine terms, (c) with period 1 and (d) with period 2.
Would there be any difference between the values of the last two series at (i) $x = 0$, (ii) $x = 1$?

The function and its four continuations are shown as (a)–(d) in Figure 4.1. Note that in the range $0 \le x \le 1$, all four graphs are identical.

Where a continued function has a discontinuity at the ends of its basic period, the series will yield a value at those end-points that is the average of the function's values on the

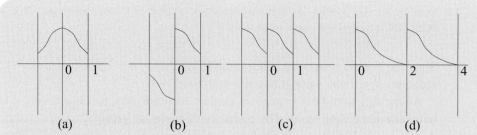

Figure 4.1 The continuations of $\exp(-x^2)$ in $0 \le x \le 1$ to give: (a) cosine terms only; (b) sine terms only; (c) period 1; (d) period 2.

two sides of the discontinuity. Thus for continuation (c) both (i) $x = 0$ and (ii) $x = 1$ are end-points, and the value of the series there will be $(1 + e^{-1})/2$. For continuation (d), $x = 0$ is an end-point, and the series will have value $\frac{1}{2}(1 + e^{-4})$. However, $x = 1$ is not a point of discontinuity, and the series will have the expected value of e^{-1}.

4.13 Consider the representation as a Fourier series of the displacement of a string lying in the interval $0 \le x \le L$ and fixed at its ends, when it is pulled aside by y_0 at the point $x = L/4$. Sketch the continuations for the region outside the interval that will

(a) produce a series of period L,
(b) produce a series that is antisymmetric about $x = 0$, and
(c) produce a series that will contain only cosine terms.
(d) What are (i) the periods of the series in (b) and (c) and (ii) the value of the "a_0-term" in (c)?
(e) Show that a typical term of the series obtained in (b) is

$$\frac{32 y_0}{3 n^2 \pi^2} \sin \frac{n \pi}{4} \sin \frac{n \pi x}{L}.$$

Parts (a), (b) and (c) of Figure 4.2 show the three required continuations. Condition (b) will result in a series containing only sine terms, whilst condition (c) requires the continued function to be symmetric about $x = 0$.

(d) (i) The period in both cases, (b) and (c), is clearly $2L$.

(ii) The average value of the displacement is found from "the area under the triangular curve" to be $(\frac{1}{2} L y_0)/L = \frac{1}{2} y_0$, and this is the value of the "a_0-term".

(e) For the antisymmetric continuation there will be no cosine terms. The sine term coefficients (for a period of $2L$) are given by

$$b_n = 2 \frac{2}{2L} \int_0^L f(x) \sin(nkx) \, dx, \quad \text{where } k = 2\pi/2L = \pi/L.$$

Here use has been made of the antisymmetry about $x = 0$ of both $f(x)$ and the sine functions. However, because of the change in analytic form of $f(x)$ between $x < L/4$ and

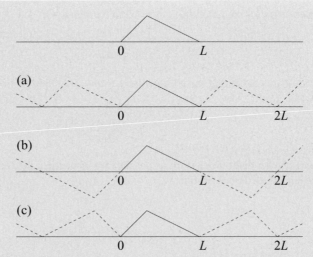

Figure 4.2 Plucked string with fixed ends: (a)–(c) show possible mathematical continuations; (b) is antisymmetric about 0 and (c) is symmetric.

$x > L/4$, the integral will have to be split into two parts. Thus

$$
b_n = \frac{2y_0}{L}\left[\int_0^{L/4} \frac{4x}{L}\sin(nkx)\,dx + \int_{L/4}^{L}\left(\frac{4}{3}-\frac{4x}{3L}\right)\sin(nkx)\,dx\right]
$$

$$
= \frac{8y_0}{3L^2}\left[\int_0^{L/4} 3x\sin(nkx)\,dx + \int_{L/4}^{L}(L-x)\sin(nkx)\,dx\right]
$$

$$
= \frac{8y_0}{3L^2}\left\{\left[-\frac{3x\cos(nkx)}{nk}\right]_0^{L/4} + \int_0^{L/4}\frac{3\cos(nkx)}{nk}\,dx\right.
$$

$$
\left. + \left[-\frac{L\cos(nkx)}{nk}\right]_{L/4}^{L} + \left[\frac{x\cos(nkx)}{nk}\right]_{L/4}^{L} - \int_{L/4}^{L}\frac{\cos(nkx)}{nk}\,dx\right\}.
$$

Evaluation of the remaining integrals then yields

$$
b_n = \frac{8y_0}{3L^2}\left\{-\frac{3L\cos(n\pi/4)}{4n(\pi/L)} - 0 + \left[\frac{3\sin(nkx)}{n^2k^2}\right]_0^{L/4} - \frac{L\cos(n\pi)}{n(\pi/L)}\right.
$$

$$
\left. + \frac{L\cos(n\pi/4)}{n(\pi/L)} + \frac{L\cos(n\pi)}{n(\pi/L)} - \frac{L\cos(n\pi/4)}{4n(\pi/L)} - \left[\frac{\sin(nkx)}{n^2k^2}\right]_{L/4}^{L}\right\}
$$

$$
= \frac{8y_0}{3L^2}\left[\frac{3L^2\sin(n\pi/4)}{n^2\pi^2} - \frac{L^2\sin(n\pi)}{n^2\pi^2} + \frac{L^2\sin(n\pi/4)}{n^2\pi^2}\right] = \frac{32y_0}{3n^2\pi^2}\sin\left(\frac{n\pi}{4}\right).
$$

A typical term is therefore

$$
\frac{32y_0}{3n^2\pi^2}\sin\left(\frac{n\pi}{4}\right)\sin\left(\frac{n\pi x}{L}\right).
$$

We note that every fourth term ($n = 4m$ with m an integer) will be missing.

4.15 The Fourier series for the function $y(x) = |x|$ in the range $-\pi \le x < \pi$ is

$$y(x) = \frac{\pi}{2} - \frac{4}{\pi} \sum_{m=0}^{\infty} \frac{\cos(2m+1)x}{(2m+1)^2}.$$

By integrating this equation term by term from 0 to x, find the function $g(x)$ whose Fourier series is

$$\frac{4}{\pi} \sum_{m=0}^{\infty} \frac{\sin(2m+1)x}{(2m+1)^3}.$$

Using these results, determine, as far as possible by inspection, the form of the functions of which the following are the Fourier series:

(a)

$$\cos\theta + \frac{1}{9}\cos 3\theta + \frac{1}{25}\cos 5\theta + \cdots;$$

(b)

$$\sin\theta + \frac{1}{27}\sin 3\theta + \frac{1}{125}\sin 5\theta + \cdots;$$

(c)

$$\frac{L^2}{3} - \frac{4L^2}{\pi^2}\left[\cos\frac{\pi x}{L} - \frac{1}{4}\cos\frac{2\pi x}{L} + \frac{1}{9}\cos\frac{3\pi x}{L} - \cdots\right].$$

[You may find it helpful to first set $x = 0$ in the quoted result and so obtain values for $S_o = \sum(2m+1)^{-2}$ and other sums derivable from it.]

First, define

$$S = \sum_{\text{all } n \ne 0} n^{-2}, \quad S_o = \sum_{\text{odd } n} n^{-2}, \quad S_e = \sum_{\text{even } n \ne 0} n^{-2}.$$

Clearly, $S_e = \frac{1}{4}S$.

Now set $x = 0$ in the quoted result to obtain

$$0 = \frac{\pi}{2} - \frac{4}{\pi} \sum_{m=0}^{\infty} \frac{1}{(2m+1)^2} = \frac{\pi}{2} - \frac{4}{\pi}S_o.$$

Thus, $S_o = \pi^2/8$. Further, $S = S_o + S_e = S_o + \frac{1}{4}S$; it follows that $S = \pi^2/6$ and, by subtraction, that $S_e = \pi^2/24$.

We now consider the integral of $y(x) = |x|$ from 0 to x.

(i) For $x < 0$, $\displaystyle\int_0^x |x|\,dx = \int_0^x (-x)\,dx = -\frac{1}{2}x^2.$

(ii) For $x > 0$, $\displaystyle\int_0^x |x|\,dx = \int_0^x x\,dx = \frac{1}{2}x^2.$

Integrating the series term by term gives

$$\frac{\pi x}{2} - \frac{4}{\pi} \sum_{m=0}^{\infty} \frac{\sin(2m+1)x}{(2m+1)^3}.$$

Equating these two results and isolating the series gives

$$\frac{4}{\pi} \sum_{m=0}^{\infty} \frac{\sin(2m+1)x}{(2m+1)^3} = \tfrac{1}{2}x(\pi - x) \text{ for } x \geq 0,$$

$$= \tfrac{1}{2}x(\pi + x) \text{ for } x \leq 0.$$

Questions (a)–(c) are to be solved largely through inspection and so detailed working is not (cannot be) given.

(a) Straightforward substitution of θ for x and rearrangement of the original Fourier series give $g_1(\theta) = \tfrac{1}{4}\pi(\tfrac{1}{2}\pi - |\theta|)$.

(b) Straightforward substitution of θ for x and rearrangement of the integrated Fourier series give $g_2(\theta) = \tfrac{1}{8}\pi\theta(\pi - |\theta|)$.

(c) This contains only cosine terms and is therefore an even function of x. Its average value (given by the a_0 term) is $\tfrac{1}{3}L^2$. Setting $x = 0$ gives

$$f(0) = \frac{L^2}{3} - \frac{4L^2}{\pi^2}\left(1 - \frac{1}{4} + \frac{1}{9} - \cdots\right)$$

$$= \frac{L^2}{3} - \frac{4L^2}{\pi^2}(S_o - S_e)$$

$$= \frac{L^2}{3} - \frac{4L^2}{\pi^2}\left(\frac{\pi^2}{8} - \frac{\pi^2}{24}\right) = 0.$$

Setting $x = L$ gives

$$f(L) = \frac{L^2}{3} - \frac{4L^2}{\pi^2}\left(-1 - \frac{1}{4} - \frac{1}{9} - \cdots\right)$$

$$= \frac{L^2}{3} - \frac{4L^2}{\pi^2}(-S) = L^2.$$

All of this evidence suggests that $f(x) = x^2$ (which it is).

4.17 Find the (real) Fourier series of period 2 for $f(x) = \cosh x$ and $g(x) = x^2$ in the range $-1 \leq x \leq 1$. By integrating the series for $f(x)$ twice, prove that

$$\sum_{n=1}^{\infty} \frac{(-1)^{n+1}}{n^2\pi^2(n^2\pi^2 + 1)} = \frac{1}{2}\left(\frac{1}{\sinh 1} - \frac{5}{6}\right).$$

Since both functions are even, we need consider only constants and cosine terms. The series for x^2 can be calculated directly or, more easily, by using the result of the final part of Problem 4.15 with L set equal to 1:

$$g(x) = x^2 = \frac{1}{3} + \frac{4}{\pi^2} \sum_{n=1}^{\infty} \frac{(-1)^n}{n^2} \cos \pi nx \text{ for } -1 \leq x \leq 1.$$

For $f(x) = \cosh x$,

$$a_0 = \frac{2}{2} 2 \int_0^1 \cosh x \, dx = 2 \sinh(1),$$

$$a_n = \frac{2}{2} 2 \int_0^1 \cosh x \cos(n\pi x) \, dx$$

$$= 2 \left[\frac{\cosh x \sin(n\pi x)}{n\pi} \right]_0^1 - 2 \int_0^1 \frac{\sinh x \sin(n\pi x)}{n\pi} \, dx$$

$$= 0 + 2 \left[\frac{\sinh x \cos(n\pi x)}{n^2\pi^2} \right]_0^1 - \frac{a_n}{n^2\pi^2}.$$

Rearranging this gives

$$a_n = \frac{(-1)^n 2 \sinh(1)}{1 + n^2\pi^2}.$$

Thus,

$$\cosh x = \sinh(1) \left(1 + 2 \sum_{n=1}^{\infty} \frac{(-1)^n}{1 + n^2\pi^2} \cos n\pi x \right).$$

We now integrate this expansion twice from 0 to x (anticipating that we will recover a hyperbolic cosine function plus some additional terms). Since $\sinh(0) = \sin(m\pi 0) = 0$, the first integration yields

$$\sinh x = \sinh(1) \left(x + 2 \sum_{n=1}^{\infty} \frac{(-1)^n}{n\pi(1 + n^2\pi^2)} \sin n\pi x \right).$$

For the second integration we use $\cosh(0) = \cos(m\pi 0) = 1$ to obtain

$$\cosh(x) - 1 = \sinh(1) \left(\frac{1}{2} x^2 + 2 \sum_{n=1}^{\infty} \frac{(-1)^{n+1}}{n^2\pi^2(1 + n^2\pi^2)} [\cos(n\pi x) - 1] \right).$$

However, this expansion must be the same as the original expansion for $\cosh(x)$ after a Fourier series has been substituted for the $\frac{1}{2} \sinh(1) x^2$ term. The coefficients of $\cos n\pi x$ in the two expressions must be equal; in particular, the equality of the constant terms (formally $\cos n\pi x$ with $n = 0$) requires that

$$\sinh(1) - 1 = \frac{1}{2} \sinh(1) \frac{1}{3} + 2 \sinh(1) \sum_{n=1}^{\infty} \frac{(-1)^{n+2}}{n^2\pi^2(1 + n^2\pi^2)},$$

i.e.

$$\sum_{n=1}^{\infty} \frac{(-1)^{n+1}}{n^2\pi^2(n^2\pi^2 + 1)} = \frac{1}{2} \left(\frac{1}{\sinh 1} - \frac{5}{6} \right),$$

as stated in the question.

4.19 Demonstrate explicitly for the odd (about $x = 0$) square-wave function that Parseval's theorem is valid. You will need to use the relationship

$$\sum_{m=0}^{\infty} \frac{1}{(2m+1)^2} = \frac{\pi^2}{8}.$$

Show that a filter that transmits frequencies only up to $8\pi/T$ will still transmit more than 90% of the power in a square-wave voltage signal of period T.

As stated in the solution to Problem 4.7, and in virtually every textbook, the odd square-wave function has only the odd harmonics present in its Fourier sine series representation. The coefficient of the $\sin(2m+1)\pi x$ term is

$$b_{2m+1} = \frac{4}{(2m+1)\pi}.$$

For a periodic function of period L whose complex Fourier coefficients are c_r, or whose cosine and sine coefficients are a_r and b_r, respectively, Parseval's theorem for one function states that

$$\frac{1}{L} \int_{x_0}^{x_0+L} |f(x)|^2 dx = \sum_{r=-\infty}^{\infty} |c_r|^2$$

$$= \left(\tfrac{1}{2}a_0\right)^2 + \frac{1}{2}\sum_{r=1}^{\infty}(a_r^2 + b_r^2),$$

and therefore requires in this particular case, in which all the a_r are zero and $L = 2$, that

$$\frac{1}{2}\sum_{m=0}^{\infty} \frac{16}{(2m+1)^2\pi^2} = \frac{1}{2}\sum_{n=1}^{\infty} b_n^2 = \frac{1}{2}\int_{-1}^{1} |\pm 1|^2 dx = 1.$$

Since

$$\sum_{m=0}^{\infty} \frac{1}{(2m+1)^2} = \frac{\pi^2}{8},$$

this reduces to the identity

$$\frac{1}{2}\frac{16}{\pi^2}\frac{\pi^2}{8} = 1.$$

The power at any particular frequency in an electrical signal is proportional to the square of the amplitude at that frequency, i.e. to $|b_n|^2$ in the present case. If the filter passes only frequencies up to $8\pi/T = 4\omega$, then only the $n = 1$ and the $n = 3$ components will be passed. They contribute a fraction

$$\left(\frac{1}{1} + \frac{1}{9}\right) \div \frac{\pi^2}{8} = 0.901$$

of the total, i.e. more than 90%.

4.21 Find the complex Fourier series for the periodic function of period 2π defined in the range $-\pi \le x \le \pi$ by $y(x) = \cosh x$. By setting $x = 0$ prove that

$$\sum_{n=1}^{\infty} \frac{(-1)^n}{n^2 + 1} = \frac{1}{2}\left(\frac{\pi}{\sinh \pi} - 1\right).$$

We first note that, although $\cosh x$ is an even function of x, e^{-inx} is neither even nor odd. Consequently it will not be possible to convert the integral into one over the range $0 \le x \le \pi$. The complex Fourier coefficients c_n ($-\infty < n < \infty$) are therefore calculated as

$$c_n = \frac{1}{2\pi} \int_{-\pi}^{\pi} \cosh x \, e^{-inx} \, dx$$

$$= \frac{1}{2\pi} \int_{-\pi}^{\pi} \frac{1}{2} \left(e^{-inx+x} + e^{-inx-x}\right) dx$$

$$= \frac{1}{4\pi} \left[\frac{e^{(1-in)x}}{1-in}\right]_{-\pi}^{\pi} + \frac{1}{4\pi} \left[\frac{e^{(-1-in)x}}{-1-in}\right]_{-\pi}^{\pi}$$

$$= \frac{1}{4\pi} \frac{(1+in)(-1)^n(2\sinh \pi) - (1-in)(-1)^n(-2\sinh \pi)}{1+n^2}$$

$$= \frac{(-1)^n 4 \sinh(\pi)}{4\pi(1+n^2)}.$$

Thus,

$$\cosh x = \sum_{n=-\infty}^{\infty} \frac{(-1)^n \sinh \pi}{\pi(1+n^2)} e^{inx}.$$

We now set $x = 0$ on both sides of the equation:

$$1 = \sum_{n=-\infty}^{\infty} \frac{(-1)^n \sinh \pi}{\pi(1+n^2)},$$

$$\Rightarrow \quad \sum_{n=-\infty}^{\infty} \frac{(-1)^n}{1+n^2} = \frac{\pi}{\sinh \pi}.$$

Separating out the $n = 0$ term, and noting that $(-1)^n = (-1)^{-n}$, now gives

$$1 + 2 \sum_{n=1}^{\infty} \frac{(-1)^n}{1+n^2} = \frac{\pi}{\sinh \pi}$$

and hence the stated result.

4.23 The complex Fourier series for the periodic function generated by $f(t) = \sin t$ for $0 \le t \le \pi/2$, and repeated in every subsequent interval of $\pi/2$, is

$$\sin(t) = \frac{2}{\pi} \sum_{n=-\infty}^{\infty} \frac{4ni - 1}{16n^2 - 1} e^{i4nt}.$$

Apply Parseval's theorem to this series and so derive a value for the sum of the series

$$\frac{17}{(15)^2} + \frac{65}{(63)^2} + \frac{145}{(143)^2} + \cdots + \frac{16n^2 + 1}{(16n^2 - 1)^2} + \cdots.$$

Applying Parseval's theorem (see Solution 4.19) in a straightforward manner to the given equation:

$$\frac{2}{\pi} \int_0^{\pi/2} \sin^2(t)\, dt = \frac{4}{\pi^2} \sum_{n=-\infty}^{\infty} \frac{4ni - 1}{16n^2 - 1} \frac{-4ni - 1}{16n^2 - 1},$$

$$\frac{2}{\pi} \frac{1}{2} \frac{\pi}{2} = \frac{4}{\pi^2} \sum_{n=-\infty}^{\infty} \frac{16n^2 + 1}{(16n^2 - 1)^2},$$

$$\frac{\pi^2}{8} = 1 + 2\sum_{n=1}^{\infty} \frac{16n^2 + 1}{(16n^2 - 1)^2},$$

$$\Rightarrow \quad \sum_{n=1}^{\infty} \frac{16n^2 + 1}{(16n^2 - 1)^2} = \frac{\pi^2 - 8}{16}.$$

To obtain the second line we have used the standard result that the average value of the square of a sinusoid is $1/2$.

4.25 Show that Parseval's theorem for two real functions whose Fourier expansions have cosine and sine coefficients a_n, b_n and α_n, β_n takes the form

$$\frac{1}{L} \int_0^L f(x)g(x)\, dx = \frac{1}{4} a_0 \alpha_0 + \frac{1}{2} \sum_{n=1}^{\infty} (a_n \alpha_n + b_n \beta_n).$$

(a) Demonstrate that for $g(x) = \sin mx$ or $\cos mx$ this reduces to the definition of the Fourier coefficients.

(b) Explicitly verify the above result for the case in which $f(x) = x$ and $g(x)$ is the square-wave function, both in the interval $-1 \le x \le 1$.

[Note that $g = g^*$, and it is the integral of fg^* that will have to be formally evaluated using the complex Fourier series representations of the two functions.]

If c_n and γ_n are the complex Fourier coefficients for the real functions $f(x)$ and $g(x)$ that have real Fourier coefficients a_n, b_n and α_n, β_n, respectively, then

$$c_n = \tfrac{1}{2}(a_n - ib_n) \quad \text{and} \quad \gamma_n = \alpha_n - i\beta_n,$$

$$c_{-n} = \tfrac{1}{2}(a_n + ib_n) \quad \text{and} \quad \gamma_{-n} = \alpha_n + i\beta_n.$$

The two functions can be written as

$$f(x) = \sum_{n=-\infty}^{\infty} c_n \exp\left(\frac{2\pi inx}{L}\right),$$

$$g(x) = \sum_{n=-\infty}^{\infty} \gamma_n \exp\left(\frac{2\pi inx}{L}\right). \qquad (*)$$

Thus,

$$f(x)g^*(x) = \sum_{n=-\infty}^{\infty} c_n g^*(x) \exp\left(\frac{2\pi inx}{L}\right).$$

Integrating this equation with respect to x over the interval $(0, L)$ and dividing by L, we find

$$\frac{1}{L}\int_0^L f(x)g^*(x)\,dx = \sum_{n=-\infty}^{\infty} c_n \frac{1}{L}\int_0^L g^*(x) \exp\left(\frac{2\pi inx}{L}\right)dx$$

$$= \sum_{n=-\infty}^{\infty} c_n \left[\frac{1}{L}\int_0^L g(x) \exp\left(\frac{-2\pi inx}{L}\right)dx\right]^*$$

$$= \sum_{n=-\infty}^{\infty} c_n \gamma_n^*.$$

To obtain the last line we have used the inverse of relationship $(*)$.

Dividing up the sum over all n into a sum over positive n, a sum over negative n and the $n = 0$ term, and then substituting for c_n and γ_n, gives

$$\frac{1}{L}\int_0^L f(x)g^*(x)\,dx = \frac{1}{4}\sum_{n=1}^{\infty}(a_n - ib_n)(\alpha_n + i\beta_n)$$

$$+ \frac{1}{4}\sum_{n=1}^{\infty}(a_n + ib_n)(\alpha_n - i\beta_n) + \frac{1}{4}a_0\alpha_0$$

$$= \frac{1}{4}\sum_{n=1}^{\infty}(2a_n\alpha_n + 2b_n\beta_n) + \frac{1}{4}a_0\alpha_0$$

$$= \frac{1}{2}\sum_{n=1}^{\infty}(a_n\alpha_n + b_n\beta_n) + \frac{1}{4}a_0\alpha_0,$$

i.e. the stated result.

(a) For $g(x) = \sin mx$, $\beta_m = 1$ and all other α_n and β_n are zero. The above equation then reduces to

$$\frac{1}{L}\int_0^L f(x)\sin(mx)\,dx = \frac{1}{2}b_n,$$

which is the normal definition of b_n. Similarly, setting $g(x) = \cos mx$ leads to the normal definition of a_n.

(b) For the function $f(x) = x$ in the interval $-1 < x \le 1$, the sine coefficients are

$$b_n = \frac{2}{2} \int_{-1}^{1} x \sin n\pi x \, dx$$

$$= 2 \int_{0}^{1} x \sin n\pi x \, dx$$

$$= 2 \left\{ \left[\frac{-x \cos n\pi x}{n\pi} \right]_{0}^{1} + \int_{0}^{1} \frac{\cos n\pi x}{n\pi} \, dx \right\}$$

$$= 2 \left\{ \frac{(-1)^{n+1}}{n\pi} + \left[\frac{\sin n\pi x}{n^2 \pi^2} \right]_{0}^{1} \right\}$$

$$= \frac{2(-1)^{n+1}}{n\pi}.$$

As stated in Problem 4.19, for the (antisymmetric) square-wave function $\beta_n = 4/(n\pi)$ for odd n and $\beta_n = 0$ for even n.

Now the integral

$$\frac{1}{L} \int_{0}^{L} f(x) g^*(x) \, dx = \frac{1}{2} \left[\int_{-1}^{0} (-1)x \, dx + \int_{0}^{1} (+1)x \, dx \right] = \frac{1}{2},$$

whilst

$$\frac{1}{2} \sum_{n=1}^{\infty} b_n \beta_n = \frac{1}{2} \sum_{n \text{ odd}} \frac{4}{n\pi} \frac{2(-1)^{n+1}}{n\pi} = \frac{4}{\pi^2} \sum_{n \text{ odd}} \frac{1}{n^2} = \frac{4}{\pi^2} \frac{\pi^2}{8} = \frac{1}{2}.$$

The value of the sum $\sum n^{-2}$ for odd n is taken from S_o in the solution to Problem 4.15. Thus, the two sides of the equation agree, verifying the validity of Parseval's theorem in this case.

5 Integral transforms

5.1 Find the Fourier transform of the function $f(t) = \exp(-|t|)$.

(a) By applying Fourier's inversion theorem prove that

$$\frac{\pi}{2}\exp(-|t|) = \int_0^\infty \frac{\cos\omega t}{1+\omega^2}\,d\omega.$$

(b) By making the substitution $\omega = \tan\theta$, demonstrate the validity of Parseval's theorem for this function.

As the function $|t|$ is not representable by the same integrable function throughout the integration range, we must divide the range into two sections and use different explicit expressions for the integrand in each:

$$\tilde{f}(\omega) = \frac{1}{\sqrt{2\pi}}\int_{-\infty}^\infty e^{-|t|}e^{-i\omega t}\,dt$$

$$= \frac{1}{\sqrt{2\pi}}\int_0^\infty e^{-(1+i\omega)t}\,dt + \frac{1}{\sqrt{2\pi}}\int_{-\infty}^0 e^{(1-i\omega)t}\,dt$$

$$= \frac{1}{\sqrt{2\pi}}\left(\frac{1}{1+i\omega} + \frac{1}{1-i\omega}\right)$$

$$= \frac{1}{\sqrt{2\pi}}\frac{2}{1+\omega^2}.$$

(a) Substituting this result into the inversion theorem gives

$$\exp^{-|t|} = \frac{1}{\sqrt{2\pi}}\int_{-\infty}^\infty \frac{2}{\sqrt{2\pi}(1+\omega^2)}e^{i\omega t}\,d\omega.$$

Equating the real parts on the two sides of this equation and noting that the resulting integrand is symmetric in ω, shows that

$$\exp^{-|t|} = \frac{2}{\pi}\int_0^\infty \frac{\cos\omega t}{(1+\omega^2)}\,d\omega,$$

as given in the question.

(b) For Parseval's theorem, which states that

$$\int_{-\infty}^\infty |f(t)|^2\,dt = \int_{-\infty}^\infty |\tilde{f}(\omega)|^2\,d\omega,$$

we first evaluate

$$\int_{-\infty}^{\infty} |f(t)|^2 \, dt = \int_{-\infty}^{0} e^{2t} \, dt + \int_{0}^{\infty} e^{-2t} \, dt$$

$$= 2 \int_{0}^{\infty} e^{-2t} \, dt$$

$$= 2 \left[\frac{e^{-2t}}{-2} \right]_{0}^{\infty} = 1.$$

The second integral, over ω, is

$$\int_{-\infty}^{\infty} |\tilde{f}(\omega)|^2 \, d\omega = 2 \int_{0}^{\infty} \frac{2}{\pi(1 + \omega^2)^2} \, d\omega, \quad \text{set } \omega \text{ equal to } \tan\theta,$$

$$= \frac{4}{\pi} \int_{0}^{\pi/2} \frac{1}{\sec^4\theta} \sec^2\theta \, d\theta$$

$$= \frac{4}{\pi} \int_{0}^{\pi/2} \cos^2\theta \, d\theta = \frac{4}{\pi} \frac{1}{2} \frac{\pi}{2} = 1,$$

i.e. the same as the first one, thus verifying the theorem for this function.

5.3 Find the Fourier transform of $H(x - a)e^{-bx}$, where $H(x)$ is the Heaviside function.

The Heaviside function $H(x)$ has value 0 for $x < 0$ and value 1 for $x \geq 0$. Write $H(x - a)e^{-bx} = h(x)$ with b assumed > 0. Then,

$$\tilde{h}(k) = \frac{1}{\sqrt{2\pi}} \int_{-\infty}^{\infty} H(x - a)e^{-bx} e^{-ikx} \, dx$$

$$= \frac{1}{\sqrt{2\pi}} \int_{a}^{\infty} e^{-bx-ikx} \, dx$$

$$= \frac{1}{\sqrt{2\pi}} \left[\frac{e^{-bx-ikx}}{-b - ik} \right]_{a}^{\infty}$$

$$= \frac{1}{\sqrt{2\pi}} \frac{e^{-ba}e^{-ika}}{b + ik} = e^{-ika} \frac{e^{-ba}}{\sqrt{2\pi}} \frac{b - ik}{b^2 + k^2}.$$

This same result could be obtained by setting $y = x - a$, finding the transform of $e^{-ba}e^{-by}$, and then using the translation property of Fourier transforms.

5.5 By taking the Fourier transform of the equation

$$\frac{d^2\phi}{dx^2} - K^2\phi = f(x),$$

show that its solution, $\phi(x)$, can be written as

$$\phi(x) = \frac{-1}{\sqrt{2\pi}} \int_{-\infty}^{\infty} \frac{e^{ikx} \tilde{f}(k)}{k^2 + K^2} \, dk,$$

where $\tilde{f}(k)$ is the Fourier transform of $f(x)$.

We take the Fourier transform of each term of

$$\frac{d^2\phi}{dx^2} - K^2\phi = f(x)$$

to give

$$\frac{1}{\sqrt{2\pi}} \int_{-\infty}^{\infty} \frac{d^2\phi}{dx^2} e^{-ikx}\, dx - K^2\tilde{\phi}(k) = \frac{1}{\sqrt{2\pi}} \int_{-\infty}^{\infty} f(x)e^{-ikx}\, dx.$$

Since ϕ must vanish at $\pm\infty$, the first term can be integrated twice by parts with no contributions at the end-points. This gives the full equation as

$$-k^2\tilde{\phi}(k) - K^2\tilde{\phi}(k) = \tilde{f}(k).$$

Now, by the Fourier inversion theorem,

$$\phi(x) = \frac{1}{\sqrt{2\pi}} \int_{-\infty}^{\infty} \tilde{\phi}(k)e^{ikx}\, dk$$

$$= -\frac{1}{\sqrt{2\pi}} \int_{-\infty}^{\infty} \frac{\tilde{f}(k)e^{ikx}}{k^2 + K^2}\, dk.$$

Note

The principal advantage of this Fourier approach to a set of one or more linear differential equations is that the differential operators act only on exponential functions whose exponents are linear in x. This means that the derivatives are no more than multiples of the original function and what were originally differential equations are turned into algebraic ones. As the differential equations are linear the algebraic equations can be solved explicitly for the transforms of their solutions, and the solutions themselves may then be found using the inversion theorem. The "price" to be paid for this great simplification is that the inversion integral may not be tractable analytically, but, as a last resort, numerical integration can always be employed.

5.7 Find the Fourier transform of the unit rectangular distribution

$$f(t) = \begin{cases} 1 & |t| < 1, \\ 0 & \text{otherwise.} \end{cases}$$

Determine the convolution of f with itself and, without further integration, determine its transform. Deduce that

$$\int_{-\infty}^{\infty} \frac{\sin^2\omega}{\omega^2}\, d\omega = \pi, \qquad \int_{-\infty}^{\infty} \frac{\sin^4\omega}{\omega^4}\, d\omega = \frac{2\pi}{3}.$$

The function to be transformed is unity in the range $-1 \le t \le 1$ and so

$$\tilde{f}(\omega) = \frac{1}{\sqrt{2\pi}} \int_{-1}^{1} 1\, e^{-i\omega t}\, dt = \frac{1}{\sqrt{2\pi}}\left[\frac{e^{-i\omega} - e^{i\omega}}{-i\omega}\right] = \frac{2\sin\omega}{\sqrt{2\pi}\,\omega}.$$

Denote by $p(t)$ the convolution of f with itself and, in the second line of the calculation below, change the integration variable from s to $u = t - s$:

$$p(t) \equiv \int_{-\infty}^{\infty} f(t-s)f(s)\,ds = \int_{-1}^{1} f(t-s)\,1\,ds$$
$$= \int_{t+1}^{t-1} f(u)(-du) = \int_{t-1}^{t+1} f(u)\,du.$$

It follows that

$$p(t) = \begin{cases} (t+1) - (-1) & 0 > t > -2 \\ 1 - (t-1) & 2 > t > 0 \end{cases} = \begin{cases} 2 - |t| & 0 < |t| < 2, \\ 0 & \text{otherwise.} \end{cases}$$

The transform of p is given directly by the convolution theorem [which states that if $h(t)$, given by $h = f * g$, is the convolution of f and g, then $\tilde{h} = \sqrt{2\pi}\,\tilde{f}\,\tilde{g}$] as

$$\tilde{p}(\omega) = \sqrt{2\pi}\,\frac{2\sin\omega}{\sqrt{2\pi}\,\omega}\frac{2\sin\omega}{\sqrt{2\pi}\,\omega} = \frac{4}{\sqrt{2\pi}}\frac{\sin^2\omega}{\omega^2}.$$

Noting that the two integrals to be evaluated have as integrands the squares of functions that are essentially the known transforms of simple functions, we are led to apply Parseval's theorem to each. Applying the theorem to $f(t)$ and $p(t)$ yields

$$\int_{-\infty}^{\infty} \frac{4\sin^2\omega}{2\pi\,\omega^2}\,d\omega = \int_{-\infty}^{\infty} |f(t)|^2\,dt = 2 \quad \Rightarrow \quad \int_{-\infty}^{\infty} \frac{\sin^2\omega}{\omega^2} = \pi,$$

and

$$\int_{-\infty}^{\infty} \frac{16}{2\pi}\frac{\sin^4\omega}{\omega^4}\,d\omega = \int_{-2}^{0} (2+t)^2\,dt + \int_{0}^{2} (2-t)^2\,dt$$
$$= \left[\frac{(2+t)^3}{3}\right]_{-2}^{0} - \left[\frac{(2-t)^3}{3}\right]_{0}^{2}$$
$$= \frac{8}{3} + \frac{8}{3},$$
$$\Rightarrow \quad \int_{-\infty}^{\infty} \frac{\sin^4\omega}{\omega^4}\,d\omega = \frac{2\pi}{3}.$$

5.9 By finding the complex Fourier *series* for its LHS show that either side of the equation

$$\sum_{n=-\infty}^{\infty} \delta(t+nT) = \frac{1}{T}\sum_{n=-\infty}^{\infty} e^{-2\pi nit/T}$$

can represent a periodic train of impulses. By expressing the function $f(t+nX)$, in which X is a constant, in terms of the Fourier *transform* $\tilde{f}(\omega)$ of $f(t)$, show that

$$\sum_{n=-\infty}^{\infty} f(t+nX) = \frac{\sqrt{2\pi}}{X}\sum_{n=-\infty}^{\infty} \tilde{f}\left(\frac{2n\pi}{X}\right) e^{2\pi nit/X}.$$

This result is known as the *Poisson summation formula*.

Denote by $g(t)$ the periodic function $\sum_{n=-\infty}^{\infty} \delta(t + nT)$ with $2\pi/T = \omega$. Its complex Fourier coefficients are given by

$$c_n = \frac{1}{T} \int_0^T g(t) e^{-in\omega t} \, dt = \frac{1}{T} \int_0^T \delta(t) e^{-in\omega t} \, dt = \frac{1}{T}.$$

Thus, by the inversion theorem, its Fourier *series* representation is

$$g(t) = \sum_{n=-\infty}^{\infty} \frac{1}{T} e^{in\omega t} = \sum_{n=-\infty}^{\infty} \frac{1}{T} e^{-in\omega t} = \sum_{n=-\infty}^{\infty} \frac{1}{T} e^{-i2n\pi t/T},$$

showing that both this sum and the original one are representations of a periodic train of impulses.

In this result,

$$\sum_{n=-\infty}^{\infty} \delta(t + nT) = \frac{1}{T} \sum_{n=-\infty}^{\infty} e^{-2\pi nit/T},$$

we now make the changes of variable $t \to \omega$, $n \to -n$ and $T \to 2\pi/X$ and obtain

$$\sum_{n=-\infty}^{\infty} \delta\left(\omega - \frac{2\pi n}{X}\right) = \frac{X}{2\pi} \sum_{n=-\infty}^{\infty} e^{inX\omega}. \quad (*)$$

If we denote $f(t + nX)$ by $f_{nX}(t)$ then, by the translation theorem, we have $\tilde{f}_{nX}(\omega) = e^{inX\omega} \tilde{f}(\omega)$ and

$$f(t + nX) = \frac{1}{\sqrt{2\pi}} \int_{-\infty}^{\infty} \tilde{f}_{nX}(\omega) e^{i\omega t} \, d\omega$$

$$= \frac{1}{\sqrt{2\pi}} \int_{-\infty}^{\infty} e^{inX\omega} \tilde{f}(\omega) e^{i\omega t} \, d\omega,$$

$$\sum_{n=-\infty}^{\infty} f(t + nX) = \frac{1}{\sqrt{2\pi}} \int_{-\infty}^{\infty} \tilde{f}(\omega) e^{i\omega t} \sum_{n=-\infty}^{\infty} e^{inX\omega} \, d\omega, \quad \text{use } (*) \text{ above,}$$

$$= \frac{1}{\sqrt{2\pi}} \int_{-\infty}^{\infty} \tilde{f}(\omega) e^{i\omega t} \frac{2\pi}{X} \sum_{n=-\infty}^{\infty} \delta\left(\omega - \frac{2\pi n}{X}\right) d\omega$$

$$= \frac{\sqrt{2\pi}}{X} \sum_{n=-\infty}^{\infty} \tilde{f}\left(\frac{2\pi n}{X}\right) e^{i2\pi nt/X}.$$

In the final line we have made use of the properties of a δ-function when it appears as a factor in an integrand.

5.11 For a function $f(t)$ that is non-zero only in the range $|t| < T/2$, the full frequency spectrum $\tilde{f}(\omega)$ can be constructed, in principle exactly, from values at discrete sample points $\omega = n(2\pi/T)$. Prove this as follows.

(a) Show that the coefficients of a complex Fourier *series* representation of $f(t)$ with period T can be written as

$$c_n = \frac{\sqrt{2\pi}}{T} \tilde{f}\left(\frac{2\pi n}{T}\right).$$

(b) Use this result to represent $f(t)$ as an infinite sum in the defining integral for $\tilde{f}(\omega)$, and hence show that

$$\tilde{f}(\omega) = \sum_{n=-\infty}^{\infty} \tilde{f}\left(\frac{2\pi n}{T}\right) \operatorname{sinc}\left(n\pi - \frac{\omega T}{2}\right),$$

where sinc x is defined as $(\sin x)/x$.

(a) The complex coefficients for the Fourier series for $f(t)$ are given by

$$c_n = \frac{1}{T} \int_{-T/2}^{T/2} f(t) e^{-i2\pi nt/T} \, dt.$$

But, we also know that the Fourier transform of $f(t)$ is given by

$$\tilde{f}(\omega) = \frac{1}{\sqrt{2\pi}} \int_{-\infty}^{\infty} f(t) e^{-i\omega t} \, dt = \frac{1}{\sqrt{2\pi}} \int_{-T/2}^{T/2} f(t) e^{-i\omega t} \, dt.$$

Comparison of these two equations shows that $c_n = \frac{1}{T}\sqrt{2\pi}\, \tilde{f}\left(\frac{2\pi n}{T}\right)$.

(b) Using the Fourier series representation of $f(t)$, the frequency spectrum at a general frequency ω can now be constructed as

$$\tilde{f}(\omega) = \frac{1}{\sqrt{2\pi}} \int_{-T/2}^{T/2} f(t) e^{-i\omega t} \, dt$$

$$= \frac{1}{\sqrt{2\pi}} \int_{-T/2}^{T/2} \left[\sum_{n=-\infty}^{\infty} \frac{1}{T}\sqrt{2\pi}\, \tilde{f}\left(\frac{2\pi n}{T}\right) e^{i2\pi nt/T} \right] e^{-i\omega t} \, dt$$

$$= \frac{1}{T} \sum_{n=-\infty}^{\infty} \tilde{f}\left(\frac{2\pi n}{T}\right) \frac{2\sin\left(\dfrac{2\pi n}{2} - \dfrac{\omega T}{2}\right)}{\dfrac{2\pi n}{T} - \omega} = \sum_{n=-\infty}^{\infty} \tilde{f}\left(\frac{2\pi n}{T}\right) \operatorname{sinc}\left(n\pi - \frac{\omega T}{2}\right).$$

This final formula gives a prescription for calculating the frequency spectrum $\tilde{f}(\omega)$ of $f(t)$ for *any* ω, given the spectrum at the (admittedly infinite number of) discrete values $\omega = 2\pi n/T$. The sinc functions give the weights to be assigned to the known discrete values; of course, the weights vary as ω is varied, with, as expected, the largest weights for the nth contribution occurring when ω is close to $2\pi n/T$.

5.13 Find the Fourier transform specified in part (a) and then use it to answer part (b).

(a) Find the Fourier transform of

$$f(\gamma, p, t) = \begin{cases} e^{-\gamma t} \sin pt & t > 0, \\ 0 & t < 0, \end{cases}$$

where γ (> 0) and p are constant parameters.

(b) The current $I(t)$ flowing through a certain system is related to the applied voltage $V(t)$ by the equation

$$I(t) = \int_{-\infty}^{\infty} K(t - u)V(u)\,du,$$

where

$$K(\tau) = a_1 f(\gamma_1, p_1, \tau) + a_2 f(\gamma_2, p_2, \tau).$$

The function $f(\gamma, p, t)$ is as given in part (a) and all the a_i, γ_i (> 0) and p_i are fixed parameters. By considering the Fourier transform of $I(t)$, find the relationship that must hold between a_1 and a_2 if the total net charge Q passed through the system (over a very long time) is to be zero for an arbitrary applied voltage.

(a) Write the given sine function in terms of exponential functions. Its Fourier transform is then easily calculated as

$$\tilde{f}(\omega, \gamma, p) = \frac{1}{\sqrt{2\pi}} \int_0^\infty \frac{e^{(-\gamma - i\omega + ip)t} - e^{(-\gamma - i\omega - ip)t}}{2i}\,dt$$

$$= \frac{1}{\sqrt{2\pi}} \frac{1}{2i} \left(\frac{-1}{-\gamma - i\omega + ip} + \frac{1}{-\gamma - i\omega - ip} \right)$$

$$= \frac{1}{\sqrt{2\pi}} \frac{p}{(\gamma + i\omega)^2 + p^2}.$$

(b) Since the current is given by the convolution

$$I(t) = \int_{-\infty}^{\infty} K(t - u)V(u)\,du,$$

the convolution theorem implies that the Fourier transforms of I, K and V are related by $\tilde{I}(\omega) = \sqrt{2\pi}\,\tilde{K}(\omega)\,\tilde{V}(\omega)$ with, from part (a),

$$\tilde{K}(\omega) = \frac{1}{\sqrt{2\pi}} \left[\frac{a_1 p_1}{(\gamma_1 + i\omega)^2 + p_1^2} + \frac{a_2 p_2}{(\gamma_2 + i\omega)^2 + p_2^2} \right].$$

Now, by expressing $I(t')$ in its Fourier integral form, we can write

$$Q(\infty) = \int_{-\infty}^{\infty} I(t')\,dt' = \int_{-\infty}^{\infty} dt' \int_{-\infty}^{\infty} \frac{1}{\sqrt{2\pi}}\,\tilde{I}(\omega)\,e^{i\omega t'}\,d\omega.$$

But $\int_{-\infty}^{\infty} e^{i\omega t'} \, dt' = 2\pi \, \delta(\omega)$ and so

$$Q(\infty) = \int_{-\infty}^{\infty} \frac{1}{\sqrt{2\pi}} \tilde{I}(\omega) \, 2\pi \, \delta(\omega) \, d\omega$$

$$= \frac{2\pi}{\sqrt{2\pi}} \tilde{I}(0) = \sqrt{2\pi} \sqrt{2\pi} \, \tilde{K}(0) \, \tilde{V}(0)$$

$$= 2\pi \frac{1}{\sqrt{2\pi}} \left[\frac{a_1 p_1}{\gamma_1^2 + p_1^2} + \frac{a_2 p_2}{\gamma_2^2 + p_2^2} \right] \tilde{V}(0).$$

For $Q(\infty)$ to be zero for an arbitrary $V(t)$, we must have

$$\frac{a_1 p_1}{\gamma_1^2 + p_1^2} + \frac{a_2 p_2}{\gamma_2^2 + p_2^2} = 0,$$

and so this is the required relationship.

5.15 A linear amplifier produces an output that is the convolution of its input and its response function. The Fourier transform of the response function for a particular amplifier is

$$\tilde{K}(\omega) = \frac{i\omega}{\sqrt{2\pi}(\alpha + i\omega)^2}.$$

Determine the time variation of its output $g(t)$ when its input is the Heaviside step function.

This result is immediate, since differentiating the definition of a Fourier transform (under the integral sign) gives

$$i \frac{d\tilde{f}(\omega)}{d\omega} = \frac{i}{\sqrt{2\pi}} \frac{\partial}{\partial \omega} \left(\int_{-\infty}^{\infty} f(t) \, e^{-i\omega t} \, dt \right) = \frac{-i^2}{\sqrt{2\pi}} \int_{-\infty}^{\infty} t f(t) \, e^{-i\omega t} \, dt,$$

i.e. the transform of $t f(t)$.

Since the amplifier's output is the convolution of its input and response function, we will need the Fourier transforms of both to determine that of its output (using the convolution theorem). We already have that of its response function.

The input Heaviside step function $H(t)$ has a Fourier transform

$$\tilde{H}(\omega) = \frac{1}{\sqrt{2\pi}} \int_{-\infty}^{\infty} H(t) \, e^{-i\omega t} \, dt = \frac{1}{\sqrt{2\pi}} \int_{0}^{\infty} e^{-i\omega t} \, dt = \frac{1}{\sqrt{2\pi}} \frac{1}{i\omega}.$$

Thus, using the convolution theorem,

$$\tilde{g}(\omega) = \sqrt{2\pi} \frac{i\omega}{\sqrt{2\pi}(\alpha + i\omega)^2} \frac{1}{\sqrt{2\pi}} \frac{1}{i\omega}$$

$$= \frac{1}{\sqrt{2\pi}} \frac{1}{(\alpha + i\omega)^2}$$

$$= \frac{i}{\sqrt{2\pi}} \frac{\partial}{\partial \omega} \left(\frac{1}{\alpha + i\omega} \right)$$

$$= i \frac{\partial}{\partial \omega} \left\{ \mathcal{F} \left[e^{-\alpha t} H(t) \right] \right\}$$

$$= \mathcal{F} \left[t e^{-\alpha t} H(t) \right],$$

where we have used the "library" result to recognize the transform of a decaying exponential in the penultimate line and the result proved above in the final step. The output of the amplifier is therefore of the form $g(t) = te^{-\alpha t}$ for $t > 0$ when its input takes the form of the Heaviside step function.

5.17 [This problem can only be attempted if Problem 5.16 of the main text has been studied.]

For some ion–atom scattering processes, the spherically symmetric potential $V(\mathbf{r})$ of the previous problem may be approximated by $V = |\mathbf{r}_1 - \mathbf{r}_2|^{-1} \exp(-\mu |\mathbf{r}_1 - \mathbf{r}_2|)$. Show, using the result of the worked example in Subsection 5.1.9, that the probability that the ion will scatter from, say, $\mathbf{p}_1$ to $\mathbf{p}_1'$ is proportional to $(\mu^2 + k^2)^{-2}$, where $k = |\mathbf{k}|$ and $\mathbf{k}$ is as given in part (c) of the previous problem.

As shown in Problem 5.16, the Fourier transform of a spherically symmetric potential $V(r)$ is given by

$$\tilde{V}(\mathbf{k}) = \frac{1}{(2\pi)^{3/2}k} \int_0^\infty 4\pi V(r) r \sin kr \, dr.$$

The ion–atom interaction potential in this particular example is $V(r) = r^{-1} \exp(-\mu r)$. As this is spherically symmetric, we may apply the result to it. Substituting for $V(r)$ gives

$$M \propto \tilde{V}(\mathbf{k}) \propto \frac{1}{k} \int_0^\infty \frac{e^{-\mu r}}{r} r \sin kr \, dr$$

$$= \frac{1}{k} \operatorname{Im} \int_0^\infty e^{-\mu r + ikr} \, dr$$

$$= \frac{1}{k} \operatorname{Im} \left[\frac{-1}{-\mu + ik} \right]$$

$$= \frac{1}{k} \frac{k}{\mu^2 + k^2}.$$

Since the probability of the ion scattering from $\mathbf{p}_1$ to $\mathbf{p}_1'$ is proportional to the modulus squared of M, the probability is $\propto |M|^2 \propto (\mu^2 + k^2)^{-2}$.

5.19 Find the Laplace transforms of $t^{-1/2}$ and $t^{1/2}$, by setting $x^2 = ts$ in the result

$$\int_0^\infty \exp(-x^2) \, dx = \tfrac{1}{2}\sqrt{\pi}.$$

Setting $x^2 = st$, and hence $2x \, dx = s \, dt$ and $dx = s \, dt/(2\sqrt{st})$, we obtain

$$\int_0^\infty e^{-st} \frac{\sqrt{s}}{2} t^{-1/2} \, dt = \frac{\sqrt{\pi}}{2},$$

$$\Rightarrow \quad \mathcal{L}\left[t^{-1/2}\right] \equiv \int_0^\infty t^{-1/2} e^{-st} \, dt = \sqrt{\frac{\pi}{s}}.$$

Integrating the LHS of this result by parts yields

$$\left[e^{-st}\, 2t^{1/2}\right]_0^{\infty} - \int_0^{\infty} (-s)\, e^{-st}\, 2t^{1/2}\, dt = \sqrt{\frac{\pi}{s}}.$$

The first term vanishes at both limits, whilst the second is a multiple of the required Laplace transform of $t^{1/2}$. Hence,

$$\mathcal{L}\left[t^{1/2}\right] \equiv \int_0^{\infty} e^{-st}\, t^{1/2}\, dt = \frac{1}{2s}\sqrt{\frac{\pi}{s}}.$$

5.21 Use the properties of Laplace transforms to prove the following without evaluating any Laplace integrals explicitly:

(a) $\mathcal{L}\left[t^{5/2}\right] = \frac{15}{8}\sqrt{\pi}\, s^{-7/2}$;

(b) $\mathcal{L}\left[(\sinh at)/t\right] = \frac{1}{2}\ln\left[(s+a)/(s-a)\right]$, $s > |a|$;

(c) $\mathcal{L}\left[\sinh at \cos bt\right] = a(s^2 - a^2 + b^2)[(s-a)^2 + b^2]^{-1}[(s+a)^2 + b^2]^{-1}$.

(a) We use the general result for Laplace transforms that

$$\mathcal{L}\left[t^n f(t)\right] = (-1)^n \frac{d^n \bar{f}(s)}{ds^n}, \qquad \text{for } n = 1, 2, 3, \dots.$$

If we take $n = 2$, then $f(t)$ becomes $t^{1/2}$, for which we found the Laplace transform in Problem 5.19:

$$\mathcal{L}\left[t^{5/2}\right] = \mathcal{L}\left[t^2\, t^{1/2}\right] = (-1)^2 \frac{d^2}{ds^2}\left(\frac{\sqrt{\pi}\, s^{-3/2}}{2}\right)$$

$$= \frac{\sqrt{\pi}}{2}\left(-\frac{3}{2}\right)\left(-\frac{5}{2}\right) s^{-7/2} = \frac{15\sqrt{\pi}}{8} s^{-7/2}.$$

(b) Here we apply a second general result for Laplace transforms which states that

$$\mathcal{L}\left[\frac{f(t)}{t}\right] = \int_s^{\infty} \bar{f}(u)\, du,$$

provided $\lim_{t \to 0}[f(t)/t]$ exists, which it does in this case.

$$\mathcal{L}\left[\frac{\sinh(at)}{t}\right] = \int_s^{\infty} \frac{a}{u^2 - a^2}\, du, \quad u > |a|,$$

$$= \frac{1}{2}\int_s^{\infty} \left(\frac{1}{u-a} - \frac{1}{u+a}\right) du$$

$$= \frac{1}{2}\ln\left(\frac{s+a}{s-a}\right), \quad s > |a|.$$

(c) The translation property of Laplace transforms can be used here to deal with the sinh(at) factor, as it can be expressed in terms of exponential functions:

$$\mathcal{L}\left[\sinh(at)\cos(bt)\right] = \mathcal{L}\left[\tfrac{1}{2}e^{at}\cos(bt)\right] - \mathcal{L}\left[\tfrac{1}{2}e^{-at}\cos(bt)\right]$$

$$= \frac{1}{2}\frac{s-a}{(s-a)^2+b^2} - \frac{1}{2}\frac{s+a}{(s+a)^2+b^2}$$

$$= \frac{1}{2}\frac{(s^2-a^2)2a+2ab^2}{[(s-a)^2+b^2][(s+a)^2+b^2]}$$

$$= \frac{a(s^2-a^2+b^2)}{[(s-a)^2+b^2][(s+a)^2+b^2]}.$$

The result is valid for $s > |a|$.

5.23 This problem is concerned with the limiting behavior of Laplace transforms.

(a) If $f(t) = A + g(t)$, where A is a constant and the indefinite integral of $g(t)$ is bounded as its upper limit tends to ∞, show that

$$\lim_{s\to 0} s\bar{f}(s) = A.$$

(b) For $t > 0$, the function $y(t)$ obeys the differential equation

$$\frac{d^2 y}{dt^2} + a\frac{dy}{dt} + by = c\cos^2\omega t,$$

where a, b and c are positive constants. Find $\bar{y}(s)$ and show that $s\bar{y}(s) \to c/2b$ as $s \to 0$. Interpret the result in the t-domain.

(a) From the definition,

$$\bar{f}(s) = \int_0^\infty [A + g(t)]\, e^{-st}\, dt$$

$$= \left[\frac{A\,e^{-st}}{-s}\right]_0^\infty + \lim_{T\to\infty}\int_0^T g(t)\, e^{-st}\, dt,$$

$$s\bar{f}(s) = A + s\lim_{T\to\infty}\int_0^T g(t)\, e^{-st}\, dt.$$

Now, for $s \geq 0$,

$$\left|\lim_{T\to\infty}\int_0^T g(t)\, e^{-st}\, dt\right| \leq \left|\lim_{T\to\infty}\int_0^T g(t)\, dt\right| < B,\text{ say.}$$

Thus, taking the limit $s \to 0$,

$$\lim_{s\to 0} s\bar{f}(s) = A \pm \lim_{s\to 0} sB = A.$$

(b) We will need

$$\mathcal{L}\left[\cos^2\omega t\right] = \mathcal{L}\left[\tfrac{1}{2}\cos 2\omega + \tfrac{1}{2}\right] = \frac{s}{2(s^2+4\omega^2)} + \frac{1}{2s}.$$

Taking the transform of the differential equation yields

$$-y'(0) - sy(0) + s^2\bar{y} + a[-y(0) + s\bar{y}] + b\bar{y} = c\left[\frac{s}{2(s^2 + 4\omega^2)} + \frac{1}{2s}\right].$$

This can be rearranged as

$$s\bar{y} = \frac{c\left(\dfrac{s^2}{2(s^2 + 4\omega^2)} + \dfrac{1}{2}\right) + sy'(0) + asy(0) + s^2y(0)}{s^2 + as + b}.$$

In the limit $s \to 0$, this tends to $(c/2)/b = c/(2b)$, a value independent of that of a and the initial values of y and y'.

The $s = 0$ component of the transform corresponds to long-term values, when a steady state has been reached and rates of change are negligible. With the first two terms of the differential equation ignored, it reduces to $by = c\cos^2 \omega t$, and, as the average value of $\cos^2 \omega t$ is $\frac{1}{2}$, the solution is the more or less steady value of $y = \frac{1}{2}c/b$.

5.25 The function $f_a(x)$ is defined as unity for $0 < x < a$ and zero otherwise. Find its Laplace transform $\bar{f}_a(s)$ and deduce that the transform of $xf_a(x)$ is

$$\frac{1}{s^2}\left[1 - (1 + as)e^{-sa}\right].$$

Write $f_a(x)$ in terms of Heaviside functions and hence obtain an explicit expression for

$$g_a(x) = \int_0^x f_a(y)f_a(x - y)\,dy.$$

Use the expression to write $\bar{g}_a(s)$ in terms of the functions $\bar{f}_a(s)$ and $\bar{f}_{2a}(s)$, and their derivatives, and hence show that $\bar{g}_a(s)$ is equal to the square of $\bar{f}_a(s)$, in accordance with the convolution theorem.

From their definitions,

$$\bar{f}_a(s) = \int_0^a 1\,e^{-sx}\,dx = \frac{1}{s}(1 - e^{-sa}),$$

$$\int_0^a x\,f_a(x)\,e^{-sx}\,dx = -\frac{d\bar{f}_a}{ds} = \frac{1}{s^2}(1 - e^{-sa}) - \frac{a}{s}e^{-sa}$$

$$= \frac{1}{s^2}\left[1 - (1 + as)e^{-sa}\right]. \qquad (*)$$

In terms of Heaviside functions,

$$f(x) = H(x) - H(x - a),$$

and so the expression for $g_a(x) = \int_0^x f_a(y)f_a(x - y)\,dy$ is

$$\int_{-\infty}^{\infty} [H(y) - H(y - a)]\,[H(x - y) - H(x - y - a)]\,dy.$$

This can be expanded as the sum of four integrals, each of which contains the common factors $H(y)$ and $H(x - y)$, implying that, in all cases, unless x is positive and greater than y, the integral has zero value. The other factors in the four integrands are generated analogously to the terms of the expansion $(a - b)(c - d) = ac - ad - bc + bd$:

$$\int_{-\infty}^{\infty} H(y)H(x - y)\,dy$$

$$- \int_{-\infty}^{\infty} H(y)H(x - y - a)\,dy$$

$$- \int_{-\infty}^{\infty} H(y - a)H(x - y)\,dy$$

$$+ \int_{-\infty}^{\infty} H(y - a)H(x - y - a)\,dy.$$

In all four integrals the integrand is either 0 or 1 and the value of each integral is equal to the length of the y-interval in which the integrand is non-zero.

- The first integral requires $0 < y < x$ and therefore has value x for $x > 0$.
- The second integral requires $0 < y < x - a$ and therefore has value $x - a$ for $x > a$ and 0 for $x < a$.
- The third integral requires $a < y < x$ and therefore has value $x - a$ for $x > a$ and 0 for $x < a$.
- The final integral requires $a < y < x - a$ and therefore has value $x - 2a$ for $x > 2a$ and 0 for $x < 2a$.

Collecting these together:

$$
\begin{aligned}
x < 0 \qquad & g_a(x) = 0 - 0 - 0 + 0 = 0, \\
0 < x < a \qquad & g_a(x) = x - 0 - 0 + 0 = x, \\
a < x < 2a \qquad & g_a(x) = x - (x - a) - (x - a) + 0 = 2a - x, \\
2a < x \qquad & g_a(x) = x - (x - a) - (x - a) + (x - 2a) = 0.
\end{aligned}
$$

Consequently, the transform of $g_a(x)$ is given by

$$
\begin{aligned}
\bar{g}_a(s) &= \int_0^a x e^{-sx}\,dx + \int_a^{2a} (2a - x)e^{-sx}\,dx \\
&= -\int_0^{2a} x e^{-sx}\,dx + 2 \int_0^a x e^{-sx}\,dx + 2a \int_a^{2a} e^{-sx}\,dx \\
&= -\frac{1}{s^2}\left[1 - (1 + 2as)e^{-2sa}\right] + \frac{2}{s^2}\left[1 - (1 + as)e^{-sa}\right] \\
&\quad + \frac{2a}{s}(e^{-sa} - e^{-2sa}) \\
&= \frac{1}{s^2}(1 - 2e^{-sa} + e^{-2sa}) \\
&= \frac{1}{s^2}(1 - e^{-as})^2 = [\bar{f}_a(s)]^2,
\end{aligned}
$$

which is as expected. In order to adjust the integral limits in the second line, we both added and subtracted

$$\int_0^a (-x)e^{-sx}\,dx.$$

In the third line we used the result (∗) twice, once as it stands and once with a replaced by $2a$.

6 Higher-order ordinary differential equations

6.1 A simple harmonic oscillator, of mass m and natural frequency ω_0, experiences an oscillating driving force $f(t) = ma \cos \omega t$. Therefore, its equation of motion is

$$\frac{d^2x}{dt^2} + \omega_0^2 x = a \cos \omega t,$$

where x is its position. Given that at $t = 0$ we have $x = dx/dt = 0$, find the function $x(t)$. Describe the solution if ω is approximately, but not exactly, equal to ω_0.

To find the full solution given the initial conditions, we need the complete general solution made up of a complementary function (CF) and a particular integral (PI). The CF is clearly of the form $A \cos \omega_0 t + B \sin \omega_0 t$ and, in view of the form of the RHS, we try $x(t) = C \cos \omega t + D \sin \omega t$ as a PI. Substituting this gives

$$-\omega^2 C \cos \omega t - \omega^2 D \sin \omega t + \omega_0^2 C \cos \omega t + \omega_0^2 D \sin \omega t = a \cos \omega t.$$

Equating coefficients of the independent functions $\cos \omega t$ and $\sin \omega t$ requires that

$$-\omega^2 C + \omega_0^2 C = a \quad \Rightarrow \quad C = \frac{a}{\omega_0^2 - \omega^2},$$

$$-\omega^2 D + \omega_0^2 D = 0 \quad \Rightarrow \quad D = 0.$$

Thus, the general solution is

$$x(t) = A \cos \omega_0 t + B \sin \omega_0 t + \frac{a}{\omega_0^2 - \omega^2} \cos \omega t.$$

The initial conditions impose the requirements

$$x(0) = 0 \quad \Rightarrow \quad 0 = A + \frac{a}{\omega_0^2 - \omega^2},$$

$$\text{and } \dot{x}(0) = 0 \quad \Rightarrow \quad 0 = \omega_0 B.$$

Incorporating the implications of these into the general solution gives

$$x(t) = \frac{a}{\omega_0^2 - \omega^2} (\cos \omega t - \cos \omega_0 t)$$

$$= \frac{2a \sin[\frac{1}{2}(\omega + \omega_0)t] \sin[\frac{1}{2}(\omega_0 - \omega)t]}{(\omega_0 + \omega)(\omega_0 - \omega)}.$$

For $\omega_0 - \omega = \epsilon$ with $|\epsilon| t \ll 1$,

$$x(t) \approx \frac{2a \sin \omega_0 t \, \frac{1}{2}\epsilon t}{2\omega_0 \, \epsilon} = \frac{at}{2\omega_0} \sin \omega_0 t.$$

Thus, for moderate t, $x(t)$ is a sine wave of linearly increasing amplitude.

Over a long time, $x(t)$ will vary between $\pm 2a/(\omega_0^2 - \omega^2)$ with sizeable intervals between the two extremes, i.e. it will show beats of amplitude $2a/(\omega_0^2 - \omega^2)$.

6.3 The theory of bent beams shows that at any point in the beam the "bending moment" is given by K/ρ, where K is a constant (that depends upon the beam material and cross-sectional shape) and ρ is the radius of curvature at that point. Consider a light beam of length L whose ends, $x = 0$ and $x = L$, are supported at the same vertical height and which has a weight W suspended from its center. Verify that at any point x ($0 \le x \le L/2$ for definiteness) the net magnitude of the bending moment (bending moment = force $\times$ perpendicular distance) due to the weight and support reactions, evaluated on either side of x, is $Wx/2$.

If the beam is only slightly bent, so that $(dy/dx)^2 \ll 1$, where $y = y(x)$ is the downward displacement of the beam at x, show that the beam profile satisfies the approximate equation

$$\frac{d^2 y}{dx^2} = -\frac{Wx}{2K}.$$

By integrating this equation twice and using physically imposed conditions on your solution at $x = 0$ and $x = L/2$, show that the downward displacement at the center of the beam is $WL^3/(48K)$.

The upward reaction of the support at each end of the beam is $\frac{1}{2}W$.

At the position x the moment on the left is due to

(i) the support at $x = 0$ providing a clockwise moment of $\frac{1}{2}Wx$.

The moment on the right is due to

(ii) the support at $x = L$ providing an anticlockwise moment of $\frac{1}{2}W(L - x)$;

(iii) the weight at $x = \frac{1}{2}L$ providing a clockwise moment of $W(\frac{1}{2}L - x)$.

The net clockwise moment on the right is therefore $W(\frac{1}{2}L - x) - \frac{1}{2}W(L - x) = -\frac{1}{2}Wx$, i.e. equal in magnitude, but opposite in sign, to that on the left.

The radius of curvature of the beam is $\rho = [\,1 + (-y')^2\,]^{3/2}/(-y'')$, but if $|y'| \ll 1$ this simplifies to $-1/y''$ and the equation of the beam profile satisfies

$$\frac{Wx}{2} = M = \frac{K}{\rho} = -K\frac{d^2 y}{dx^2}.$$

We now need to integrate this, taking into account the boundary conditions $y(0) = 0$ and, on symmetry grounds, $y'(\frac{1}{2}L) = 0$:

$$y' = -\frac{Wx^2}{4K} + A, \text{ with } y'(\tfrac{1}{2}L) = 0 \quad \Rightarrow \quad A = \frac{WL^2}{16K},$$

$$y' = \frac{W}{4K}\left(\frac{L^2}{4} - x^2\right),$$

$$y = \frac{W}{4K}\left(\frac{L^2 x}{4} - \frac{x^3}{3} + B\right), \text{ with } y(0) = 0 \quad \Rightarrow \quad B = 0.$$

The center is lowered by

$$y(\tfrac{1}{2}L) = \frac{W}{4K}\left(\frac{L^2}{4}\frac{L}{2} - \frac{1}{3}\frac{L^3}{8}\right) = \frac{WL^3}{48K}.$$

Note that the derived analytic form for $y(x)$ is not applicable in the range $\frac{1}{2}L \le x \le L$; the beam profile is symmetrical about $x = \frac{1}{2}L$, but the expression $\frac{1}{4}L^2 x - \frac{1}{3}x^3$ is not invariant under the substitution $x \to L - x$.

6.5 The function $f(t)$ satisfies the differential equation

$$\frac{d^2 f}{dt^2} + 8\frac{df}{dt} + 12f = 12e^{-4t}.$$

For the following sets of boundary conditions determine whether it has solutions, and, if so, find them:

(a) $f(0) = 0$, $f'(0) = 0$, $f(\ln \sqrt{2}) = 0$;
(b) $f(0) = 0$, $f'(0) = -2$, $f(\ln \sqrt{2}) = 0$.

Three boundary conditions have been given, and, as this is a second-order linear equation for which only two independent conditions are needed, they may be inconsistent. The plan is to solve it using two of the conditions and then test whether the third one is compatible. The auxiliary equation for obtaining the CF is

$$m^2 + 8m + 12 = 0 \quad \Rightarrow \quad m = -2 \text{ or } m = -6$$
$$\Rightarrow \quad f(t) = Ae^{-6t} + Be^{-2t}.$$

Since the form of the RHS, Ce^{-4t}, is not included in the CF, we can try it as the particular integral:

$$16C - 32C + 12C = 12 \quad \Rightarrow \quad C = -3.$$

The general solution is therefore

$$f(t) = Ae^{-6t} + Be^{-2t} - 3e^{-4t}.$$

(a) For boundary conditions $f(0) = 0$, $f'(0) = 0$, $f(\ln \sqrt{2}) = 0$:

$$f(0) = 0 \quad \Rightarrow \quad A + B - 3 = 0,$$
$$f'(0) = 0 \quad \Rightarrow \quad -6A - 2B + 12 = 0,$$
$$\Rightarrow \quad A = \tfrac{3}{2}, \quad B = \tfrac{3}{2}.$$
$$\text{Hence, } f(t) \quad = \quad \tfrac{3}{2}e^{-6t} + \tfrac{3}{2}e^{-2t} - 3e^{-4t}.$$

Recalling that $e^{-(\ln \sqrt{2})} = 1/\sqrt{2}$, we evaluate

$$f(\ln \sqrt{2}) = \frac{3}{2}\frac{1}{8} + \frac{3}{2}\frac{1}{2} - 3\frac{1}{4} = \frac{3}{16} \ne 0.$$

Thus the boundary conditions are inconsistent and there is no solution.

(b) For boundary conditions $f(0) = 0$, $f'(0) = -2$, $f(\ln \sqrt{2}) = 0$, we proceed as before:

$$f(0) = 0 \quad \Rightarrow \quad A + B - 3 = 0,$$
$$f'(0) = 0 \quad \Rightarrow \quad -6A - 2B + 12 = -2,$$
$$\Rightarrow \quad A = 2, \quad B = 1.$$
$$\text{Hence, } f(t) \quad = \quad 2e^{-6t} + e^{-2t} - 3e^{-4t}.$$

We again evaluate

$$f(\ln \sqrt{2}) = 2\frac{1}{8} + \frac{1}{2} - 3\frac{1}{4} = 0.$$

This time the boundary conditions are consistent and there is a unique solution as given above.

6.7 A solution of the differential equation

$$\frac{d^2y}{dx^2} + 2\frac{dy}{dx} + y = 4e^{-x}$$

takes the value 1 when $x = 0$ and the value e^{-1} when $x = 1$. What is its value when $x = 2$?

The auxiliary equation, $m^2 + 2m + 1 = 0$, has repeated roots $m = -1$, and so the general CF has the special form $y(x) = (A + Bx)e^{-x}$.

Turning to the PI, we note that the form of the RHS of the original equation is contained in the CF, and (to make matters worse) so is x times the RHS. We therefore need to take x^2 times the RHS as a trial PI:

$$y(x) = Cx^2e^{-x}, \quad y' = C(2x - x^2)e^{-x}, \quad y'' = C(2 - 4x + x^2)e^{-x}.$$

Substituting these into the original equation shows that

$$2Ce^{-x} = 4e^{-x} \quad \Rightarrow \quad C = 2$$

and that the full general solution is given by

$$y(x) = (A + Bx)e^{-x} + 2x^2e^{-x}.$$

We now determine the unknown constants using the information given about the solution. Since $y(0) = 1$, $A = 1$. Further, $y(1) = e^{-1}$ requires

$$e^{-1} = (1 + B)e^{-1} + 2e^{-1} \quad \Rightarrow \quad B = -2.$$

Finally, we conclude that $y(x) = (1 - 2x + 2x^2)e^{-x}$ and, therefore, that $y(2) = 5e^{-2}$.

6.9 Find the general solutions of

(a) $\dfrac{d^3y}{dx^3} - 12\dfrac{dy}{dx} + 16y = 32x - 8,$

(b) $\dfrac{d}{dx}\left(\dfrac{1}{y}\dfrac{dy}{dx}\right) + (2a \coth 2ax)\left(\dfrac{1}{y}\dfrac{dy}{dx}\right) = 2a^2,$

where a is a constant.

(a) As this is a third-order equation, we expect three terms in the CF.

Since it is linear with constant coefficients, we can make use of the auxiliary equation, which is

$$m^3 - 12m + 16 = 0.$$

By inspection, $m = 2$ is one root; the other two can be found by factorization:

$$m^3 - 12m + 16 = (m - 2)(m^2 + 2m - 8) = (m - 2)(m + 4)(m - 2) = 0.$$

Thus we have one repeated root ($m = 2$) and one other ($m = -4$) leading to a CF of the form

$$y(x) = (A + Bx)e^{2x} + Ce^{-4x}.$$

As the RHS contains no exponentials, we try $y(x) = Dx + E$ for the PI. We then need $16D = 32$ and $-12D + 16E = -8$, giving $D = 2$ and $E = 1$.

The general solution is therefore

$$y(x) = (A + Bx)e^{2x} + Ce^{-4x} + 2x + 1.$$

(b) The equation is already arranged in the form

$$\frac{dg(y)}{dx} + h(x)g(y) = j(x)$$

and so needs only an integrating factor to allow the first integration step to be made. For this equation the IF is

$$\exp\left\{\int 2a \coth 2ax \, dx\right\} = \exp(\ln \sinh 2ax) = \sinh 2ax.$$

After multiplication through by this factor, the equation can be written

$$\sinh 2ax \frac{d}{dx}\left(\frac{1}{y}\frac{dy}{dx}\right) + (2a \cosh 2ax)\left(\frac{1}{y}\frac{dy}{dx}\right) = 2a^2 \sinh 2ax,$$

$$\frac{d}{dx}\left(\sinh 2ax \frac{1}{y}\frac{dy}{dx}\right) = 2a^2 \sinh 2ax.$$

Integrating this gives

$$\sinh 2ax \frac{1}{y}\frac{dy}{dx} = \frac{2a^2}{2a} \cosh 2ax + A,$$

$$\Rightarrow \quad \frac{1}{y}\frac{dy}{dx} = a \coth 2ax + \frac{A}{\sinh 2ax}.$$

Integrating again, $\quad \ln y = \frac{1}{2}\ln(\sinh 2ax) + \int \frac{A}{\sinh 2ax} \, dx + B$

$$= \frac{1}{2}\ln(\sinh 2ax) + \frac{A}{2a}\ln(|\tanh ax|) + B,$$

$$\Rightarrow \quad y = C(\sinh 2ax)^{1/2} (|\tanh ax|)^D.$$

The indefinite integral of $(\sinh 2ax)^{-1}$ appearing in the fourth line can be verified by differentiating $y = \ln |\tanh ax|$ in the form $y = \frac{1}{2} \ln(\tanh^2 ax)$ and recalling that

$$\cosh ax \sinh ax = \frac{1}{2} \sinh 2ax.$$

6.11 The quantities $x(t)$, $y(t)$ satisfy the simultaneous equations

$$\ddot{x} + 2n\dot{x} + n^2 x = 0,$$
$$\ddot{y} + 2n\dot{y} + n^2 y = \mu\dot{x},$$

where $x(0) = y(0) = \dot{y}(0) = 0$ and $\dot{x}(0) = \lambda$. Show that

$$y(t) = \tfrac{1}{2}\mu\lambda t^2 \left(1 - \tfrac{1}{3}nt\right) \exp(-nt).$$

For these two coupled equations, in which an "output" from the first acts as the "driving input" for the second, we take Laplace transforms and incorporate the boundary conditions:

$$(s^2\bar{x} - 0 - \lambda) + 2n(s\bar{x} - 0) + n^2\bar{x} = 0,$$
$$(s^2\bar{y} - 0 - 0) + 2n(s\bar{y} - 0) + n^2\bar{y} = \mu(s\bar{x} - 0).$$

From the first transformed equation,

$$\bar{x} = \frac{\lambda}{s^2 + 2ns + n^2}.$$

Substituting this into the second transformed equation gives

$$\bar{y} = \frac{\mu s\bar{x}}{(s+n)^2} = \frac{\mu\lambda s}{(s+n)^2(s+n)^2}$$
$$= \frac{\mu\lambda}{(s+n)^3} - \frac{\mu\lambda n}{(s+n)^4},$$
$$\Rightarrow \quad y(t) = \mu\lambda \left(\frac{t^2}{2!} e^{-nt} - \frac{nt^3}{3!} e^{-nt}\right), \text{ from the look-up table,}$$
$$= \frac{1}{2}\mu\lambda t^2 \left(1 - \frac{nt}{3}\right) e^{-nt},$$

i.e. as stated in the question.

6.13 Two unstable isotopes A and B and a stable isotope C have the following decay rates per atom present: $A \rightarrow B$, $3\,\text{s}^{-1}$; $A \rightarrow C$, $1\,\text{s}^{-1}$; $B \rightarrow C$, $2\,\text{s}^{-1}$. Initially a quantity x_0 of A is present but there are no atoms of the other two types. Using Laplace transforms, find the amount of C present at a later time t.

Using the name symbol to represent the corresponding number of atoms and taking Laplace transforms, we have

$$\frac{dA}{dt} = -(3+1)A \quad \Rightarrow \quad s\bar{A} - x_0 = -4\bar{A}$$

$$\Rightarrow \quad \bar{A} = \frac{x_0}{s+4},$$

$$\frac{dB}{dt} = 3A - 2B \quad \Rightarrow \quad s\bar{B} = 3\bar{A} - 2\bar{B}$$

$$\Rightarrow \quad \bar{B} = \frac{3x_0}{(s+2)(s+4)},$$

$$\frac{dC}{dt} = A + 2B \quad \Rightarrow \quad s\bar{C} = \bar{A} + 2\bar{B}$$

$$\Rightarrow \quad \bar{C} = \frac{x_0(s+2) + 6x_0}{s(s+2)(s+4)}.$$

Using the "cover-up" method for finding the coefficients of a partial fraction expansion without repeated factors, e.g. the coefficient of $(s+2)^{-1}$ is $[(-2+8)x_0]/[(-2)(-2+4)] = -6x_0/4$, we have

$$\bar{C} = \frac{x_0(s+8)}{s(s+2)(s+4)} = \frac{x_0}{s} - \frac{6x_0}{4(s+2)} + \frac{4x_0}{8(s+4)}$$

$$\Rightarrow \quad C(t) = x_0 \left(1 - \tfrac{3}{2}e^{-2t} + \tfrac{1}{2}e^{-4t} \right).$$

This is the required expression.

6.15 The "golden mean", which is said to describe the most aesthetically pleasing proportions for the sides of a rectangle (e.g. the ideal picture frame), is given by the limiting value of the ratio of successive terms of the Fibonacci series u_n, which is generated by

$$u_{n+2} = u_{n+1} + u_n,$$

with $u_0 = 0$ and $u_1 = 1$. Find an expression for the general term of the series and verify that the golden mean is equal to the larger root of the recurrence relation's characteristic equation.

The recurrence relation is second order and its characteristic equation, obtained by setting $u_n = A\lambda^n$, is

$$\lambda^2 - \lambda - 1 = 0 \quad \Rightarrow \quad \lambda = \tfrac{1}{2}(1 \pm \sqrt{5}).$$

The general solution is therefore

$$u_n = A \left(\frac{1+\sqrt{5}}{2} \right)^n + B \left(\frac{1-\sqrt{5}}{2} \right)^n.$$

The initial values (boundary conditions) determine A and B:

$$u_0 = 0 \quad \Rightarrow \quad B = -A,$$

$$u_1 = 1 \quad \Rightarrow \quad A\left(\frac{1+\sqrt{5}}{2} - \frac{1-\sqrt{5}}{2}\right) = 1 \quad \Rightarrow \quad A = \frac{1}{\sqrt{5}}.$$

Hence, $\quad u_n \quad = \quad \dfrac{1}{\sqrt{5}}\left[\left(\dfrac{1+\sqrt{5}}{2}\right)^n - \left(\dfrac{1-\sqrt{5}}{2}\right)^n\right].$

If we write $(1 - \sqrt{5})/(1 + \sqrt{5}) = r < 1$, the ratio of successive terms in the series is

$$\frac{u_{n+1}}{u_n} = \frac{\frac{1}{2}[(1+\sqrt{5})^{n+1} - (1-\sqrt{5})^{n+1}]}{(1+\sqrt{5})^n - (1-\sqrt{5})^n}$$

$$= \frac{\frac{1}{2}[1+\sqrt{5} - (1-\sqrt{5})r^n]}{1 - r^n}$$

$$\rightarrow \frac{1+\sqrt{5}}{2} \quad \text{as } n \rightarrow \infty;$$

i.e. the limiting ratio is the same as the larger value of λ.

This result is a particular example of the more general one that the ratio of successive terms in a series generated by a recurrence relation tends to the largest (in absolute magnitude) of the roots of the characteristic equation. Here there are only two roots, but for an Nth-order relation there will be N roots.

6.17 The first few terms of a series u_n, starting with u_0, are 1, 2, 2, 1, 6, −3. The series is generated by a recurrence relation of the form

$$u_n = Pu_{n-2} + Qu_{n-4},$$

where P and Q are constants. Find an expression for the general term of the series and show that, in fact, the series consists of two interleaved series given by

$$u_{2m} = \tfrac{2}{3} + \tfrac{1}{3}4^m,$$

$$u_{2m+1} = \tfrac{7}{3} - \tfrac{1}{3}4^m,$$

for $m = 0, 1, 2, \ldots$.

We first find P and Q using

$$n = 4 \qquad 6 = 2P + Q,$$

$$n = 5 \qquad -3 = P + 2Q, \qquad \Rightarrow \qquad Q = -4 \text{ and } P = 5.$$

The recurrence relation is thus

$$u_n = 5u_{n-2} - 4u_{n-4}.$$

To solve this we try $u_n = A + B\lambda^n$ for arbitrary constants A and B and obtain

$$A + B\lambda^n = 5A + 5B\lambda^{n-2} - 4A - 4B\lambda^{n-4},$$
$$\Rightarrow \quad 0 = \lambda^4 - 5\lambda^2 + 4$$
$$= (\lambda^2 - 1)(\lambda^2 - 4) \quad \Rightarrow \quad \lambda = \pm 1, \pm 2.$$

The general solution is $u_n = A + B(-1)^n + C2^n + D(-2)^n$.

We now need to solve the simultaneous equations for A, B, C and D provided by the values of $u_0, \ldots, u_3$:

$$1 = A + B + C + D,$$
$$2 = A - B + 2C - 2D,$$
$$2 = A + B + 4C + 4D,$$
$$1 = A - B + 8C - 8D.$$

These have the straightforward solution

$$A = \frac{3}{2}, \qquad B = -\frac{5}{6}, \qquad C = \frac{1}{12}, \qquad D = \frac{1}{4},$$

and so

$$u_n = \frac{3}{2} - \frac{5}{6}(-1)^n + \frac{1}{12}2^n + \frac{1}{4}(-2)^n.$$

When n is even and equal to $2m$,

$$u_{2m} = \frac{3}{2} - \frac{5}{6} + \frac{4^m}{12} + \frac{4^m}{4} = \frac{2}{3} + \frac{4^m}{3}.$$

When n is odd and equal to $2m + 1$,

$$u_{2m+1} = \frac{3}{2} + \frac{5}{6} + \frac{4^m}{6} - \frac{4^m}{2} = \frac{7}{3} - \frac{4^m}{3}.$$

In passing, we note that the fact that both P and Q, and all of the given values $u_0, \ldots, u_4$, are integers, and hence that all terms in the series are integers, provides an indirect proof that $4^m + 2$ is divisible by 3 (without remainder) for all non-negative integers m. This can be more easily proved by induction, as the reader may like to verify.

6.19 Find the general expression for the u_n satisfying

$$u_{n+1} = 2u_{n-2} - u_n$$

with $u_0 = u_1 = 0$ and $u_2 = 1$, and show that they can be written in the form

$$u_n = \frac{1}{5} - \frac{2^{n/2}}{\sqrt{5}} \cos\left(\frac{3\pi n}{4} - \phi\right),$$

where $\tan \phi = 2$.

The characteristic equation (which will be a cubic since the recurrence relation is third order) and its solution are given by

$$\lambda^{n+1} = 2\lambda^{n-2} - \lambda^n,$$

$$\lambda^3 + \lambda^2 - 2 = 0,$$

$$(\lambda - 1)(\lambda^2 + 2\lambda + 2) = 0 \quad \Rightarrow \quad \lambda = 1 \text{ or } \lambda = -1 \pm i.$$

Thus the general solution of the recurrence relation, which has the generic form $A\lambda_1^n + B\lambda_2^n + C\lambda_3^n$, is

$$u_n = A + B(-1+i)^n + C(-1-i)^n$$

$$= A + B\, 2^{n/2} e^{i3\pi n/4} + C\, 2^{n/2} e^{i5\pi n/4}.$$

To determine A, B and C we use

$$u_0 = 0, \qquad 0 = A + B + C,$$

$$u_1 = 0, \qquad 0 = A + B\, 2^{1/2} e^{i3\pi/4} + C\, 2^{1/2} e^{i5\pi/4}$$

$$= A + B(-1+i) + C(-1-i),$$

$$u_2 = 1, \qquad 1 = A + B\, 2e^{i6\pi/4} + C\, 2e^{i10\pi/4} = A + 2B(-i) + 2C(i).$$

Adding twice each of the first two equations to the last one gives $5A = 1$. Substituting this into the first and last equations then leads to

$$B + C = -\frac{1}{5} \qquad \text{and} \qquad -B + C = \frac{2}{5i},$$

from which it follows that

$$B = \frac{-1+2i}{10} = \frac{\sqrt{5}}{10} e^{i(\pi-\phi)}$$

$$\text{and} \quad C = \frac{-1-2i}{10} = \frac{\sqrt{5}}{10} e^{i(\pi+\phi)},$$

where $\tan\phi = 2/1 = 2$.
   Thus, collecting these results together, we have

$$u_n = \frac{1}{5} + \frac{2^{n/2}\sqrt{5}}{10} (e^{i3\pi n/4} e^{i(\pi-\phi)} + e^{i5\pi n/4} e^{i(\pi+\phi)})$$

$$= \frac{1}{5} - \frac{2^{n/2}\sqrt{5}}{10} (e^{i3\pi n/4} e^{-i\phi} + e^{-i3\pi n/4} e^{i\phi})$$

$$= \frac{1}{5} - \frac{2^{n/2}\sqrt{5}}{10} \left[2\cos\left(\frac{3\pi n}{4} - \phi\right) \right]$$

$$= \frac{1}{5} - \frac{2^{n/2}}{\sqrt{5}} \cos\left(\frac{3\pi n}{4} - \phi\right),$$

i.e. the form of solution given in the question.

6.21 Find the general solution of

$$x^2\frac{d^2y}{dx^2} - x\frac{dy}{dx} + y = x,$$

given that $y(1) = 1$ and $y(e) = 2e$.

This is Euler's equation and can be solved either by a change of variables, $x = e^t$, or by trying $y = x^\lambda$; we will adopt the second approach. Doing so in the homogeneous equation (RHS set to zero) gives

$$x^2\,\lambda(\lambda - 1)x^{\lambda-2} - x\,\lambda x^{\lambda-1} + x^\lambda = 0.$$

The CF is therefore obtained when λ satisfies

$$\lambda(\lambda - 1) - \lambda + 1 = 0 \quad \Rightarrow \quad (\lambda - 1)^2 = 0 \quad \Rightarrow \quad \lambda = 1 \text{ (repeated)}.$$

Thus, one solution is $y = x$; the other linearly independent solution implied by the repeated root is $x \ln x$ (see a textbook if this is not known).

There is now a further complication as the RHS of the original equation (x) is contained in the CF. We therefore need an extra factor of $\ln x$ in the trial PI, beyond those already in the CF. (This corresponds to the extra power of t needed in the PI if the transformation to a linear equation with constant coefficients is made via the $x = e^t$ change of variable.) As a consequence, the PI to be tried is $y = Cx(\ln x)^2$:

$$x^2\left[2C\frac{\ln x}{x} + \frac{2C}{x}\right] - x\left[Cx\frac{2\ln x}{x} + C(\ln x)^2\right] + Cx(\ln x)^2 = x.$$

This implies that $C = \frac{1}{2}$ and gives the general solution as

$$y(x) = Ax + Bx\ln x + \tfrac{1}{2}x(\ln x)^2.$$

It remains only to determine the unknown constants A and B; this is done using the two given values of $y(x)$. The boundary condition $y(1) = 1$ requires that $A = 1$, and $y(e) = 2e$ implies that $B = \frac{1}{2}$; the solution is now completely determined as

$$y(x) = x + \tfrac{1}{2}x\ln x(1 + \ln x).$$

6.23 Prove that the general solution of

$$(x - 2)\frac{d^2y}{dx^2} + 3\frac{dy}{dx} + \frac{4y}{x^2} = 0$$

is given by

$$y(x) = \frac{1}{(x-2)^2}\left[k\left(\frac{2}{3x} - \frac{1}{2}\right) + cx^2\right].$$

This equation is not of any plausible standard form, and the only solution method is to try to make it into an exact equation. If this is possible the order of the equation will be reduced by one.

We first multiply through by x^2 and then note that the resulting factor $3x^2$ in the second term can be written as $[x^2(x-2)]' + 4x$, i.e. as the derivative of the function multiplying y'' together with another simple function. This latter can be combined with the undifferentiated term and allow the whole equation to be written as an exact equation:

$$\frac{d}{dx}\left[x^2(x-2)\frac{dy}{dx}\right] + 4x\frac{dy}{dx} + 4y = 0,$$

$$\frac{d}{dx}\left[x^2(x-2)\frac{dy}{dx}\right] + \frac{d(4xy)}{dx} = 0,$$

$$\Rightarrow \quad x^2(x-2)\frac{dy}{dx} + 4xy = k.$$

Either by inspection or by use of the standard formula, the IF is $(x-2)/x^4$ and leads to

$$\frac{d}{dx}\left[\frac{(x-2)^2}{x^2}y\right] = \frac{k(x-2)}{x^4},$$

$$\Rightarrow \quad \frac{(x-2)^2}{x^2}y = k\left(-\frac{1}{2x^2} + \frac{2}{3x^3}\right) + c,$$

$$\Rightarrow \quad y = \frac{1}{(x-2)^2}\left(-\frac{k}{2} + \frac{2k}{3x} + cx^2\right).$$

6.25 Find the Green's function that satisfies

$$\frac{d^2G(x,\xi)}{dx^2} - G(x,\xi) = \delta(x-\xi) \qquad \text{with} \qquad G(0,\xi) = G(1,\xi) = 0.$$

It is clear from inspection that the CF has solutions of the form $e^{\pm x}$. The other pair of solutions that may suggest themselves are $\sinh x$ and $\cosh x$, but these are merely independent linear combinations of the same two functions.

As both boundary conditions are given at finite values of x (rather than at $x \to \pm\infty$) and both are of the form $y(x) = 0$, it is more convenient to work with those particular linear combinations of e^x and e^{-x} that vanish at the boundary points. The only common linear combination of these two functions that vanishes at a finite value of x is a sinh function. To construct one that vanishes at $x = x_0$ the argument of the sinh function must be made to be $x - x_0$. For the present case the appropriate combinations are

$$\sinh x = \frac{1}{2}(e^x - e^{-x}) \quad \text{and} \quad \sinh(1-x) = \left(\frac{e}{2}\right)e^{-x} - \left(\frac{1}{2e}\right)e^x.$$

Thus, with $0 \le \xi \le 1$, we take

$$G(x,\xi) = \begin{cases} A(\xi)\sinh x & x < \xi, \\ B(\xi)\sinh(1-x) & x > \xi. \end{cases}$$

The continuity requirement on $G(x, \xi)$ at $x = \xi$ and the unit discontinuity requirement on its derivative at the same point give

$$A \sinh \xi - B \sinh(1 - \xi) = 0$$

and $\qquad - B \cosh(1 - \xi) - A \cosh \xi = 1,$

leading to

$$A \sinh \xi \cosh(1 - \xi) + A \cosh \xi \sinh(1 - \xi) = - \sinh(1 - \xi),$$
$$A[\sinh(\xi + 1 - \xi)] = - \sinh(1 - \xi).$$

Hence,

$$A = - \frac{\sinh(1 - \xi)}{\sinh 1} \qquad \text{and} \qquad B = - \frac{\sinh \xi}{\sinh 1},$$

giving the full Green's function as

$$G(x, \xi) = \begin{cases} - \dfrac{\sinh(1 - \xi)}{\sinh 1} \sinh x & x < \xi, \\[3mm] - \dfrac{\sinh \xi}{\sinh 1} \sinh(1 - x) & x > \xi. \end{cases}$$

6.27 Show generally that if $y_1(x)$ and $y_2(x)$ are linearly independent solutions of

$$\frac{d^2 y}{dx^2} + p(x)\frac{dy}{dx} + q(x)y = 0,$$

with $y_1(0) = 0$ and $y_2(1) = 0$, then the Green's function $G(x, \xi)$ for the interval $0 \le x, \xi \le 1$ and with $G(0, \xi) = G(1, \xi) = 0$ can be written in the form

$$G(x, \xi) = \begin{cases} y_1(x)y_2(\xi)/W(\xi) & 0 < x < \xi, \\ y_2(x)y_1(\xi)/W(\xi) & \xi < x < 1, \end{cases}$$

where $W(x) = W[y_1(x), y_2(x)]$ is the Wronskian of $y_1(x)$ and $y_2(x)$.

As usual, we start by writing the general solution as a weighted sum of the linearly independent solutions, whilst leaving the possibility that the weights may be different for different x-ranges:

$$G(x, \xi) = \begin{cases} A(\xi)y_1(x) + B(\xi)y_2(x) & 0 < x < \xi, \\ C(\xi)y_1(x) + D(\xi)y_2(x) & \xi < x < 1. \end{cases}$$

Imposing the boundary conditions and using $y_1(0) = y_2(1) = 0,$

$$0 = G(0, \xi) = A(\xi)y_1(0) + B(\xi)y_2(0) \quad \Rightarrow \quad B(\xi) = 0,$$
$$0 = G(1, \xi) = C(\xi)y_1(1) + D(\xi)y_2(1) \quad \Rightarrow \quad C(\xi) = 0.$$

The continuity requirement on $G(x, \xi)$ at $x = \xi$ and the unit discontinuity requirement on its derivative at the same point give

$$A(\xi)y_1(\xi) - D(\xi)y_2(\xi) = 0,$$
$$A(\xi)y_1'(\xi) - D(\xi)y_2'(\xi) = -1,$$

leading to

$$A(\xi)[\, y_1 y_2' - y_2 y_1' \,] = y_2 \quad \Rightarrow \quad A(\xi) = \frac{y_2(\xi)}{W(\xi)},$$

$$D(\xi) = \frac{y_1(\xi)}{y_2(\xi)} A(\xi) = \frac{y_1(\xi)}{W(\xi)}.$$

Thus,

$$G(x, \xi) = \begin{cases} y_1(x)y_2(\xi)/W(\xi) & 0 < x < \xi, \\ y_2(x)y_1(\xi)/W(\xi) & \xi < x < 1. \end{cases}$$

This result is perfectly general for linear second-order equations of the type stated and can be a quick way to find the corresponding Green's function, provided the solutions that vanish at the end-points can be identified easily. Problem 6.25 is a particular example of this general result.

6.29 The equation of motion for a driven damped harmonic oscillator can be written

$$\ddot{x} + 2\dot{x} + (1 + \kappa^2)x = f(t),$$

with $\kappa \neq 0$. If it starts from rest with $x(0) = 0$ and $\dot{x}(0) = 0$, find the corresponding Green's function $G(t, \tau)$ and verify that it can be written as a function of $t - \tau$ only. Find the explicit solution when the driving force is the unit step function, i.e. $f(t) = H(t)$. Confirm your solution by taking the Laplace transforms of both it and the original equation.

The auxiliary equation is

$$m^2 + 2m + (1 + \kappa^2) = 0 \quad \Rightarrow \quad m = -1 \pm i\kappa,$$

and the CF is $x(t) = Ae^{-t}\cos\kappa t + Be^{-t}\sin\kappa t$.

Let

$$G(t, \tau) = \begin{cases} A(\tau)e^{-t}\cos\kappa t + B(\tau)e^{-t}\sin\kappa t & 0 < t < \tau, \\ C(\tau)e^{-t}\cos\kappa t + D(\tau)e^{-t}\sin\kappa t & t > \tau. \end{cases}$$

The boundary condition $x(0) = 0$ implies that $A = 0$, and

$$\dot{x}(0) = 0 \quad \Rightarrow \quad B(-e^{-t}\sin\kappa t + \kappa e^{-t}\cos\kappa t) = 0 \quad \Rightarrow \quad B = 0.$$

Thus $G(t, \tau) = 0$ for $t < \tau$.

The continuity of G at $t = \tau$ gives

$$Ce^{-\tau}\cos\kappa\tau + De^{-\tau}\sin\kappa\tau = 0 \quad \Rightarrow \quad D = -\frac{C\cos\kappa\tau}{\sin\kappa\tau}.$$

The unit discontinuity in the derivative of G at $t = \tau$ requires (using $s = \sin \kappa \tau$ and $c = \cos \kappa \tau$ as shorthand)

$$Ce^{-\tau}(-c - \kappa s) + De^{-\tau}(-s + \kappa c) - 0 = 1,$$

$$C\left[-c - \kappa s - \frac{c}{s}(-s + \kappa c) \right] = e^{\tau},$$

$$C(-sc - \kappa s^2 + cs - \kappa c^2) = se^{\tau},$$

giving

$$C = -\frac{e^{\tau} \sin \kappa \tau}{\kappa} \quad \text{and} \quad D = \frac{e^{\tau} \cos \kappa \tau}{\kappa}.$$

Thus, for $t > \tau$,

$$G(t, \tau) = \frac{e^{\tau}}{\kappa}(-\sin \kappa \tau \cos \kappa t + \cos \kappa \tau \sin \kappa t)e^{-t}$$

$$= \frac{e^{-(t-\tau)}}{\kappa} \sin \kappa(t - \tau).$$

This form verifies that the Green's function is a function only of the difference $t - \tau$ and not of t and τ separately.

The explicit solution to the given equation when $f(t) = H(t)$ is thus

$$x(t) = \int_0^{\infty} G(t, \tau)f(\tau)\,d\tau$$

$$= \int_0^t G(t, \tau)H(\tau)\,d\tau, \text{ since } G(t, \tau) = 0 \text{ for } \tau > t,$$

$$= \frac{1}{\kappa} \int_0^t e^{-(t-\tau)} \sin \kappa(t - \tau)\,d\tau$$

$$= \frac{e^{-t}}{\kappa} \text{Im} \int_0^t e^{\tau + i\kappa(t-\tau)}\,d\tau$$

$$= \frac{e^{-t}}{\kappa} \text{Im} \left[\frac{e^{i\kappa t} e^{\tau - i\kappa \tau}}{1 - i\kappa} \right]_{\tau=0}^{\tau=t}$$

$$= \frac{e^{-t}}{\kappa} \text{Im} \left[\frac{e^t - e^{i\kappa t}}{1 - i\kappa} \right].$$

Now multiplying both numerator and denominator by $1 + i\kappa$ to make the latter real gives

$$x(t) = \frac{e^{-t}}{\kappa(1 + \kappa^2)} \text{Im} \left[(e^t - e^{i\kappa t})(1 + i\kappa) \right]$$

$$= \frac{e^{-t}}{\kappa(1 + \kappa^2)} [\kappa(e^t - \cos \kappa t) - \sin \kappa t]$$

$$= \frac{1}{1 + \kappa^2} \left(1 - e^{-t} \cos \kappa t - \frac{1}{\kappa} e^{-t} \sin \kappa t \right).$$

The Laplace transform of this solution is given by

$$\bar{x} = \frac{1}{1+\kappa^2}\left(\frac{1}{s} - \frac{s+1}{(s+1)^2+\kappa^2} - \frac{1}{\kappa}\frac{\kappa}{(s+1)^2+\kappa^2}\right)$$

$$= \frac{(s+1)^2+\kappa^2 - s(s+1) - s}{(1+\kappa^2)s[(s+1)^2+\kappa^2]}$$

$$= \frac{1}{s[(s+1)^2+\kappa^2]}.$$

The Laplace transform of the original equation with the given initial conditions reads

$$[s^2\bar{x} - 0s - 0] + 2[s\bar{x} - 0] + (1+\kappa^2)\bar{x} = \frac{1}{s},$$

again showing that

$$\bar{x} = \frac{1}{s[s^2+2s+1+\kappa^2]} = \frac{1}{s[(s+1)^2+\kappa^2]},$$

and so confirming the solution reached using the Green's function approach.

6.31 Find the Green's function $x = G(t, t_0)$ that solves

$$\frac{d^2x}{dt^2} + \alpha\frac{dx}{dt} = \delta(t - t_0)$$

under the initial conditions $x = dx/dt = 0$ at $t = 0$. Hence solve

$$\frac{d^2x}{dt^2} + \alpha\frac{dx}{dt} = f(t),$$

where $f(t) = 0$ for $t < 0$. Evaluate your answer explicitly for $f(t) = Ae^{-\beta t}$ $(t > 0)$.

It is clear that one solution, $x(t)$, to the homogeneous equation has $\ddot{x} = -\alpha\dot{x}$ and is therefore $x(t) = Ae^{-\alpha t}$. The equation is of second order and therefore has a second solution; this is the trivial (but perfectly valid) x is a constant. The CF is thus $x(t) = Ae^{-\alpha t} + B$.

Let

$$G(t, t_0) = \begin{cases} Ae^{-\alpha t} + B, & 0 \le t \le t_0, \\ Ce^{-\alpha t} + D, & t > t_0. \end{cases}$$

Now, the initial conditions give

$$x(0) = 0 \quad \Rightarrow \quad A + B = 0,$$
$$\dot{x}(0) = 0 \quad \Rightarrow \quad -\alpha A = 0 \quad \Rightarrow \quad A = B = 0.$$

Thus $G(t, t_0) = 0$ for $0 \le t \le t_0$.

The continuity/discontinuity conditions determine C and D through

$$Ce^{-\alpha t_0} + D - 0 = 0,$$

$$-\alpha C e^{-\alpha t_0} - 0 = 1, \quad \Rightarrow \quad C = -\frac{e^{\alpha t_0}}{\alpha} \text{ and } D = \frac{1}{\alpha}.$$

It follows that $\qquad G(t, t_0) = \frac{1}{\alpha}[1 - e^{-\alpha(t - t_0)}]$ for $t > t_0$.

The general formalism now gives the solution of

$$\frac{d^2 x}{dt^2} + \alpha \frac{dx}{dt} = f(t)$$

as

$$x(t) = \int_0^t \frac{1}{\alpha}[1 - e^{-\alpha(t - \tau)}] f(\tau) \, d\tau.$$

With $f(t) = Ae^{-\beta t}$ this becomes

$$x(t) = \int_0^t \frac{1}{\alpha}[1 - e^{-\alpha(t - \tau)}] A e^{-\beta \tau} \, d\tau$$

$$= \frac{A}{\alpha} \int_0^t (e^{-\beta \tau} - e^{-\alpha t} e^{(\alpha - \beta)\tau}) \, d\tau$$

$$= A \left[\frac{1 - e^{-\beta t}}{\alpha \beta} - \frac{e^{-\beta t} - e^{-\alpha t}}{\alpha(\alpha - \beta)} \right]$$

$$= A \left[\frac{\alpha - \beta - \alpha e^{-\beta t} + \beta e^{-\alpha t}}{\beta \alpha(\alpha - \beta)} \right]$$

$$= A \left[\frac{\alpha(1 - e^{-\beta t}) - \beta(1 - e^{-\alpha t})}{\beta \alpha(\alpha - \beta)} \right].$$

This is the required explicit solution.

6.33 Solve

$$2y \frac{d^3 y}{dx^3} + 2\left(y + 3\frac{dy}{dx}\right)\frac{d^2 y}{dx^2} + 2\left(\frac{dy}{dx}\right)^2 = \sin x.$$

The only realistic hope for this non-linear equation is to try to arrange it as an exact equation! We note that the second and fourth terms can be written as the derivative of a product, and that adding and subtracting $2y' y''$ will enable the first term to be written in a similar way. We therefore rewrite the equation as

$$\frac{d}{dx}\left(2y\frac{d^2 y}{dx^2}\right) + \frac{d}{dx}\left(2y\frac{dy}{dx}\right) + (6 - 2)\frac{dy}{dx}\frac{d^2 y}{dx^2} = \sin x,$$

$$\frac{d}{dx}\left(2y\frac{d^2 y}{dx^2}\right) + \frac{d}{dx}\left(2y\frac{dy}{dx}\right) + \frac{d}{dx}\left[2\left(\frac{dy}{dx}\right)^2\right] = \sin x.$$

This second form is obtained by noting that the final term on the LHS of the first equation happens to be an exact differential. Thus the whole of the LHS is an exact differential and one stage of integration can be carried out:

$$2y\frac{d^2y}{dx^2} + 2y\frac{dy}{dx} + 2\left(\frac{dy}{dx}\right)^2 = -\cos x + A.$$

We now note that the first and third terms of this integrated equation can be combined as the derivative of a product, whilst the second term is the derivative of y^2. This allows a further step of integration:

$$\frac{d}{dx}\left(2y\frac{dy}{dx}\right) + 2y\frac{dy}{dx} = -\cos x + A,$$

$$\frac{d}{dx}\left(2y\frac{dy}{dx}\right) + \frac{d(y^2)}{dx} = -\cos x + A,$$

$$\Rightarrow \quad 2y\frac{dy}{dx} + y^2 = -\sin x + Ax + B,$$

$$\frac{d(y^2)}{dx} + y^2 = -\sin x + Ax + B.$$

At this stage an integrating factor is needed. However, as the LHS consists of the sum of the differentiated and undifferentiated forms of the same function, the required IF is simply e^x. After multiplying through by this, we obtain

$$\frac{d}{dx}\left(e^x y^2\right) = -e^x \sin x + Axe^x + Be^x,$$

$$\Rightarrow \quad y^2 = e^{-x}\left[C + \int^x (B + Au - \sin u)e^u\, du\right]$$

$$= Ce^{-x} + B + A(x-1) - \tfrac{1}{2}(\sin x - \cos x).$$

The last term in this final solution is obtained by considering

$$\int^x e^u \sin u\, du = \text{Im}\int^x e^{(1+i)u}\, du$$

$$= \text{Im}\left[\frac{e^{(1+i)u}}{1+i}\right]^x$$

$$= \text{Im}\left[\tfrac{1}{2}(1-i)e^{(1+i)x}\right]$$

$$= \tfrac{1}{2}e^x(\sin x - \cos x).$$

6.35 Confirm that the equation

$$2x^2 y \frac{d^2y}{dx^2} + y^2 = x^2 \left(\frac{dy}{dx}\right)^2 \qquad (*)$$

is homogeneous in both x and y separately. Make two successive transformations that exploit this fact, starting with a substitution for x, to obtain an equation of the form

$$2\frac{d^2v}{dt^2} + \left(\frac{dv}{dt}\right)^2 - 2\frac{dv}{dt} + 1 = 0.$$

By writing $dv/dt = p$, solve this equation for $v = v(t)$ and hence find the solution to $(*)$.

The "net-power" (weight) of x in each of the terms is $2 - 2 = 0$, 0, $2 - 2 = 0$, i.e. they are all equal (to zero) and the equation is homogeneous in x. Similarly for y, the weights are 2, 2 and 2; and so the equation is also (separately) homogeneous in y.

Using the homogeneity in x, set $x = e^t$ with $\frac{d}{dx} = e^{-t}\frac{d}{dt}$. The equation then reads

$$2e^{2t} y e^{-t} \frac{d}{dt}\left(e^{-t}\frac{dy}{dt}\right) + y^2 = e^{2t}\left(e^{-t}\frac{dy}{dt}\right)^2,$$

$$2y e^t \left(e^{-t}\frac{d^2y}{dt^2} - e^{-t}\frac{dy}{dt}\right) + y^2 = e^{2t}e^{-2t}\left(\frac{dy}{dt}\right)^2,$$

$$2y\frac{d^2y}{dt^2} - 2y\frac{dy}{dt} + y^2 = \left(\frac{dy}{dt}\right)^2.$$

Now using the homogeneity in y, set $y = e^v$ with

$$\frac{dy}{dt} = \frac{dy}{dv}\frac{dv}{dt} = e^v\frac{dv}{dt} \quad \text{and} \quad \frac{d^2y}{dt^2} = \frac{d}{dt}\left(e^v\frac{dv}{dt}\right) = e^v\frac{dv}{dt}\frac{dv}{dt} + e^v\frac{d^2v}{dt^2}$$

and obtain

$$2e^v\left[e^v\left(\frac{dv}{dt}\right)^2 + e^v\frac{d^2v}{dt^2}\right] - 2e^v e^v\frac{dv}{dt} + e^{2v} = e^{2v}\left(\frac{dv}{dt}\right)^2.$$

Canceling e^{2v} all through, this reduces to

$$2\frac{d^2v}{dt^2} + \left(\frac{dv}{dt}\right)^2 - 2\frac{dv}{dt} + 1 = 0.$$

Since v does not appear in this equation undifferentiated, write $dv/dt = p$, obtaining

$$2\frac{dp}{dt} + p^2 - 2p + 1 = 0 \qquad \Rightarrow \qquad \frac{dp}{dt} = -\tfrac{1}{2}(p-1)^2.$$

This is separable:

$$\frac{dp}{(p-1)^2} = -\tfrac{1}{2}dt \qquad \Rightarrow \qquad (p-1)^{-1} = +\tfrac{1}{2}t + A.$$

Rewriting p as dv/dt, we now have

$$\frac{dv}{dt} - 1 = \frac{1}{\tfrac{1}{2}t + A} \qquad \Rightarrow \qquad v - t = 2\ln\left(A + \tfrac{1}{2}t\right) + B.$$

Resubstituting for v and t now gives

$$\ln y = 2\ln(A + \tfrac{1}{2}\ln x) + B + \ln x,$$
$$\Rightarrow \qquad y = x(C + D\ln x)^2,$$

where $C = Ae^{B/2}$ and $D = \tfrac{1}{2}e^{B/2}$.

7 Series solutions of ordinary differential equations

7.1 Find two power series solutions about $z = 0$ of the differential equation

$$(1 - z^2)y'' - 3zy' + \lambda y = 0.$$

Deduce that the value of λ for which the corresponding power series becomes an Nth-degree polynomial $U_N(z)$ is $N(N + 2)$. Construct $U_2(z)$ and $U_3(z)$.

If the equation is imagined divided through by $(1 - z^2)$ it is straightforward to see that, although $z = \pm 1$ are singular points of the equation, the point $z = 0$ is an *ordinary* point. We therefore expect two (uncomplicated!) series solutions with indicial values $\sigma = 0$ and $\sigma = 1$.

(a) $\sigma = 0$ and $y(z) = \sum_{n=0}^{\infty} a_n z^n$ with $a_0 \neq 0$.

Substituting and equating the coefficients of z^m,

$$(1 - z^2) \sum_{n=0}^{\infty} n(n - 1)a_n z^{n-2} - 3 \sum_{n=0}^{\infty} n a_n z^n + \lambda \sum_{n=0}^{\infty} a_n z^n = 0,$$

$$(m + 2)(m + 1)a_{m+2} - m(m - 1)a_m - 3m a_m + \lambda a_m = 0,$$

gives as the recurrence relation

$$a_{m+2} = \frac{m(m - 1) + 3m - \lambda}{(m + 2)(m + 1)} a_m = \frac{m(m + 2) - \lambda}{(m + 1)(m + 2)} a_m.$$

Since this recurrence relation connects alternate coefficients a_m, and $a_0 \neq 0$, only the coefficients with even indices are generated. All such coefficients with index higher than m will become zero, and the series will become an Nth-degree polynomial $U_N(z)$, if $\lambda = m(m + 2) = N(N + 2)$ for some (even) m appearing in the series; here, this means any positive even integer N.

To construct $U_2(z)$ we need to take $\lambda = 2(2 + 2) = 8$. The recurrence relation gives a_2 as

$$a_2 = \frac{0 - 8}{(0 + 1)(0 + 2)} a_0 = -4a_0 \quad \Rightarrow \quad U_2(z) = a_0(1 - 4z^2).$$

(b) $\sigma = 1$ and $y(z) = z \sum_{n=0}^{\infty} a_n z^n$ with $a_0 \neq 0$.

Substituting and equating the coefficients of z^{m+1},

$$(1 - z^2) \sum_{n=0}^{\infty} (n + 1)n a_n z^{n-1} - 3 \sum_{n=0}^{\infty} (n + 1)a_n z^{n+1} + \lambda \sum_{n=0}^{\infty} a_n z^{n+1} = 0,$$

$$(m + 3)(m + 2)a_{m+2} - (m + 1)m a_m - 3(m + 1)a_m + \lambda a_m = 0,$$

gives as the recurrence relation

$$a_{m+2} = \frac{m(m+1) + 3(m+1) - \lambda}{(m+2)(m+3)} a_m = \frac{(m+1)(m+3) - \lambda}{(m+2)(m+3)} a_m.$$

Again, all coefficients with index higher than m will become zero, and the series will become an Nth-degree polynomial $U_N(z)$, if $\lambda = (m+1)(m+3) = N(N+2)$ for some (even) m appearing in the series; here, this means any positive odd integer N.

To construct $U_3(z)$ we need to take $\lambda = 3(3+2) = 15$. The recurrence relation gives a_2 as

$$a_2 = \frac{3 - 15}{(0+2)(0+3)} a_0 = -2a_0.$$

Thus,

$$U_3(z) = a_0(z - 2z^3).$$

7.3 Find power series solutions in z of the differential equation

$$zy'' - 2y' + 9z^5 y = 0.$$

Identify closed forms for the two series, calculate their Wronskian, and verify that they are linearly independent. Compare the Wronskian with that calculated from the differential equation.

Putting the equation in its standard form shows that $z = 0$ is a singular point of the equation but, as $-2z/z$ and $9z^7/z$ are finite as $z \to 0$, it is a regular singular point. We therefore substitute a Frobenius-type solution,

$$y(z) = z^\sigma \sum_{n=0}^{\infty} a_n z^n \text{ with } a_0 \neq 0,$$

and obtain

$$\sum_{n=0}^{\infty} (n+\sigma)(n+\sigma-1) a_n z^{n+\sigma-1}$$

$$- 2\sum_{n=0}^{\infty} (n+\sigma) a_n z^{n+\sigma-1} + 9\sum_{n=0}^{\infty} a_n z^{n+\sigma+5} = 0.$$

Equating the coefficient of $z^{\sigma-1}$ to zero gives the indicial equation as

$$\sigma(\sigma-1)a_0 - 2\sigma a_0 = 0 \quad \Rightarrow \quad \sigma = 0, \ 3.$$

These differ by an integer and may or may not yield two independent solutions. The larger root, $\sigma = 3$, will give a solution; the smaller one, $\sigma = 0$, may not.

(a) $\sigma = 3$.

Equating the general coefficient of z^{m+2} to zero (with $\sigma = 3$) gives

$$(m+3)(m+2)a_m - 2(m+3)a_m + 9a_{m-6} = 0.$$

Hence the recurrence relation is

$$a_m = -\frac{9a_{m-6}}{m(m+3)},$$

$$\Rightarrow \quad a_{6p} = -\frac{9}{6p(6p+3)}a_{6p-6} = -\frac{a_{6p-6}}{2p(2p+1)} = \frac{(-1)^p a_0}{(2p+1)!}.$$

The first solution is therefore given by

$$y_1(x) = a_0 z^3 \sum_{n=0}^{\infty} \frac{(-1)^n}{(2n+1)!} z^{6n} = a_0 \sum_{n=0}^{\infty} \frac{(-1)^n}{(2n+1)!} z^{3(2n+1)} = a_0 \sin z^3.$$

(b) $\sigma = 0$.

Equating the general coefficient of z^{m-1} to zero (with $\sigma = 0$) gives

$$m(m-1)a_m - 2ma_m + 9a_{m-6} = 0.$$

Hence the recurrence relation is

$$a_m = -\frac{9a_{m-6}}{m(m-3)},$$

$$\Rightarrow \quad a_{6p} = -\frac{9}{6p(6p-3)}a_{6p-6} = -\frac{a_{6p-6}}{2p(2p-1)} = \frac{(-1)^p a_0}{(2p)!}.$$

A second solution is thus

$$y_2(x) = a_0 \sum_{n=0}^{\infty} \frac{(-1)^n}{(2n)!} z^{6n} = a_0 \cos z^3.$$

We see that $\sigma = 0$ does, in fact, produce a (different) series solution. This is because the recurrence relation relates a_n to a_{n+6} and does not involve a_{n+3}; the relevance here of considering the subscripted index "$m + 3$" is that "3" is the difference between the two indicial values.

We now calculate the Wronskian of the two solutions, $y_1 = a_0 \sin z^3$ and $y_2 = b_0 \cos z^3$:

$$W(y_1, y_2) = y_1 y_2' - y_2 y_1'$$
$$= a_0 \sin z^3 (-3b_0 z^2 \sin z^3) - b_0 \cos z^3 (3a_0 z^2 \cos z^3)$$
$$= -3a_0 b_0 z^2 \neq 0.$$

The fact that the Wronskian is non-zero shows that the two solutions are linearly independent.

We can also calculate the Wronskian from the original equation in its standard form,

$$y'' - \frac{2}{z} y' + 9z^4 y = 0,$$

as

$$W = C \exp\left\{-\int^z \frac{-2}{u} \, du\right\} = C \exp(2 \ln z) = C z^2.$$

This is in agreement with the Wronskian calculated from the solutions, as it must be.

7.5 Investigate solutions of Legendre's equation at one of its singular points as follows.

(a) Verify that $z = 1$ is a regular singular point of Legendre's equation and that the indicial equation for a series solution in powers of $(z - 1)$ has a double root $\sigma = 0$.

(b) Obtain the corresponding recurrence relation and show that a polynomial solution is obtained if ℓ is a positive integer.

(c) Determine the radius of convergence R of the $\sigma = 0$ series and relate it to the positions of the singularities of Legendre's equation.

(a) In standard form, Legendre's equation reads

$$y'' - \frac{2z}{1 - z^2} y' + \frac{\ell(\ell + 1)}{1 - z^2} y = 0.$$

This has a singularity at $z = 1$, but, since

$$\frac{-2z(z - 1)}{1 - z^2} \to 1 \text{ and } \frac{\ell(\ell + 1)(z - 1)^2}{1 - z^2} \to 0 \text{ as } z \to 1,$$

i.e. both limits are finite, the point is a regular singular point.

We next change the origin to the point $z = 1$ by writing $u = z - 1$ and $y(z) = f(u)$. The transformed equation is

$$f'' - \frac{2(u + 1)}{-u(u + 2)} f' + \frac{\ell(\ell + 1)}{-u(u + 2)} y = 0$$

or $- u(u + 2) f'' - 2(u + 1) f' + \ell(\ell + 1) f = 0.$

The point $u = 0$ is a regular singular point of this equation and so we set $f(u) = u^{\sigma} \sum_{n=0}^{\infty} a_n u^n$ and obtain

$$-u(u + 2) \sum_{n=0}^{\infty} (\sigma + n)(\sigma + n - 1) a_n u^{\sigma+n-2}$$

$$- 2(u + 1) \sum_{n=0}^{\infty} (\sigma + n) a_n u^{\sigma+n-1} + \ell(\ell + 1) \sum_{n=0}^{\infty} a_n u^{\sigma+n} = 0.$$

Equating to zero the coefficient of $u^{\sigma-1}$ gives

$$-2\sigma(\sigma - 1) a_0 - 2\sigma a_0 = 0 \quad \Rightarrow \quad \sigma^2 = 0;$$

i.e. the indicial equation has a double root $\sigma = 0$.

(b) To obtain the recurrence relation we set the coefficient of u^m equal to zero for general m:

$$-m(m - 1)a_m - 2(m + 1)m a_{m+1} - 2m a_m - 2(m + 1) a_{m+1} + \ell(\ell + 1)a_m = 0.$$

Tidying this up gives

$$2(m + 1)(m + 1)a_{m+1} = [\ell(\ell + 1) - m^2 + m - 2m]a_m,$$

$$\Rightarrow \quad a_{m+1} = \frac{\ell(\ell + 1) - m(m + 1)}{2(m + 1)^2} a_m.$$

From this it is clear that, if ℓ is a positive integer, then $a_{\ell+1}$ and all further a_n are zero and that the solution is a polynomial (of degree ℓ).

(c) The limit of the ratio of successive terms in the series is given by

$$\left| \frac{a_{n+1} u^{n+1}}{a_n u^n} \right| = \left| \frac{u[\ell(\ell+1) - m(m+1)]}{2(m+1)^2} \right| \rightarrow \frac{|u|}{2} \quad \text{as } m \rightarrow \infty.$$

For convergence this limit needs to be < 1, i.e. $|u| < 2$. Thus the series converges in a circle of radius 2 centered on $u = 0$, i.e. on $z = 1$. The value 2 is to be expected, as it is the distance from $z = 1$ of the next nearest (actually the only other) singularity of the equation (at $z = -1$), excluding $z = 1$ itself.

7.7 The first solution of Bessel's equation for $\nu = 0$ is

$$J_0(z) = \sum_{n=0}^{\infty} \frac{(-1)^n}{n! \Gamma(n+1)} \left(\frac{z}{2} \right)^{2n}.$$

Use the derivative method to show that

$$J_0(z) \ln z - \sum_{n=1}^{\infty} \frac{(-1)^n}{(n!)^2} \left(\sum_{r=1}^{n} \frac{1}{r} \right) \left(\frac{z}{2} \right)^{2n}$$

is a second (independent) solution.

Bessel's equation with $\nu = 0$ reads

$$zy'' + y' + zy = 0.$$

The recurrence relations that gave rise to the first solution, $J_0(z)$, were $(\sigma + 1)^2 a_1 = 0$ and $(\sigma + n)^2 a_n + a_{n-2} = 0$ for $n \geq 2$. Thus, in a general form as a function of σ, the solution is given by

$$y(\sigma, z) = a_0 z^\sigma \left\{ 1 - \frac{z^2}{(\sigma+2)^2} + \frac{z^4}{(\sigma+2)^2(\sigma+4)^2} - \cdots \right.$$
$$\left. + \frac{(-1)^n z^{2n}}{[(\sigma+2)(\sigma+4)\ldots(\sigma+2n)]^2} + \cdots \right\}.$$

Setting $\sigma = 0$ reproduces the first solution given above.

To obtain a second independent solution, we must differentiate the above expression with respect to σ, before setting σ equal to 0:

$$\frac{\partial y}{\partial \sigma} = \ln z \, J_0(z) + \sum_{n=1}^{\infty} \frac{da_{2n}(\sigma)}{d\sigma} z^{\sigma+2n} \quad \text{at } \sigma = 0.$$

Now

$$\frac{da_{2n}(\sigma)}{d\sigma}\bigg|_{\sigma=0} = \frac{d}{d\sigma}\left\{\frac{(-1)^n}{[(\sigma+2)(\sigma+4)\ldots(\sigma+2n)]^2}\right\}_{\sigma=0}$$

$$= \frac{(-1)^n(-2)}{[\ldots]^3}\left(\frac{[\ldots]}{\sigma+2} + \frac{[\ldots]}{\sigma+4} + \cdots + \frac{[\ldots]}{\sigma+2n}\right)$$

$$= \frac{(-2)(-1)^n}{[\ldots]^2}\sum_{r=1}^{n}\frac{1}{\sigma+2r}$$

$$= \frac{-2(-1)^n}{2^{2n}(n!)^2}\sum_{r=1}^{n}\frac{1}{2r}, \qquad \text{at } \sigma = 0.$$

Substituting this result, we obtain the second series as

$$J_0(z)\ln z - \sum_{n=1}^{\infty}\frac{(-1)^n}{(n!)^2}\left(\sum_{r=1}^{n}\frac{1}{r}\right)\left(\frac{z}{2}\right)^{2n}.$$

This is the form given in the question.

7.9 Find series solutions of the equation $y'' - 2zy' - 2y = 0$. Identify one of the series as $y_1(z) = \exp z^2$ and verify this by direct substitution. By setting $y_2(z) = u(z)y_1(z)$ and solving the resulting equation for $u(z)$, find an explicit form for $y_2(z)$ and deduce that

$$\int_0^x e^{-v^2}\,dv = e^{-x^2}\sum_{n=0}^{\infty}\frac{n!}{2(2n+1)!}(2x)^{2n+1}.$$

(a) The origin is an ordinary point of the equation and so power series solutions will be possible. Substituting $y(z) = \sum_{n=0}^{\infty}a_n z^n$ gives

$$\sum_{n=0}^{\infty}n(n-1)a_n z^{n-2} - 2\sum_{n=0}^{\infty}na_n z^n - 2\sum_{n=0}^{\infty}a_n z^n = 0.$$

Equating to zero the coefficient of z^{m-2} yields the recurrence relation

$$a_m = \frac{2m-2}{m(m-1)}a_{m-2} = \frac{2}{m}a_{m-2}.$$

The solution with $a_0 = 1$ and $a_1 = 0$ is therefore

$$y_1(z) = 1 + \frac{2z^2}{2} + \frac{2^2 z^4}{(2)(4)} + \cdots + \frac{2^n z^{2n}}{2^n n!} + \cdots$$

$$= \sum_{n=0}^{\infty}\frac{z^{2n}}{n!} = \exp z^2.$$

Putting this result into the original equation,

$$(4z^2 + 2)\exp z^2 - 2z\,2z\exp z^2 - 2\exp z^2 = 0,$$

shows directly that it is a valid solution.

The solution with $a_0 = 0$ and $a_1 = 1$ takes the form

$$y_2(z) = z + \frac{2z^3}{3} + \frac{2^2 z^5}{(3)(5)} + \cdots + \frac{2^n\, 2^n\, n!\, z^{2n+1}}{(2n+1)!} + \cdots$$

$$= \sum_{n=0}^{\infty} \frac{n!\,(2z)^{2n+1}}{2(2n+1)!}.$$

(b) We now set $y_2(z) = u(z)y_1(z)$ and substitute it into the original equation. As they must, the terms in which u is undifferentiated cancel and leave

$$u'' \exp z^2 + 2u'(2z \exp z^2) - 2zu' \exp z^2 = 0.$$

It follows that

$$\frac{u''}{u'} = -2z \quad \Rightarrow \quad u' = A e^{-z^2} \quad \Rightarrow \quad u(x) = A \int^x e^{-v^2}\, dv.$$

Hence, setting the two derived forms for a second solution equal to each other, we have

$$\sum_{n=0}^{\infty} \frac{n!\,(2x)^{2n+1}}{2(2n+1)!} = y_2(x) = y_1(x)u(x) = e^{x^2} A \int^x e^{-v^2}\, dv.$$

For arbitrary small x, only the $n = 0$ term in the series is significant and takes the value $2x/2 = x$, whilst the integral is $A \int^x 1\, dv = Ax$. Thus $A = 1$ and the equality

$$\int_0^x e^{-v^2}\, dv = e^{-x^2} \sum_{n=0}^{\infty} \frac{n!\,(2x)^{2n+1}}{2(2n+1)!}$$

holds for all x.

7.11 For the equation $y'' + z^{-3}y = 0$, show that the origin becomes a regular singular point if the independent variable is changed from z to $x = 1/z$. Hence find a series solution of the form $y_1(z) = \sum_0^\infty a_n z^{-n}$. By setting $y_2(z) = u(z)y_1(z)$ and expanding the resulting expression for du/dz in powers of z^{-1}, show that $y_2(z)$ is a second solution with asymptotic form

$$y_2(z) = c \left[z + \ln z - \tfrac{1}{2} + O\left(\frac{\ln z}{z} \right) \right],$$

where c is an arbitrary constant.

With the equation in its original form, it is clear that, since $z^2/z^3 \to \infty$ as $z \to 0$, the origin is an irregular singular point. However, if we set $1/z = \xi$ and $y(z) = Y(\xi)$, with

$$\frac{d\xi}{dz} = -\frac{1}{z^2} = -\xi^2 \quad \Rightarrow \quad \frac{d}{dz} = -\xi^2 \frac{d}{d\xi},$$

then

$$-\xi^2 \frac{d}{d\xi}\left(-\xi^2 \frac{dY}{d\xi}\right) + \xi^3 Y = 0,$$

$$\xi^2 \frac{d^2 Y}{d\xi^2} + 2\xi \frac{dY}{d\xi} + \xi Y = 0,$$

$$Y'' + \frac{2}{\xi} Y' + \frac{1}{\xi} Y = 0.$$

By inspection, $\xi = 0$ is a regular singular point of this equation, and its indicial equation is

$$\sigma(\sigma - 1) + 2\sigma = 0 \quad \Rightarrow \quad \sigma = 0, \, -1.$$

We start with the larger root, $\sigma = 0$, as this is "guaranteed" to give a valid series solution and assume a solution of the form $Y(\xi) = \sum_{n=0}^{\infty} a_n \xi^n$, leading to

$$\sum_{n=0}^{\infty} n(n-1) a_n \xi^{n-1} + 2 \sum_{n=0}^{\infty} n a_n \xi^{n-1} + \sum_{n=0}^{\infty} a_n \xi^n = 0.$$

Equating to zero the coefficient of ξ^{m-1} gives the recurrence relation

$$a_m = \frac{-a_{m-1}}{m(m+1)} \quad \Rightarrow \quad a_m = \frac{(-1)^m}{(m+1)(m!)^2} a_0$$

and the series solution in inverse powers of z,

$$y_1(z) = a_0 \sum_{n=0}^{\infty} \frac{(-1)^n}{(n+1)(n!)^2 z^n}.$$

To find the second solution we set $y_2(z) = f(z) y_1(z)$. As usual (and as intended), all terms with f undifferentiated vanish when this is substituted in the original equation. What is left is

$$0 = f''(z) y_1(z) + 2 f'(z) y_1'(z),$$

which on rearrangement yields

$$\frac{f''}{f'} = -\frac{2 y_1'}{y_1}.$$

This equation, although it contains a second derivative, is in fact only a first-order equation (for f'). It can be integrated directly to give

$$\ln f' = -2 \ln y_1 + c.$$

After exponentiation, this equation can be written as

$$\frac{df}{dz} = \frac{A}{y_1^2(z)} = \frac{A}{a_0^2}\left(1 - \frac{1}{2 \times 1^2 \, z} + \frac{1}{3 \times 2^2 \, z^2} - \cdots\right)^{-2}$$

$$= \frac{A}{a_0^2}\left[1 + \frac{1}{z} + O\left(\frac{1}{z^2}\right)\right],$$

where $A = e^c$.

Hence, on integrating a second time, one obtains

$$f(z) = \frac{A}{a_0^2}\left(z + \ln z + \mathrm{O}\left(\frac{1}{z}\right)\right),$$

which in turn implies

$$y_2(z) = \frac{A}{a_0^2}\left[z + \ln z + \mathrm{O}\left(\frac{1}{z}\right)\right]a_0\left(1 - \frac{1}{2z} + \frac{1}{12z^2} - \cdots\right)$$

$$= c\left[z + \ln z - \frac{1}{2} + \mathrm{O}\left(\frac{\ln z}{z}\right)\right].$$

This establishes the asymptotic form of the second solution.

7.13 The origin is an ordinary point of the Chebyshev equation,

$$(1 - z^2)y'' - zy' + m^2 y = 0,$$

which therefore has series solutions of the form $z^\sigma \sum_0^\infty a_n z^n$ for $\sigma = 0$ and $\sigma = 1$.

(a) Find the recurrence relationships for the a_n in the two cases and show that there exist polynomial solutions $T_m(z)$:
   (i) for $\sigma = 0$, when m is an even integer, the polynomial having $\frac{1}{2}(m+2)$ terms;
   (ii) for $\sigma = 1$, when m is an odd integer, the polynomial having $\frac{1}{2}(m+1)$ terms.
(b) $T_m(z)$ is normalized so as to have $T_m(1) = 1$. Find explicit forms for $T_m(z)$ for $m = 0, 1, 2, 3$.
(c) Show that the corresponding non-terminating series solutions $S_m(z)$ have as their first few terms

$$S_0(z) = a_0\left(z + \frac{1}{3!}z^3 + \frac{9}{5!}z^5 + \cdots\right),$$

$$S_1(z) = a_0\left(1 - \frac{1}{2!}z^2 - \frac{3}{4!}z^4 - \cdots\right),$$

$$S_2(z) = a_0\left(z - \frac{3}{3!}z^3 - \frac{15}{5!}z^5 - \cdots\right),$$

$$S_3(z) = a_0\left(1 - \frac{9}{2!}z^2 + \frac{45}{4!}z^4 + \cdots\right).$$

(a) (i) If, for $\sigma = 0$, $y(z) = \sum_{n=0}^\infty a_n z^n$ with $a_0 \neq 0$, the condition for the coefficient of z^r in

$$(1 - z^2)\sum_{n=0}^\infty n(n-1)a_n z^{n-2} - z\sum_{n=0}^\infty n a_n z^{n-1} + m^2\sum_{n=0}^\infty a_n z^n$$

to be zero is that

$$(r+2)(r+1)a_{r+2} - r(r-1)a_r - ra_r + m^2 a_r = 0,$$

$$\Rightarrow \quad a_{r+2} = \frac{r^2 - m^2}{(r+2)(r+1)}a_r.$$

This relation relates a_{r+2} to a_r and so to a_0 if r is even. For a_{r+2} to vanish, in this case, requires that $r = m$, which must therefore be an even integer. The non-vanishing coefficients will be $a_0, a_2, \ldots, a_m$, i.e. $\frac{1}{2}(m+2)$ of them in all.

(ii) If, for $\sigma = 1$, $y(z) = \sum_{n=0}^{\infty} a_n z^{n+1}$ with $a_0 \neq 0$, the condition for the coefficient of z^{r+1} in

$$(1 - z^2) \sum_{n=0}^{\infty} (n+1)na_n z^{n-1} - z \sum_{n=0}^{\infty} (n+1)a_n z^n + m^2 \sum_{n=0}^{\infty} a_n z^{n+1}$$

to be zero is that

$$(r+3)(r+2)a_{r+2} - (r+1)ra_r - (r+1)a_r + m^2 a_r = 0,$$

$$\Rightarrow \quad a_{r+2} = \frac{(r+1)^2 - m^2}{(r+3)(r+2)} a_r.$$

This relation relates a_{r+2} to a_r and so to a_0 if r is even. For a_{r+2} to vanish, in this case, requires that $r + 1 = m$, which must therefore be an odd integer. The non-vanishing coefficients will be, as before, $a_0, a_2, \ldots, a_{m-1}$, i.e. $\frac{1}{2}(m+1)$ of them in all.

(b) For $m = 0$, $T_0(z) = a_0$. With the given normalization, $a_0 = 1$ and $T_0(z) = 1$.

For $m = 1$, $T_1(z) = a_0 z$. The required normalization implies that $a_0 = 1$ and so $T_0(z) = z$.

For $m = 2$, we need the recurrence relation in (a)(i). This shows that

$$a_2 = \frac{0^2 - 2^2}{(2)(1)} a_0 = -2a_0 \quad \Rightarrow \quad T_2(z) = a_0(1 - 2z^2).$$

With the given normalization, $a_0 = -1$ and $T_2(z) = 2z^2 - 1$.

For $m = 3$, we use the recurrence relation in (a)(ii) and obtain

$$a_2 = \frac{1^2 - 3^2}{(3)(2)} a_0 = -\frac{4}{3} a_0 \quad \Rightarrow \quad T_3(z) = a_0 \left(z - \frac{4z^3}{3} \right).$$

For the required normalization, we must have $a_0 = -\frac{1}{3}$ and consequently that $T_3(z) = 4z^3 - 3z$.

(c) The non-terminating series solutions $S_m(z)$ arise when $\sigma = 0$ but m is an odd integer and when $\sigma = 1$ with m an even integer. We take each in turn and apply the appropriate recurrence relation to generate the coefficients.

(i) $\sigma = 0$, $m = 1$, using the (a)(i) recurrence relation:

$$a_2 = \frac{0 - 1}{(2)(1)} a_0 = -\frac{1}{2!} a_0, \quad a_4 = \frac{4 - 1}{(4)(3)} a_2 = -\frac{3}{4!} a_0.$$

Hence,

$$S_1(z) = a_0 \left(1 - \frac{1}{2!} z^2 - \frac{3}{4!} z^4 - \cdots \right).$$

(ii) $\sigma = 0$, $m = 3$, using the (a)(i) recurrence relation:

$$a_2 = \frac{0 - 9}{(2)(1)} a_0 = -\frac{9}{2!} a_0, \quad a_4 = \frac{4 - 9}{(4)(3)} a_2 = \frac{45}{4!} a_0.$$

Hence,

$$S_3(z) = a_0 \left(1 - \frac{9}{2!} z^2 + \frac{45}{4!} z^4 + \cdots \right).$$

(iii) $\sigma = 1$, $m = 0$, using the (a)(ii) recurrence relation:

$$a_2 = \frac{1-0}{(3)(2)} a_0 = \frac{1}{3!} a_0, \quad a_4 = \frac{9-0}{(5)(4)} a_2 = \frac{9}{5!} a_0.$$

Hence,

$$S_0(z) = a_0 \left(z + \frac{1}{3!} z^3 + \frac{9}{5!} z^5 + \cdots \right).$$

(iv) $\sigma = 1$, $m = 2$, using the (a)(ii) recurrence relation:

$$a_2 = \frac{1-4}{(3)(2)} a_0 = -\frac{3}{3!} a_0, \quad a_4 = \frac{9-4}{(5)(4)} a_2 = -\frac{15}{5!} a_0.$$

Hence,

$$S_2(z) = a_0 \left(z - \frac{3}{3!} z^3 - \frac{15}{5!} z^5 - \cdots \right).$$

8

Eigenfunction methods for differential equations

8.1 By considering $\langle h|h \rangle$, where $h = f + \lambda g$ with λ real, prove that, for two functions f and g,

$$\langle f|f \rangle \langle g|g \rangle \geq \tfrac{1}{4}[\langle f|g \rangle + \langle g|f \rangle]^2.$$

The function $y(x)$ is real and positive for all x. Its Fourier cosine transform $\tilde{y}_c(k)$ is defined by

$$\tilde{y}_c(k) = \int_{-\infty}^{\infty} y(x)\cos(kx)\,dx,$$

and it is given that $\tilde{y}_c(0) = 1$. Prove that

$$\tilde{y}_c(2k) \geq 2[\tilde{y}_c(k)]^2 - 1.$$

For any $|h\rangle$ we have that $\langle h|h \rangle \geq 0$, with equality only if $|h\rangle = |0\rangle$. Hence, noting that λ is real, we have

$$0 \leq \langle h|h \rangle = \langle f + \lambda g|f + \lambda g \rangle = \langle f|f \rangle + \lambda \langle g|f \rangle + \lambda \langle f|g \rangle + \lambda^2 \langle g|g \rangle.$$

This equation, considered as a quadratic inequality in λ, states that the corresponding quadratic equation has no real roots. The condition for this ("$b^2 < 4ac$") is given by

$$[\langle g|f \rangle + \langle f|g \rangle]^2 \leq 4\langle f|f \rangle \langle g|g \rangle, \qquad (*)$$

from which the stated result follows immediately. Note that $\langle g|f \rangle + \langle f|g \rangle$ is real and its square is therefore non-negative.

The given datum is equivalent to

$$1 = \tilde{y}_c(0) = \int_{-\infty}^{\infty} y(x)\cos(0x)\,dx = \int_{-\infty}^{\infty} y(x)\,dx.$$

Now consider

$$\tilde{y}_c(2k) = \int_{-\infty}^{\infty} y(x)\cos(2kx)\,dx$$

$$= 2\int_{-\infty}^{\infty} y(x)\cos^2 kx - \int_{-\infty}^{\infty} y(x)\,dx,$$

$$\Rightarrow \quad \tilde{y}_c(2k) + 1 = 2\int_{-\infty}^{\infty} y(x)\cos^2 kx.$$

In order to use $(*)$, we need to choose for $f(x)$ and $g(x)$ functions whose product will form the integrand defining $\tilde{y}_c(k)$. With this in mind, we take $f(x) = y^{1/2}(x)\cos kx$ and

$g(x) = y^{1/2}(x)$; we may do this since $y(x) > 0$ for all x. Making these choices gives

$$\left(\int_{-\infty}^{\infty} y \cos kx \, dx + \int_{-\infty}^{\infty} y \cos kx \, dx \right)^2 \leq 4 \int_{-\infty}^{\infty} y \cos^2 kx \, dx \int_{-\infty}^{\infty} y \, dx,$$

$$\left(\int_{-\infty}^{\infty} 2y \cos kx \, dx \right)^2 \leq 4 \int_{-\infty}^{\infty} y \cos^2 kx \, dx \times 1,$$

$$4\tilde{y}_c^2(k) \leq 4 \int_{-\infty}^{\infty} y \cos^2 kx \, dx.$$

Thus,

$$\tilde{y}_c(2k) + 1 = 2 \int_{-\infty}^{\infty} y(x) \cos^2 kx \geq 2[\tilde{y}_c(k)]^2$$

and hence the stated result.

8.3 Consider the real eigenfunctions $y_n(x)$ of a Sturm–Liouville equation

$$(py')' + qy + \lambda \rho y = 0, \qquad a \leq x \leq b,$$

in which $p(x)$, $q(x)$ and $\rho(x)$ are continuously differentiable real functions and $p(x)$ does not change sign in $a \leq x \leq b$. Take $p(x)$ as positive throughout the interval, if necessary by changing the signs of all eigenvalues. For $a \leq x_1 \leq x_2 \leq b$, establish the identity

$$(\lambda_n - \lambda_m) \int_{x_1}^{x_2} \rho y_n y_m \, dx = \left[y_n \, p \, y'_m - y_m \, p \, y'_n \right]_{x_1}^{x_2}.$$

Deduce that if $\lambda_n > \lambda_m$ then $y_n(x)$ must change sign between two successive zeros of $y_m(x)$.

[The reader may find it helpful to illustrate this result by sketching the first few eigenfunctions of the system $y'' + \lambda y = 0$, with $y(0) = y(\pi) = 0$, and the Legendre polynomials $P_n(z)$ for $n = 2, 3, 4, 5$.]

The function $p(x)$ does not change sign in the interval $a \leq x \leq b$; we take it as positive, multiplying the equation all through by -1 if necessary. This means that the weight function ρ can still be taken as positive, but that we must consider all possible functions for $q(x)$ and eigenvalues λ of either sign.

We start with the eigenvalue equation for $y_n(x)$, multiply it through by $y_m(x)$ and then integrate from x_1 to x_2. From this result we subtract the same equation with the roles of n and m reversed, as follows. The integration limits are omitted until the explicit integration by parts is carried through:

$$\int y_m(p \, y'_n)' \, dx + \int y_m q \, y_n \, dx + \int y_m \lambda_n \rho y_n \, dx = 0,$$

$$\int y_n(p \, y'_m)' \, dx + \int y_n q \, y_m \, dx + \int y_n \lambda_m \rho y_m \, dx = 0,$$

$$\int \left[y_m(p \, y'_n)' - y_n(p \, y'_m)' \right] dx + (\lambda_n - \lambda_m) \int y_m \rho y_n \, dx = 0,$$

$$\left[y_m \, p \, y'_n \right]_{x_1}^{x_2} - \int y'_m \, p \, y'_n \, dx - \left[y_n \, p \, y'_m \right]_{x_1}^{x_2}$$

$$+ \int y'_n \, p \, y'_m \, dx + (\lambda_n - \lambda_m) \int y_m \rho y_n \, dx = 0.$$

Hence

$$(\lambda_n - \lambda_m) \int y_m \rho y_n \, dx = \left[y_n p \, y_m' - y_m p \, y_n' \right]_{x_1}^{x_2}. \qquad (*)$$

Now, in this general result, take x_1 and x_2 as successive zeros of $y_m(x)$, where m is determined by $\lambda_n > \lambda_m$ (after the signs have been changed, if that was necessary). Clearly the sign of $y_m(x)$ does not change in this interval; let it be α. It follows that the sign of $y_m'(x_1)$ is also α, whilst that of $y_m'(x_2)$ is $-\alpha$. In addition, the second term on the RHS of $(*)$ vanishes at both limits, as $y_m(x_1) = y_m(x_2) = 0$.

Let us now *suppose* that the sign of $y_n(x)$ does not change in this same interval and is always β. Then the sign of the expression on the LHS of $(*)$ is $(+1)(\alpha)(+1)\beta = \alpha\beta$. The first $(+1)$ appears because $\lambda_n > \lambda_m$.

The signs of the upper- and lower-limit contributions of the remaining term on the RHS of $(*)$ are $\beta(+1)(-\alpha)$ and $(-1)\beta(+1)\alpha$, respectively, the additional factor of (-1) in the second product arising from the fact that the contribution comes from a lower limit. The contributions at both limits have the same sign, $-\alpha\beta$, and so the sign of the total RHS must also be $-\alpha\beta$.

This contradicts, however, the sign of $+\alpha\beta$ found for the LHS. It follows that it was wrong to suppose that the sign of $y_n(x)$ does not change in the interval; in other words, a zero of $y_n(x)$ does appear between every pair of zeros of $y_m(x)$.

8.5 Use the properties of Legendre polynomials to solve the following problems.

(a) Find the solution of $(1 - x^2)y'' - 2xy' + by = f(x)$ that is valid in the range $-1 \le x \le 1$ and finite at $x = 0$, in terms of Legendre polynomials.

(b) Find the explicit solution if $b = 14$ and $f(x) = 5x^3$. Verify it by direct substitution.

[Explicit forms for the Legendre polynomials can be found in any textbook. In *Mathematical Methods for Physics and Engineering*, 3rd edition, they are given in Subsection 18.1.1.]

(a) The LHS of the given equation is the same as that of Legendre's equation and so we substitute $y(x) = \sum_{n=0}^{\infty} a_n P_n(x)$ and use the fact that $(1 - x^2)P_n'' - 2x P_n' = -n(n + 1)P_n$. This results in

$$\sum_{n=0}^{\infty} a_n[b - n(n + 1)]P_n = f(x).$$

Now, using the mutual orthogonality and normalization of the $P_n(x)$, we multiply both sides by $P_m(x)$ and integrate over x:

$$\sum_{n=0}^{\infty} a_n[b - n(n + 1)] \, \delta_{mn} \frac{2}{2m + 1} = \int_{-1}^{1} f(z) P_m(z) \, dz,$$

$$\Rightarrow \quad a_m = \frac{2m + 1}{2[b - m(m + 1)]} \int_{-1}^{1} f(z) P_m(z) \, dz.$$

This gives the coefficients in the solution $y(x)$.

(b) We now express $f(x)$ in terms of Legendre polynomials,

$$f(x) = 5x^3 = 2\left[\tfrac{1}{2}(5x^3 - 3x)\right] + 3[x] = 2P_3(x) + 3P_1(x),$$

and conclude that, because of the mutual orthogonality of the Legendre polynomials, only a_3 and a_1 in the series solution will be non-zero. To find them we need to evaluate

$$\int_{-1}^{1} f(z)P_3(z)\,dz = 2\,\frac{2}{2(3)+1} = \frac{4}{7};$$

similarly, $\int_{-1}^{1} f(z)P_1(z)\,dz = 3 \times (2/3) = 2$.
   Inserting these values gives

$$a_3 = \frac{7}{2(14-12)}\frac{4}{7} = 1 \quad \text{and} \quad a_1 = \frac{3}{2(14-2)}2 = \frac{1}{4}.$$

Thus the solution is

$$y(x) = \frac{1}{4}P_1(x) + P_3(x) = \frac{1}{4}x + \frac{1}{2}(5x^3 - 3x) = \frac{5(2x^3 - x)}{4}.$$

Check:

$$(1-x^2)\frac{60x}{4} - 2x\frac{30x^2-5}{4} + \frac{140x^3 - 70x}{4} = 5x^3,$$

$$\Rightarrow \quad 60x - 60x^3 - 60x^3 + 10x + 140x^3 - 70x = 20x^3,$$

which is satisfied.

8.7 Consider the set of functions, $\{f(x)\}$, of the real variable x defined in the interval $-\infty < x < \infty$, that $\to 0$ at least as quickly as x^{-1}, as $x \to \pm\infty$. For unit weight function, determine whether each of the following linear operators is Hermitian when acting upon $\{f(x)\}$:

$$\text{(a) } \frac{d}{dx} + x; \quad \text{(b) } -i\frac{d}{dx} + x^2; \quad \text{(c) } ix\frac{d}{dx}; \quad \text{(d) } i\frac{d^3}{dx^3}.$$

For an operator $\mathcal{L}$ to be Hermitian over the given range with respect to a unit weight function, the equation

$$\int_{-\infty}^{\infty} f^*(x)[\mathcal{L}g(x)]\,dx = \left\{\int_{-\infty}^{\infty} g^*(x)[\mathcal{L}f(x)]\,dx\right\}^* \qquad (*)$$

must be satisfied for general functions f and g.
   (a) For $\mathcal{L} = \dfrac{d}{dx} + x$, the LHS of $(*)$ is

$$\int_{-\infty}^{\infty} f^*(x)\left(\frac{dg}{dx} + xg\right)dx = \left[f^*g\right]_{-\infty}^{\infty} - \int_{-\infty}^{\infty}\frac{df^*}{dx}g\,dx + \int_{-\infty}^{\infty} f^*xg\,dx$$

$$= 0 - \int_{-\infty}^{\infty}\frac{df^*}{dx}g\,dx + \int_{-\infty}^{\infty} f^*xg\,dx.$$

The RHS of $(*)$ is

$$\int_{-\infty}^{\infty}\left\{g^*(x)\left(\frac{df}{dx}+xf\right)dx\right\}^* = \left\{\int_{-\infty}^{\infty}g^*\frac{df}{dx}dx\right\}^* + \left\{\int_{-\infty}^{\infty}g^*xf\,dx\right\}^*$$

$$= \int_{-\infty}^{\infty}g\frac{df^*}{dx}dx + \int_{-\infty}^{\infty}gxf^*\,dx.$$

Since the sign of the first term differs in the two expressions, the LHS $\neq$ RHS and $\mathcal{L}$ is *not* Hermitian. It will also be apparent that purely multiplicative terms in the operator, such as x or x^2, will always be Hermitian; thus we can ignore the x^2 term in part (b).

(b) As explained above, we need only consider

$$\int_{-\infty}^{\infty}f^*(x)\left(-i\frac{dg}{dx}\right)dx = \left[-if^*g\right]_{-\infty}^{\infty} + i\int_{-\infty}^{\infty}\frac{df^*}{dx}g\,dx$$

$$= 0 + i\int_{-\infty}^{\infty}\frac{df^*}{dx}g\,dx$$

and

$$\int_{-\infty}^{\infty}\left\{g^*(x)\left(-i\frac{df}{dx}\right)dx\right\}^* = i\int_{-\infty}^{\infty}g\frac{df^*}{dx}dx.$$

These are equal, and so $\mathcal{L} = -i\dfrac{d}{dx}$ is Hermitian, as is $\mathcal{L} = -i\dfrac{d}{dx} + x^2$.

(c) For $\mathcal{L} = ix\dfrac{d}{dx}$, the LHS of $(*)$ is

$$\int_{-\infty}^{\infty}f^*(x)\left(ix\frac{dg}{dx}\right)dx = \left[ixf^*g\right]_{-\infty}^{\infty} - i\int_{-\infty}^{\infty}x\frac{df^*}{dx}g\,dx - i\int_{-\infty}^{\infty}f^*g\,dx$$

$$= 0 - i\int_{-\infty}^{\infty}x\frac{df^*}{dx}g\,dx - i\int_{-\infty}^{\infty}f^*g\,dx.$$

The RHS of $(*)$ is given by

$$\int_{-\infty}^{\infty}\left\{g^*(x)ix\left(\frac{df}{dx}\right)dx\right\}^* = -i\int_{-\infty}^{\infty}gx\frac{df^*}{dx}dx.$$

Since, in general, $-i\int_{-\infty}^{\infty}fg^*\,dx \neq 0$, the two sides are not equal; therefore $\mathcal{L}$ is not Hermitian.

(d) Since $\mathcal{L} = i\dfrac{d^3}{dx^3}$ is the cube of the operator $-i\dfrac{d}{dx}$, which was shown in part (b) to be Hermitian, it is expected that $\mathcal{L}$ is Hermitian. This can be verified directly as follows.

The LHS of (∗) is given by

$$i \int_{-\infty}^{\infty} f^* \frac{d^3 g}{dx^3} \, dx = \left[i f^* \frac{d^2 g}{dx^2} \right]_{-\infty}^{\infty} - i \int_{-\infty}^{\infty} \frac{d f^*}{dx} \frac{d^2 g}{dx^2} \, dx$$

$$= 0 - i \left[\frac{d f^*}{dx} \frac{dg}{dx} \right]_{-\infty}^{\infty} + i \int_{-\infty}^{\infty} \frac{d^2 f^*}{dx^2} \frac{dg}{dx} \, dx$$

$$= 0 + i \left[\frac{d^2 f^*}{dx^2} g \right]_{-\infty}^{\infty} - i \int_{-\infty}^{\infty} \frac{d^3 f^*}{dx^3} g \, dx$$

$$= 0 + \left\{ \int_{-\infty}^{\infty} i g^* \frac{d^3 f}{dx^3} \, dx \right\}^* = \text{RHS of (∗)}.$$

Thus $\mathcal{L}$ is confirmed as Hermitian.

8.9 Find an eigenfunction expansion for the solution with boundary conditions $y(0) = y(\pi) = 0$ of the inhomogeneous equation

$$\frac{d^2 y}{dx^2} + \kappa y = f(x),$$

where κ is a constant and

$$f(x) = \begin{cases} x & 0 \leq x \leq \pi/2, \\ \pi - x & \pi/2 < x \leq \pi. \end{cases}$$

The eigenfunctions of the operator $\mathcal{L} = \dfrac{d^2}{dx^2} + \kappa$ are obviously

$$y_n(x) = A_n \sin nx + B_n \cos nx,$$

with corresponding eigenvalues $\lambda_n = n^2 - \kappa$.

The boundary conditions, $y(0) = y(\pi) = 0$, require that n is a positive integer and that $B_n = 0$, i.e.

$$y_n(x) = A_n \sin nx = \sqrt{\frac{2}{\pi}} \sin nx,$$

where A_n (for $n \geq 1$) has been chosen so that the eigenfunctions are normalized over the interval $x = 0$ to $x = \pi$. Since $\mathcal{L}$ is Hermitian on the range $0 \leq x \leq \pi$, the eigenfunctions are also mutually orthogonal, and so the $y_n(x)$ form an orthonormal set.

If the required solution is $y(x) = \sum_n a_n y_n(x)$, then direct substitution yields the result

$$\sum_{n=1}^{\infty} (\kappa - n^2) a_n y_n(x) = f(x).$$

Following the usual procedure for analysis using sets of orthonormal functions, this implies that

$$a_m = \frac{1}{\kappa - m^2} \int_0^{\pi} f(z) y_m(z) \, dz$$

and, consequently, that

$$y(x) = \sum_{n=1}^{\infty} \sqrt{\frac{2}{\pi}} \frac{\sin nx}{\kappa - n^2} \sqrt{\frac{2}{\pi}} \int_0^{\pi} f(z) \sin(nz) \, dz.$$

It only remains to evaluate

$$
\begin{aligned}
I_n &= \int_0^{\pi} \sin(nx) f(x) \, dx \\
&= \int_0^{\pi/2} x \sin nx \, dx + \int_{\pi/2}^{\pi} (\pi - x) \sin nx \, dx \\
&= \left[\frac{-x \cos nx}{n} \right]_0^{\pi/2} + \int_0^{\pi/2} \frac{\cos nx}{n} \, dx \\
&\quad + \left[\frac{-(\pi - x) \cos nx}{n} \right]_{\pi/2}^{\pi} + \int_{\pi/2}^{\pi} \frac{(-1) \cos nx}{n} \, dx \\
&= -\frac{\pi}{2} \frac{\cos(n\pi/2)}{n} (1 - 1) + \left[\frac{\sin nx}{n^2} \right]_0^{\pi/2} - \left[\frac{\sin nx}{n^2} \right]_{\pi/2}^{\pi} \\
&= 0 + \frac{(-1)^{(n-1)/2}}{n^2} (1 + 1) \text{ for odd } n \text{ and } = 0 \text{ for even } n.
\end{aligned}
$$

Thus,

$$y(x) = \frac{4}{\pi} \sum_{n \text{ odd}} \frac{(-1)^{(n-1)/2}}{n^2(\kappa - n^2)} \sin nx$$

is the required solution.

8.11 The differential operator $\mathcal{L}$ is defined by

$$\mathcal{L}y = -\frac{d}{dx} \left(e^x \frac{dy}{dx} \right) - \tfrac{1}{4} e^x y.$$

Determine the eigenvalues λ_n of the problem

$$\mathcal{L}y_n = \lambda_n e^x y_n \qquad 0 < x < 1,$$

with boundary conditions

$$y(0) = 0, \qquad \frac{dy}{dx} + \tfrac{1}{2} y = 0 \quad \text{at} \quad x = 1.$$

(a) Find the corresponding unnormalized y_n, and also a weight function $\rho(x)$ with respect to which the y_n are orthogonal. Hence, select a suitable normalization for the y_n.
(b) By making an eigenfunction expansion, solve the equation

$$\mathcal{L}y = -e^{x/2}, \qquad 0 < x < 1,$$

subject to the same boundary conditions as previously.

When written out explicitly, the eigenvalue equation is

$$-\frac{d}{dx} \left(e^x \frac{dy}{dx} \right) - \tfrac{1}{4} e^x y = \lambda e^x y, \qquad (*)$$

or, on differentiating out the product,

$$e^x y'' + e^x y' + \left(\lambda + \tfrac{1}{4}\right)e^x y = 0.$$

The auxiliary equation is

$$m^2 + m + \left(\lambda + \tfrac{1}{4}\right) = 0 \quad \Rightarrow \quad m = -\tfrac{1}{2} \pm i\sqrt{\lambda}.$$

The general solution is thus given by

$$y(x) = Ae^{-x/2}\cos\sqrt{\lambda}x + Be^{-x/2}\sin\sqrt{\lambda}x,$$

with the condition $y(0) = 0$ implying that $A = 0$. The other boundary condition requires that, at $x = 1$,

$$-\tfrac{1}{2}Be^{-x/2}\sin\sqrt{\lambda}x + \sqrt{\lambda}Be^{-x/2}\cos\sqrt{\lambda}x + \tfrac{1}{2}Be^{-x/2}\sin\sqrt{\lambda}x = 0,$$

i.e. that $\cos\sqrt{\lambda} = 0$ and hence that $\lambda = (n + \tfrac{1}{2})^2\pi^2$ for non-negative integral n.

(a) The unnormalized eigenfunctions are

$$y_n(x) = B_n e^{-x/2}\sin\left(n + \tfrac{1}{2}\right)\pi x$$

and (∗) is in Sturm–Liouville form. However, although $y_n(0) = 0$, the values at the upper limit in $\left[y_m' p\, y_n\right]_0^1$ are $y_n(1) = B_n e^{-1/2}(-1)^n$, $p(1) = e^1$ and $y_m'(1) = -\tfrac{1}{2}B_m e^{-1/2}(-1)^m$. Consequently, $\left[y_m' p\, y_n\right]_0^1 \neq 0$ and SL theory cannot be applied. We therefore have to find a suitable weight function $\rho(x)$ by inspection. Given the general form of the eigenfunctions, ρ has to be taken as e^x, with the orthonormality integral taking the form

$$
\begin{aligned}
I_{nm} &= \int_0^1 \rho(x) y_n(x) y_m^*(x)\, dx \\
&= B_n B_m \int_0^1 e^x e^{-x/2}\sin\left[\left(n + \tfrac{1}{2}\right)\pi x\right] e^{-x/2}\sin\left[\left(m + \tfrac{1}{2}\right)\pi x\right] dx \\
&= \begin{cases} 0 & \text{for } m \neq n, \\ \tfrac{1}{2}B_n B_m & \text{for } m = n. \end{cases}
\end{aligned}
$$

It is clear that a suitable normalization is $B_n = \sqrt{2}$ for all n.

(b) We write the solution as $y(x) = \sum_{n=0}^{\infty} a_n y_n(x)$, giving as the equation to be solved

$$-e^{x/2} = \mathcal{L}y = \mathcal{L}\sum_{n=0}^{\infty} a_n y_n(x)$$

$$= \sum_{n=0}^{\infty} a_n[\lambda_n \rho(x) y_n(x)]$$

$$= \sum_{n=0}^{\infty} a_n\left(n + \tfrac{1}{2}\right)^2\pi^2 e^x \sqrt{2}e^{-x/2}\sin\left[\left(n + \tfrac{1}{2}\right)\pi x\right]$$

$$\Rightarrow \quad -1 = \sum_{n=0}^{\infty} a_n\left(n + \tfrac{1}{2}\right)^2\pi^2\sqrt{2}\sin\left[\left(n + \tfrac{1}{2}\right)\pi x\right].$$

After multiplying both sides of this equation by $\sin\left(m+\frac{1}{2}\right)\pi x$ and integrating from 0 to 1, we obtain

$$a_m \int_0^1 \sin^2\left(m+\tfrac{1}{2}\right)\pi x\, dx = \frac{-1}{\left(m+\tfrac{1}{2}\right)^2 \pi^2 \sqrt{2}} \int_0^1 \sin\left(m+\tfrac{1}{2}\right)\pi x\, dx,$$

$$\frac{a_m}{2} = \frac{-1}{\left(m+\tfrac{1}{2}\right)^2 \pi^2 \sqrt{2}} \int_0^1 \sin\left(m+\tfrac{1}{2}\right)\pi x\, dx$$

$$= \frac{1}{\left(m+\tfrac{1}{2}\right)^2 \pi^2 \sqrt{2}} \left[\frac{\cos\left(m+\tfrac{1}{2}\right)\pi x}{\left(m+\tfrac{1}{2}\right)\pi} \right]_0^1,$$

$$a_m = -\frac{\sqrt{2}}{\left(m+\tfrac{1}{2}\right)^3 \pi^3}.$$

Substituting this result into the assumed expansion, and recalling that $B_n = \sqrt{2}$, gives as the solution

$$y(x) = -\sum_{n=0}^{\infty} \frac{2}{\left(n+\tfrac{1}{2}\right)^3 \pi^3} e^{-x/2} \sin\left(n+\tfrac{1}{2}\right)\pi x.$$

8.13 By substituting $x = \exp t$, find the normalized eigenfunctions $y_n(x)$ and the eigenvalues λ_n of the operator $\mathcal{L}$ defined by

$$\mathcal{L}y = x^2 y'' + 2xy' + \tfrac{1}{4}y, \qquad 1 \leq x \leq e,$$

with $y(1) = y(e) = 0$. Find, as a series $\sum a_n y_n(x)$, the solution of $\mathcal{L}y = x^{-1/2}$.

Putting $x = e^t$ and $y(x) = u(t)$ with $u(0) = u(1) = 0$,

$$\frac{dx}{dt} = e^t \quad \Rightarrow \quad \frac{d}{dx} = e^{-t}\frac{d}{dt}$$

and the eigenvalue equation becomes

$$e^{2t} e^{-t} \frac{d}{dt}\left(e^{-t}\frac{du}{dt}\right) + 2e^t e^{-t}\frac{du}{dt} + \frac{1}{4}u = \lambda u,$$

$$\frac{d^2 u}{dt^2} - \frac{du}{dt} + 2\frac{du}{dt} + \left(\frac{1}{4} - \lambda\right) = 0.$$

The auxiliary equation to this constant-coefficient linear equation for u is

$$m^2 + m + \left(\tfrac{1}{4} - \lambda\right) = 0 \quad \Rightarrow \quad m = -\tfrac{1}{2} \pm \sqrt{\lambda},$$

leading to

$$u(t) = e^{-t/2}\left(Ae^{\sqrt{\lambda}t} + Be^{-\sqrt{\lambda}t}\right).$$

In view of the requirement that u vanishes at two different values of t (one of which is $t = 0$), we need $\lambda < 0$ and $u(t)$ to take the form

$$u(t) = Ae^{-t/2}\sin\sqrt{-\lambda}\,t \text{ with } \sqrt{-\lambda}\,1 = n\pi, \text{ i.e. } \lambda = -n^2\pi^2,$$

where n is an integer. Thus

$$u_n(t) = A_n e^{-t/2} \sin n\pi t \quad \text{or, in terms of } x, \quad y_n(x) = \frac{A_n}{\sqrt{x}} \sin(n\pi \ln x).$$

Normalization requires that

$$1 = \int_1^e \frac{A_n^2}{x} \sin^2(n\pi \ln x)\, dx = \int_0^1 A_n^2 \sin^2(n\pi t)\, dt = \tfrac{1}{2} A_n^2 \quad \Rightarrow \quad A_n = \sqrt{2}.$$

To solve

$$\mathcal{L}y = x^2 y'' + 2xy' + \tfrac{1}{4} y = \frac{1}{\sqrt{x}},$$

we set $y(x) = \sum_{n=0}^{\infty} a_n y_n(x)$. Then the equation becomes

$$\mathcal{L}y = \sum_{n=0}^{\infty} a_n(-n^2 \pi^2) y_n(x) = \sum_{n=0}^{\infty} -n^2 \pi^2 a_n \frac{\sqrt{2}}{\sqrt{x}} \sin(n\pi \ln x) = \frac{1}{\sqrt{x}}.$$

Multiplying through by $y_m(x)$ and integrating, as with ordinary Fourier series,

$$\int_1^e \frac{2a_n}{x} \sin(n\pi \ln x) \sin(m\pi \ln x)\, dx = -\frac{1}{n^2 \pi^2} \int_1^e \frac{\sqrt{2} \sin(m\pi \ln x)}{x}\, dx.$$

The LHS of this equation is the normalization integral just considered and has the value $a_m \delta_{mn}$. Thus

$$\begin{aligned}
a_m &= -\frac{\sqrt{2}}{m^2 \pi^2} \int_1^e \frac{\sin(m\pi \ln x)}{x}\, dx \\
&= -\frac{\sqrt{2}}{m^2 \pi^2} \left[\frac{-\cos(m\pi \ln x)}{m\pi} \right]_1^e \\
&= -\frac{\sqrt{2}}{m^3 \pi^3} [1 - (-1)^m] \\
&= \begin{cases} -\dfrac{2\sqrt{2}}{m^3 \pi^3} & \text{for } m \text{ odd}, \\ 0 & \text{for } m \text{ even}. \end{cases}
\end{aligned}$$

The explicit solution is therefore

$$y(x) = -\frac{4}{\pi^3} \sum_{p=0}^{\infty} \frac{\sin[(2p+1)\pi \ln x]}{(2p+1)^3 \sqrt{x}}.$$

8.15 In the quantum mechanical study of the scattering of a particle by a potential, a Born-approximation solution can be obtained in terms of a function $y(\mathbf{r})$ that satisfies an equation of the form

$$(-\nabla^2 - K^2)y(\mathbf{r}) = F(\mathbf{r}).$$

Assuming that $y_{\mathbf{k}}(\mathbf{r}) = (2\pi)^{-3/2} \exp(i\mathbf{k} \cdot \mathbf{r})$ is a suitably normalized eigenfunction of $-\nabla^2$ corresponding to eigenvalue k^2, find a suitable Green's function $G_K(\mathbf{r}, \mathbf{r}')$. By taking the direction of the vector $\mathbf{r} - \mathbf{r}'$ as the polar axis for a $\mathbf{k}$-space integration, show that $G_K(\mathbf{r}, \mathbf{r}')$ can be reduced to

$$\frac{1}{4\pi^2 |\mathbf{r} - \mathbf{r}'|} \int_{-\infty}^{\infty} \frac{w \sin w}{w^2 - w_0^2} \, dw,$$

where $w_0 = K|\mathbf{r} - \mathbf{r}'|$.

[This integral can be evaluated using contour integration and gives the Green's function explicitly as $(4\pi|\mathbf{r} - \mathbf{r}'|)^{-1} \exp(iK|\mathbf{r} - \mathbf{r}'|)$.]

Given that $y_{\mathbf{k}}(\mathbf{r}) = (2\pi)^{-3/2} \exp(i\mathbf{k} \cdot \mathbf{r})$ satisfies

$$-\nabla^2 y_{\mathbf{k}}(\mathbf{r}) = k^2 y_{\mathbf{k}}(\mathbf{r}),$$

it follows that

$$(-\nabla^2 - K^2)y_{\mathbf{k}}(\mathbf{r}) = (k^2 - K^2)y_{\mathbf{k}}(\mathbf{r}).$$

Thus the same functions are suitable eigenfunctions for the extended operator, but with different eigenvalues.

Its Green's function is therefore (from the general expression for Green's functions in terms of eigenfunctions)

$$
\begin{aligned}
G_K(\mathbf{r}, \mathbf{r}') &= \int \frac{1}{\lambda} y_{\mathbf{k}}(\mathbf{r}) y_{\mathbf{k}}^*(\mathbf{r}') \, d\mathbf{k} \\
&= \frac{1}{(2\pi)^3} \int \frac{\exp(i\mathbf{k} \cdot \mathbf{r}) \exp(-i\mathbf{k} \cdot \mathbf{r}')}{k^2 - K^2} \, d\mathbf{k}.
\end{aligned}
$$

We carry out the three-dimensional integration in $\mathbf{k}$-space using the direction $\mathbf{r} - \mathbf{r}'$ as the polar axis (and denote $\mathbf{r} - \mathbf{r}'$ by $\mathbf{R}$). The azimuthal integral is immediate. The remaining two-dimensional integration is as follows:

$$
\begin{aligned}
G_K(\mathbf{r}, \mathbf{r}') &= \frac{1}{(2\pi)^3} \int_0^\infty \int_0^\pi \frac{\exp(i\mathbf{k} \cdot \mathbf{R})}{k^2 - K^2} 2\pi k^2 \sin\theta_k \, d\theta_k \, dk \\
&= \frac{1}{(2\pi)^2} \int_0^\infty \int_0^\pi \frac{\exp(ikR\cos\theta_k)}{k^2 - K^2} k^2 \sin\theta_k \, d\theta_k \, dk \\
&= \frac{1}{(2\pi)^2} \int_0^\infty \frac{\exp(ikR) - \exp(-ikR)}{ikR(k^2 - K^2)} k^2 \, dk \\
&= \frac{1}{2\pi^2 R} \int_0^\infty \frac{k \sin kR}{k^2 - K^2} \, dk.
\end{aligned}
$$

$$= \frac{1}{2\pi^2 R} \int_0^\infty \frac{w \sin w}{w^2 - w_0^2}\, dw, \quad \text{where } w = kR \text{ and } w_0 = kR,$$

$$= \frac{1}{4\pi^2 R} \int_{-\infty}^\infty \frac{w \sin w}{w^2 - w_0^2}\, dw.$$

Here, the final line is justified by noting that the integrand is an even function of the integration variable w.

9

Special functions

9.1 Use the explicit expressions

$$Y_0^0 = \sqrt{\tfrac{1}{4\pi}}, \qquad\qquad Y_1^0 = \sqrt{\tfrac{3}{4\pi}}\cos\theta,$$

$$Y_1^{\pm 1} = \mp\sqrt{\tfrac{3}{8\pi}}\sin\theta\exp(\pm i\phi), \qquad Y_2^0 = \sqrt{\tfrac{5}{16\pi}}(3\cos^2\theta - 1),$$

$$Y_2^{\pm 1} = \mp\sqrt{\tfrac{15}{8\pi}}\sin\theta\cos\theta\exp(\pm i\phi), \qquad Y_2^{\pm 2} = \sqrt{\tfrac{15}{32\pi}}\sin^2\theta\exp(\pm 2i\phi),$$

to verify for $\ell = 0, 1, 2$ that

$$\sum_{m=-\ell}^{\ell} \left|Y_\ell^m(\theta, \phi)\right|^2 = \frac{2\ell + 1}{4\pi}$$

and so is independent of the values of θ and ϕ. This is true for any ℓ, but a general proof is more involved. This result helps to reconcile intuition with the apparently arbitrary choice of polar axis in a general quantum mechanical system.

We first note that, since every term is the square of a modulus, factors of the form $\exp(\pm mi\phi)$ never appear in the sums. For each value of ℓ, let us denote the sum by S_ℓ. For $\ell = 0$ and $\ell = 1$, we have

$$S_0 = \sum_{m=0}^{0} \left|Y_0^m(\theta, \phi)\right|^2 = \frac{1}{4\pi},$$

$$S_1 = \sum_{m=-1}^{1} \left|Y_1^m(\theta, \phi)\right|^2 = \frac{3}{4\pi}\cos^2\theta + 2\,\frac{3}{8\pi}\sin^2\theta = \frac{3}{4\pi}.$$

For $\ell = 2$, the summation is more complicated but reads

$$S_2 = \sum_{m=-2}^{2} \left|Y_2^m(\theta, \phi)\right|^2$$

$$= \frac{5}{16\pi}(3\cos^2\theta - 1)^2 + 2\,\frac{15}{8\pi}\sin^2\theta\cos^2\theta + 2\,\frac{15}{32\pi}\sin^4\theta$$

$$= \frac{5}{16\pi}(9\cos^4\theta - 6\cos^2\theta + 1 + 12\sin^2\theta\cos^2\theta + 3\sin^4\theta)$$

$$= \frac{5}{16\pi}[6\cos^4\theta - 6\cos^2\theta + 1 + 6\sin^2\theta\cos^2\theta + 3(\cos^2\theta + \sin^2\theta)^2]$$

$$= \frac{5}{16\pi}[6\cos^2\theta(-\sin^2\theta) + 1 + 6\sin^2\theta\cos^2\theta + 3] = \frac{5}{4\pi}.$$

All three sums are independent of θ and ϕ, and are given by the general formula $(2\ell+1)/4\pi$. It will, no doubt, be noted that $2\ell+1$ is the number of terms in S_ℓ, i.e. the number of m values, and that 4π is the total solid angle subtended at the origin by all space.

9.3 Use the generating function for the Legendre polynomials $P_n(x)$ to show that

$$\int_0^1 P_{2n+1}(x)\,dx = (-1)^n \frac{(2n)!}{2^{2n+1}n!(n+1)!}$$

and that, except for the case $n=0$,

$$\int_0^1 P_{2n}(x)\,dx = 0.$$

Denote $\int_0^1 P_n(x)\,dx$ by a_n. From the generating function for the Legendre polynomials, we have

$$\frac{1}{(1-2xh+h^2)^{1/2}} = \sum_{n=0}^{\infty} P_n(x)h^n.$$

Integrating this definition with respect to x gives

$$\int_0^1 \frac{dx}{(1-2xh+h^2)^{1/2}} = \sum_{n=0}^{\infty}\left(\int_0^1 P_n(x)\,dx\right)h^n,$$

$$\left[\frac{-(1-2xh+h^2)^{1/2}}{h}\right]_0^1 = \sum_{n=0}^{\infty} a_n h^n,$$

$$\frac{1}{h}[(1+h^2)^{1/2} - 1 + h] = \sum_{n=0}^{\infty} a_n h^n.$$

Now expanding $(1+h^2)^{1/2}$ using the binomial theorem yields

$$\sum_{n=0}^{\infty} a_n h^n = \frac{1}{h}\left[1 + \sum_{m=1}^{\infty}{}^{1/2}C_m h^{2m} - 1 + h\right] = 1 + \sum_{m=1}^{\infty}{}^{1/2}C_m h^{2m-1}.$$

Comparison of the coefficients of h^n on the two sides of the equation shows that all a_{2r} are zero except for $a_0 = 1$. For $n = 2r+1$ we need $2m-1 = n = 2r+1$, i.e. $m = r+1$, and the value of a_{2r+1} is ${}^{1/2}C_{r+1}$.

Now, the binomial coefficient $^{1/2}C_m$ can be written as

$$
\begin{aligned}
^{1/2}C_m &= \frac{\frac{1}{2}\left(\frac{1}{2}-1\right)\left(\frac{1}{2}-2\right)\cdots\left(\frac{1}{2}-m+1\right)}{m!}, \\
&= \frac{1(1-2)(1-4)\cdots(1-2m+2)}{2^m\,m!} \\
&= (-1)^{m-1}\frac{(1)(1)(3)\cdots(2m-3)}{2^m\,m!} \\
&= (-1)^{m-1}\frac{(2m-2)!}{2^m\,m!\,2^{m-1}\,(m-1)!} \\
&= (-1)^{m-1}\frac{(2m-2)!}{2^{2m-1}\,m!\,(m-1)!}.
\end{aligned}
$$

Thus, setting $m = r + 1$ gives the value of the integral a_{2r+1} as

$$
a_{2r+1} = {}^{1/2}C_{r+1} = (-1)^r\frac{(2r)!}{2^{2r+1}\,(r+1)!\,r!},
$$

as stated in the question.

9.5 The Hermite polynomials $H_n(x)$ may be defined by

$$
\Phi(x, h) = \exp(2xh - h^2) = \sum_{n=0}^{\infty}\frac{1}{n!}H_n(x)h^n.
$$

Show that

$$
\frac{\partial^2\Phi}{\partial x^2} - 2x\frac{\partial\Phi}{\partial x} + 2h\frac{\partial\Phi}{\partial h} = 0,
$$

and hence that the $H_n(x)$ satisfy the Hermite equation,

$$
y'' - 2xy' + 2ny = 0,
$$

where n is an integer ≥ 0.
  Use Φ to prove that

(a) $H_n'(x) = 2n H_{n-1}(x)$,
(b) $H_{n+1}(x) - 2x H_n(x) + 2n H_{n-1}(x) = 0$.

With

$$
\Phi(x, h) = \exp(2xh - h^2) = \sum_{n=0}^{\infty}\frac{1}{n!}H_n(x)h^n,
$$

we have

$$
\frac{\partial\Phi}{\partial x} = 2h\Phi, \qquad \frac{\partial\Phi}{\partial h} = (2x - 2h)\Phi, \qquad \frac{\partial^2\Phi}{\partial x^2} = 4h^2\Phi.
$$

It then follows that

$$
\frac{\partial^2\Phi}{\partial x^2} - 2x\frac{\partial\Phi}{\partial x} + 2h\frac{\partial\Phi}{\partial h} = (4h^2 - 4hx + 4hx - 4h^2)\Phi = 0.
$$

Substituting the series form into this result gives

$$\sum_{n=0}^{\infty} \left(\frac{1}{n!} H_n'' - \frac{2x}{n!} H_n' + \frac{2n}{n!} \right) h^n = 0,$$

$$\Rightarrow \quad H_n'' - 2x H_n' + 2n H_n = 0.$$

This is the equation satisfied by $H_n(x)$, as stated in the question.

(a) From the first relationship derived above, we have that

$$\frac{\partial \Phi}{\partial x} = 2h\Phi,$$

$$\sum_{n=0}^{\infty} \frac{1}{n!} H_n'(x) h^n = 2h \sum_{n=0}^{\infty} \frac{1}{n!} H_n(x) h^n,$$

$$\Rightarrow \quad \frac{1}{m!} H_m' = \frac{2}{(m-1)!} H_{m-1}, \text{ from the coefficients of } h^m.$$

Hence,
$$H_n'(x) = 2n H_{n-1}(x).$$

(b) Differentiating result (a) and then applying it again yields

$$H_n'' = 2n H_{n-1}' = 2n \, 2(n-1) H_{n-2}.$$

Using this in the differential equation satisfied by the H_n, we obtain

$$4n(n-1)H_{n-2} - 2x \, 2n H_{n-1} + 2n H_n = 0.$$

This gives

$$H_{n+1}(x) - 2x H_n(x) + 2n H_{n-1}(x) = 0$$

after dividing through by $2n$ and changing $n \to n+1$.

9.7 For the associated Laguerre polynomials, carry out the following:

(a) Prove the Rodrigues' formula

$$L_n^m(x) = \frac{e^x x^{-m}}{n!} \frac{d^n}{dx^n} (x^{n+m} e^{-x}),$$

taking the polynomials to be defined by

$$L_n^m(x) = \sum_{k=0}^{n} (-1)^k \frac{(n+m)!}{k!(n-k)!(k+m)!} x^k.$$

(b) Prove the recurrence relations

$$(n+1)L_{n+1}^m(x) = (2n+m+1-x)L_n^m(x) - (n+m)L_{n-1}^m(x),$$

$$x(L_n^m)'(x) = nL_n^m(x) - (n+m)L_{n-1}^m(x),$$

but this time taking the polynomial as defined by

$$L_n^m(x) = (-1)^m \frac{d^m}{dx^m} L_{n+m}(x)$$

or the generating function.

(a) It is most convenient to evaluate the nth derivative directly, using Leibnitz' theorem. This gives

$$L_n^m(x) = \frac{e^x x^{-m}}{n!} \sum_{r=0}^{n} \frac{n!}{r!(n-r)!} \frac{d^r}{dx^r}(x^{n+m}) \frac{d^{n-r}}{dx^{n-r}}(e^{-x})$$

$$= e^x x^{-m} \sum_{r=0}^{n} \frac{1}{r!(n-r)!} \frac{(n+m)!}{(n+m-r)!} x^{n+m-r} (-1)^{n-r} e^{-x}$$

$$= \sum_{r=0}^{n} \frac{(-1)^{n-r}}{r!(n-r)!} \frac{(n+m)!}{(n+m-r)!} x^{n-r}.$$

Relabeling the summation using the new index $k = n - r$, we immediately obtain

$$L_n^m(x) = \sum_{k=0}^{n} (-1)^k \frac{(n+m)!}{k!(n-k)!(k+m)!} x^k,$$

which is as given in the question.

(b) The first recurrence relation can be proved using the generating function for the associated Laguerre functions:

$$G(x, h) = \frac{e^{-xh/(1-h)}}{(1-h)^{m+1}} = \sum_{n=0}^{\infty} L_n^m(x) h^n.$$

Differentiating the second equality with respect to h, we obtain

$$\frac{(m+1)(1-h) - x}{(1-h)^{m+3}} e^{-xh/(1-h)} = \sum n L_n^m h^{n-1}.$$

Using the generating function for a second time, we may rewrite this as

$$[(m+1)(1-h) - x] \sum L_n^m h^n = (1-h)^2 \sum n L_n^m h^{n-1}.$$

Equating the coefficients of h^n now yields

$$(m+1)L_n^m - (m+1)L_{n-1}^m - x L_n^m = (n+1)L_{n+1}^m - 2n L_n^m + (n-1)L_{n-1}^m,$$

which can be rearranged and simplified to give the first recurrence relation.

The second result is most easily proved by differentiating one of the standard recurrence relations satisfied by the ordinary Laguerre polynomials, but with n replaced by $n + m$. This standard equality reads

$$x L'_{n+m}(x) = (n+m)L_{n+m}(x) - (n+m)L_{n-1+m}(x).$$

We convert this into an equation for the associated polynomials,

$$L_n^m(x) = (-1)^m \frac{d^m}{dx^m} L_{n+m}(x),$$

by differentiating it m times with respect to x and multiplying through by $(-1)^m$. The result is

$$x(L_n^m)' + m L_n^m = (n+m)L_n^m - (n+m)L_{n-1}^m,$$

which immediately simplifies to give the second recurrence relation satisfied by the associated Laguerre polynomials.

9.9 By initially writing $y(x)$ as $x^{1/2} f(x)$ and then making subsequent changes of variable, reduce Stokes' equation,

$$\frac{d^2 y}{dx^2} + \lambda x y = 0,$$

to Bessel's equation. Hence show that a solution that is finite at $x = 0$ is a multiple of $x^{1/2} J_{1/3}(\frac{2}{3}\sqrt{\lambda x^3})$.

With $y(x) = x^{1/2} f(x)$,

$$y' = \frac{f}{2x^{1/2}} + x^{1/2} f' \quad \text{and} \quad y'' = -\frac{f}{4x^{3/2}} + \frac{f'}{x^{1/2}} + x^{1/2} f''$$

and the equation becomes

$$-\frac{f}{4x^{3/2}} + \frac{f'}{x^{1/2}} + x^{1/2} f'' + \lambda x^{3/2} f = 0,$$

$$x^2 f'' + x f' + (\lambda x^3 - \tfrac{1}{4}) f = 0.$$

Now, guided by the known form of Bessel's equation, change the independent variable to $u = x^{3/2}$ with $f(x) = g(u)$ and

$$\frac{du}{dx} = \frac{3}{2} x^{1/2} \quad \Rightarrow \quad \frac{d}{dx} = \frac{3}{2} u^{1/3} \frac{d}{du}.$$

This gives

$$u^{4/3} \frac{3}{2} u^{1/3} \frac{d}{du}\left(\frac{3}{2} u^{1/3} \frac{dg}{du}\right) + u^{2/3} \frac{3}{2} u^{1/3} \frac{dg}{du} + \left(\lambda u^2 - \frac{1}{4}\right) g = 0,$$

$$\frac{3}{2} u^{5/3} \left(\frac{3}{2} u^{1/3} \frac{d^2 g}{du^2} + \frac{1}{2} u^{-2/3} \frac{dg}{du}\right) + \frac{3}{2} u \frac{dg}{du} + \left(\lambda u^2 - \frac{1}{4}\right) g = 0,$$

$$\frac{9}{4} u^2 \frac{d^2 g}{du^2} + \frac{9}{4} u \frac{dg}{du} + \left(\lambda u^2 - \frac{1}{4}\right) g = 0,$$

$$u^2 \frac{d^2 g}{du^2} + u \frac{dg}{du} + \left(\frac{4}{9}\lambda u^2 - \frac{1}{9}\right) g = 0.$$

This is close to Bessel's equation but still needs a scaling of the variables. So, set $\frac{2}{3}\sqrt{\lambda} u \equiv \mu u = v$ and $g(u) = h(v)$, obtaining

$$\frac{v^2}{\mu^2} \mu^2 \frac{d^2 h}{dv^2} + \frac{v}{\mu} \mu \frac{dh}{dv} + \left(v^2 - \frac{1}{9}\right) h = 0.$$

This *is* Bessel's equation and has a general solution

$$h(v) = c_1 J_{1/3}(v) + c_2 J_{-1/3}(v),$$

$$\Rightarrow \quad g(u) = c_1 J_{1/3}\left(\tfrac{2\sqrt{\lambda}}{3} u\right) + c_2 J_{-1/3}\left(\tfrac{2\sqrt{\lambda}}{3} u\right),$$

$$\Rightarrow \quad f(x) = c_1 J_{1/3}\left(\tfrac{2\sqrt{\lambda}}{3} x^{3/2}\right) + c_2 J_{-1/3}\left(\tfrac{2\sqrt{\lambda}}{3} x^{3/2}\right).$$

For a solution that is finite at $x = 0$, only the Bessel function with a positive subscript can be accepted. Therefore the required solution is

$$y(x) = c_1 x^{1/2} J_{1/3}\left(\tfrac{2\sqrt{\lambda}}{3} x^{3/2}\right).$$

9.11 The complex function $z!$ is defined by

$$z! = \int_0^\infty u^z e^{-u}\, du \qquad \text{for Re } z > -1.$$

For Re $z \leq -1$ it is defined by

$$z! = \frac{(z+n)!}{(z+n)(z+n-1)\cdots(z+1)},$$

where n is any (positive) integer $> -\text{Re } z$. Being the ratio of two polynomials, $z!$ is analytic everywhere in the finite complex plane except at the poles that occur when z is a negative integer.

(a) Show that the definition of $z!$ for Re $z \leq -1$ is independent of the value of n chosen.
(b) Prove that the residue of $z!$ at the pole $z = -m$, where m is an integer > 0, is $(-1)^{m-1}/(m-1)!$.

(a) Let m and n be two choices of integer with $m > n > -\text{Re } z$. Denote the corresponding definitions of $z!$ by $(z!)_m$ and $(z!)_n$ and consider the ratio of these two functions:

$$\frac{(z!)_m}{(z!)_n} = \frac{(z+m)!}{(z+m)(z+m-1)\cdots(z+1)} \frac{(z+n)(z+n-1)\cdots(z+1)}{(z+n)!}$$

$$= \frac{(z+m)!}{(z+m)(z+m-1)\cdots(z+n+1) \times (z+n)!}$$

$$= \frac{(z+m)!}{(z+m)!} = 1.$$

Thus the two functions are identical for all z, i.e the definition of $z!$ is independent of the choice of n, provided that $n > -\text{Re } z$.

(b) From the given definition of $z!$ it is clear that its pole at $z = -m$ is a simple one. The residue R at the pole is therefore given by

$$R = \lim_{z \to -m} (z+m)z!$$

$$= \lim_{z \to -m} \frac{(z+m)(z+n)!}{(z+n)(z+n-1)\cdots(z+1)} \qquad (\text{integer } n \text{ is chosen} > m)$$

$$= \lim_{z \to -m} \frac{(z+n)!}{(z+n)(z+n-1)\cdots(z+m+1)(z+m-1)\cdots(z+1)}$$

$$= \frac{(-m+n)!}{(-m+n)\cdots(-m+m+1)(-m+m-1)\cdots(-m+1)}$$

$$= \frac{1}{[-1][-2]\cdots[-(m-1)]}$$

$$= (-1)^{m-1}\frac{1}{(m-1)!},$$

as stated in the question.

9.13 The integral

$$I = \int_{-\infty}^{\infty} \frac{e^{-k^2}}{k^2 + a^2}\, dk, \qquad (*)$$

in which $a > 0$, occurs in some statistical mechanics problems. By first considering the integral

$$J = \int_0^{\infty} e^{iu(k+ia)}\, du$$

and a suitable variation of it, show that $I = (\pi/a)\exp(a^2)\operatorname{erfc}(a)$, where $\operatorname{erfc}(x)$ is the complementary error function.

The fact that $a > 0$ will ensure that the improper integral J is well defined. It is

$$J = \int_0^{\infty} e^{iu(k+ia)}\, du = \left[\frac{e^{iu(k+ia)}}{i(k+ia)}\right]_0^{\infty} = \frac{i}{k+ia}.$$

We note that this result contains one of the factors that would appear as a denominator in one term of a partial fraction expansion of the integrand in $(*)$. Another term would contain a factor $(k - ia)^{-1}$, and this can be generated by

$$J' = \int_0^{\infty} e^{-iu(k-ia)}\, du = \left[\frac{e^{-iu(k-ia)}}{-i(k-ia)}\right]_0^{\infty} = \frac{-i}{k-ia}.$$

Now, actually expressing the integrand in partial fractions, using the integral expressions J and J' for the factors, and then reversing the order of integration gives

$$I = \frac{1}{2a} \int_{-\infty}^{\infty} \left(\frac{ie^{-k^2}}{k+ia} - \frac{ie^{-k^2}}{k-ia}\right) dk$$

$$= \frac{1}{2a} \int_{-\infty}^{\infty} e^{-k^2}\, dk \int_0^{\infty} e^{iu(k+ia)}\, du + \frac{1}{2a} \int_{-\infty}^{\infty} e^{-k^2}\, dk \int_0^{\infty} e^{-iu(k-ia)}\, du,$$

$$\Rightarrow \quad 2aI = \int_0^{\infty} du \int_{-\infty}^{\infty} e^{-k^2+iuk-ua}\, dk + \int_0^{\infty} du \int_{-\infty}^{\infty} e^{-k^2-iuk-ua}\, dk$$

$$= \int_0^{\infty} du \int_{-\infty}^{\infty} e^{-(k-iu/2)^2 - u^2/4 - ua}\, dk$$

$$\quad + \int_0^{\infty} du \int_{-\infty}^{\infty} e^{-(k+iu/2)^2 - u^2/4 - ua}\, dk$$

$$= 2\sqrt{\pi} \int_0^{\infty} e^{-u^2/4 - ua}\, du,$$

using the standard Gaussian result. We now complete the square in the exponent and set $2v = u + 2$, obtaining

$$2aI = 2\sqrt{\pi} \int_0^{\infty} e^{-(u+2a)^2/4 + a^2}\, du,$$

$$= 2\sqrt{\pi} \int_a^{\infty} e^{-v^2} e^{a^2}\, 2dv.$$

From this it follows that

$$I = \frac{\sqrt{\pi}}{a} 2e^{a^2} \frac{\sqrt{\pi}}{2} \mathrm{erfc}(a) = \frac{\pi}{a} e^{a^2} \mathrm{erfc}(a),$$

as stated in the question.

9.15 Prove two of the properties of the incomplete gamma function $P(a, x^2)$ as follows.

(a) By considering its form for a suitable value of a, show that the error function can be expressed as a particular case of the incomplete gamma function.

(b) The Fresnel integrals, of importance in the study of the diffraction of light, are given by

$$C(x) = \int_0^x \cos\left(\frac{\pi}{2} t^2\right) dt, \qquad S(x) = \int_0^x \sin\left(\frac{\pi}{2} t^2\right) dt.$$

Show that they can be expressed in terms of the error function by

$$C(x) + i S(x) = A \, \mathrm{erf}\left[\frac{\sqrt{\pi}}{2}(1 - i)x\right],$$

where A is a (complex) constant, which you should determine. Hence express $C(x) + i S(x)$ in terms of the incomplete gamma function.

(a) From the definition of the incomplete gamma function, we have

$$P(a, x^2) = \frac{1}{\Gamma(a)} \int_0^{x^2} e^{-t} t^{a-1} \, dt.$$

Guided by the x^2 in the upper limit, we now change the integration variable to $y = +\sqrt{t}$, with $2y \, dy = dt$, and obtain

$$P(a, x^2) = \frac{1}{\Gamma(a)} \int_0^x e^{-y^2} y^{2(a-1)} \, 2y \, dy.$$

To make the RHS into an error function we need to remove the y-term; to do this we choose a such that $2(a - 1) + 1 = 0$, i.e. $a = \frac{1}{2}$. With this choice, $\Gamma(a) = \sqrt{\pi}$ and

$$P\left(\tfrac{1}{2}, x^2\right) = \frac{2}{\sqrt{\pi}} \int_0^x e^{-y^2} \, dy,$$

i.e. a correctly normalized error function.

(b) Consider the given expression:

$$z = A \, \mathrm{erf}\left[\frac{\sqrt{\pi}}{2}(1 - i)x\right] = \frac{2A}{\sqrt{\pi}} \int_0^{\sqrt{\pi}(1-i)x/2} e^{-u^2} \, du.$$

Changing the variable of integration to s, given by $u = \frac{1}{2}\sqrt{\pi}(1-i)s$, and recalling that $(1-i)^2 = -2i$, we obtain

$$z = \frac{2A}{\sqrt{\pi}} \int_0^x e^{-s^2\pi(-2i)/4} \frac{\sqrt{\pi}}{2}(1-i)\,ds$$

$$= A(1-i) \int_0^x e^{i\pi s^2/2}\,ds$$

$$= A(1-i) \int_0^x \left[\cos\left(\frac{\pi s^2}{2}\right) + i\sin\left(\frac{\pi s^2}{2}\right) \right] ds$$

$$= A(1-i)[C(x) + iS(x)].$$

For the correct normalization we need $A(1-i) = 1$, implying that $A = (1+i)/2$.

Now, from part (a), the error function can be expressed in terms of the incomplete gamma function $P(a, x)$ by

$$\mathrm{erf}(x) = P\left(\tfrac{1}{2}, x^2\right).$$

Here the argument of the error function is $\frac{1}{2}\sqrt{\pi}(1-i)x$, whose square is $-\frac{1}{2}\pi i x^2$, and so

$$C(x) + iS(x) = \frac{1+i}{2}\,P\left(\frac{1}{2}, -\frac{i\pi}{2}x^2\right).$$

10

Partial differential equations

10.1 Determine whether the following can be written as functions of $p = x^2 + 2y$ only, and hence whether they are solutions of

$$\frac{\partial u}{\partial x} = x \frac{\partial u}{\partial y}.$$

(a) $x^2(x^2 - 4) + 4y(x^2 - 2) + 4(y^2 - 1)$;
(b) $x^4 + 2x^2 y + y^2$;
(c) $[x^4 + 4x^2 y + 4y^2 + 4]/[2x^4 + x^2(8y + 1) + 8y^2 + 2y]$.

As a first step, we verify that any function of $p = x^2 + 2y$ will satisfy the given equation. Using the chain rule, we have

$$\frac{\partial u}{\partial p}\frac{\partial p}{\partial x} = x \frac{\partial u}{\partial p}\frac{\partial p}{\partial y},$$

$$\Rightarrow \quad \frac{\partial u}{\partial p} 2x = x \frac{\partial u}{\partial p} 2.$$

This is satisfied for *any* function $u(p)$, thus completing the verification.

To test the given functions we substitute for $y = \frac{1}{2}(p - x^2)$ or for $x^2 = p - 2y$ in each of the $f(x, y)$ and then examine whether the resulting forms are independent of x or y, respectively.

(a) $f(x, y) = x^2(x^2 - 4) + 4y(x^2 - 2) + 4(y^2 - 1)$

$$= x^2(x^2 - 4) + 2(p - x^2)(x^2 - 2) + p^2 - 2p x^2 + x^4 - 4$$

$$= x^4(1 - 2 + 1) + x^2(-4 + 2p + 4 - 2p) - 4p + p^2 - 4$$

$$= p^2 - 4p - 4 = g(p).$$

This is a function of p only, and therefore the original $f(x, y)$ is a solution of the PDE.

Though not necessary for answering the question, we will repeat the verification, but this time by substituting for x rather than for y:

$$f(x, y) = x^2(x^2 - 4) + 4y(x^2 - 2) + 4(y^2 - 1)$$

$$= (p - 2y)(p - 2y - 4) + 4y(p - 2y - 2) + 4(y^2 - 1)$$

$$= p^2 - 4py + 4y^2 - 4p + 8y + 4yp - 8y^2 - 8y + 4y^2 - 4$$

$$= p^2 - 4p - 4 = g(p);$$

i.e. it is the same as before, as it must be, and again this shows that $f(x, y)$ is a solution of the PDE.

(b) $f(x, y) = x^4 + 2x^2y + y^2$
$$= (p - 2y)^2 + 2y(p - 2y) + y^2$$
$$= p^2 - 4p\,y + 4y^2 + 2p\,y - 4y^2 + y^2$$
$$= (p - y)^2 \neq g(p).$$

As this is a function of both p and y, it is not a solution of the PDE.

(c) $f(x, y) = \dfrac{x^4 + 4x^2y + 4y^2 + 4}{2x^4 + x^2(8y + 1) + 8y^2 + 2y}$

$$= \frac{p^2 - 4p\,y + 4y^2 + 4yp - 8y^2 + 4y^2 + 4}{2p^2 - 8p\,y + 8y^2 + 8yp + p - 16y^2 - 2y + 8y^2 + 2y}$$

$$= \frac{p^2 + 4}{2p^2 + p} = g(p).$$

This is a function of p only and therefore $f(x, y)$ is a solution of the PDE.

10.3 Solve the following partial differential equations for $u(x, y)$ with the boundary conditions given:

(a) $x\dfrac{\partial u}{\partial x} + xy = u, \qquad u = 2y$ on the line $x = 1$;

(b) $1 + x\dfrac{\partial u}{\partial y} = xu, \qquad u(x, 0) = x.$

(a) This can be solved as an ODE for u as a function of x, though the "constant of integration" will be a function of y. In standard form, the equation reads

$$\frac{\partial u}{\partial x} - \frac{u}{x} = -y.$$

By inspection (or formal calculation) the IF for this is x^{-1} and the equation can be rearranged as

$$\frac{\partial}{\partial x}\left(\frac{u}{x}\right) = -\frac{y}{x},$$
$$\Rightarrow \quad \frac{u}{x} = -y\ln x + f(y),$$
$$u = 2y \text{ on } x = 1 \ \Rightarrow \ f(y) = 2y,$$

and so $u(x, y) = xy(2 - \ln x)$.

(b) This equation can be written in standard form, with u as a function of y:

$$\frac{\partial u}{\partial y} - u = -\frac{1}{x},$$

for which the IF is clearly e^{-y}, leading to

$$\frac{\partial}{\partial y}\left(e^{-y}u\right) = -\frac{e^{-y}}{x},$$

$$\Rightarrow \quad e^{-y}u = \frac{e^{-y}}{x} + f(x),$$

$$u(x, 0) = x \Rightarrow f(x) = x - \frac{1}{x}.$$

Substituting this result, and multiplying through by e^y, gives $u(x, y)$ as

$$u(x, y) = \frac{1}{x} + \left(x - \frac{1}{x}\right)e^y.$$

10.5 Find solutions of

$$\frac{1}{x}\frac{\partial u}{\partial x} + \frac{1}{y}\frac{\partial u}{\partial y} = 0$$

for which (a) $u(0, y) = y$ and (b) $u(1, 1) = 1$.

As usual, we find $p\,(x, y)$ from

$$\frac{dx}{x^{-1}} = \frac{dy}{y^{-1}} \quad \Rightarrow \quad x^2 - y^2 = p.$$

(a) On $x = 0$, $p = -y^2$ and

$$u(0, y) = y = (-p)^{1/2} \quad \Rightarrow \quad u(x, y) = [-(x^2 - y^2)]^{1/2} = (y^2 - x^2)^{1/2}.$$

(b) At $(1, 1)$, $p = 0$ and

$$u(1, 1) = 1 \quad \Rightarrow \quad u(x, y) = 1 + g(x^2 - y^2),$$

where g is any function that has $g(0) = 0$.

We note that in part (a) the solution is uniquely determined because the boundary values are given along a line, whereas in part (b), where the value is fixed at only an isolated point, the solution is indeterminate to the extent of a loosely determined function. This is the normal situation, though it is modified if the boundary line in (a) happens to be one along which p has a constant value.

10.7 Solve

$$\sin x \frac{\partial u}{\partial x} + \cos x \frac{\partial u}{\partial y} = \cos x \quad (*)$$

subject to (a) $u(\pi/2, y) = 0$ and (b) $u(\pi/2, y) = y(y + 1)$.

As usual, the CF is found from

$$\frac{dx}{\sin x} = \frac{dy}{\cos x} \quad \Rightarrow \quad y - \ln \sin x = p.$$

Since the RHS of (∗) is a factor in one of the terms on the LHS, a trivial PI is any function of y only whose derivative (with respect to y) is unity, of which the simplest is $u(x, y) = y$. The general solution is therefore

$$u(x, y) = f(y - \ln \sin x) + y.$$

The actual form of the arbitrary function $f(p)$ is determined by the form that $u(x, y)$ takes on the boundary, here the line $x = \pi/2$.

(a) With $u(\pi/2, y) = 0$:

$$0 = f(y - 0) + y \quad \Rightarrow \quad f(p) = -p$$
$$\Rightarrow \quad u(x, y) = \ln \sin x - y + y = \ln \sin x.$$

(b) With $u(\pi/2, y) = y(y + 1)$:

$$y(y + 1) = f(y - 0) + y \quad \Rightarrow \quad f(p) = p^2$$
$$\Rightarrow \quad u(x, y) = (y - \ln \sin x)^2 + y.$$

10.9 If $u(x, y)$ satisfies

$$\frac{\partial^2 u}{\partial x^2} - 3\frac{\partial^2 u}{\partial x \partial y} + 2\frac{\partial^2 u}{\partial y^2} = 0$$

and $u = -x^2$ and $\partial u/\partial y = 0$ for $y = 0$ and all x, find the value of $u(0, 1)$.

If we are to find solutions to this homogeneous second-order PDE of the form $u(x, y) = f(x + \lambda y)$, then λ must satisfy

$$1 - 3\lambda + 2\lambda^2 = 0 \quad \Rightarrow \quad \lambda = \tfrac{1}{2},\ 1.$$

Thus $u(x, y) = g(x + \tfrac{1}{2}y) + f(x + y) \equiv g(p_1) + f(p_2)$.

On $y = 0$, $p_1 = p_2 = x$ and

$$-x^2 = u(x, 0) = g(x) + f(x), \qquad (\ast)$$
$$0 = \frac{\partial u}{\partial y}(x, 0) = \tfrac{1}{2}g'(x) + f'(x).$$

From (∗),
$$-2x = g'(x) + f'(x).$$

Subtracting,
$$2x = -\tfrac{1}{2}g'(x).$$

Integrating,
$$g(x) = -2x^2 + k \quad \Rightarrow \quad f(x) = x^2 - k, \quad \text{from (∗)}.$$

Hence,
$$u(x, y) = -2\left(x + \tfrac{1}{2}y\right)^2 + k + (x + y)^2 - k$$
$$= -x^2 + \tfrac{1}{2}y^2.$$

At the particular point $(0, 1)$ we have $u(0, 1) = -0^2 + \tfrac{1}{2}(1)^2 = \tfrac{1}{2}$.

10.11 In those cases in which it is possible to do so, evaluate $u(2, 2)$, where $u(x, y)$ is the solution of

$$2y\frac{\partial u}{\partial x} - x\frac{\partial u}{\partial y} = xy(2y^2 - x^2)$$

that satisfies the (separate) boundary conditions given below.

(a) $u(x, 1) = x^2$.
(b) $u(1, \sqrt{10}) = 5$.
(c) $u(\sqrt{10}, 1) = 5$.

To find the CF, $u(x, y) = f(p)$, we set

$$\frac{dx}{2y} = -\frac{dy}{x} \quad \Rightarrow \quad x^2 + 2y^2 = p.$$

The point $(2, 2)$ corresponds to $p = 2^2 + 2(2^2) = 12$.
For a PI we try $u(x, y) = Ax^n y^m$:

$$2Anx^{n-1}y^{m+1} - Amx^{n+1}y^{m-1} = 2xy^3 - x^3y,$$

which has a solution, $n = m = 2$ with $A = \frac{1}{2}$. Thus the general solution is

$$u(x, y) = f(x^2 + 2y^2) + \tfrac{1}{2}x^2y^2.$$

(a) We must find the function f that makes $u(x, 1) = x^2$. This requires f to satisfy

$$\begin{aligned} x^2 &= u(x, 1) = f(x^2 + 2) + \tfrac{1}{2}x^2 \\ \Rightarrow \quad f(p) &= \tfrac{1}{2}(p - 2) \\ \Rightarrow \quad u(x, y) &= \tfrac{1}{2}(x^2 + 2y^2 - 2) + \tfrac{1}{2}x^2y^2 \\ &= \tfrac{1}{2}(x^2 + x^2y^2 + 2y^2 - 2). \end{aligned}$$

From which it follows that $u(2, 2) = \tfrac{1}{2}(4 + 16 + 8 - 2) = 13$.

(b) With $u(1, \sqrt{10}) = 5$: At the point $(1, \sqrt{10})$ the value of p is $1 + 2(10) = 21$. As the "boundary" consists of just this one point, it is only at the points that have $p = 21$ that the value of $u(x, y)$ can be known. Since for the point $(2, 2)$ the value of p is 12, the value of $u(2, 2)$ cannot be determined.

(c) With $u(\sqrt{10}, 1) = 5$: At the point $(\sqrt{10}, 1)$ the value of p is $10 + 2(1) = 12$. Since for $(2, 2)$ it is also 12, the value of $u(2, 2)$ can be determined and is given by $f(12) + \tfrac{1}{2}(4)(4) = 5 + 8 = 13$.

10.13 Find the most general solution of $\dfrac{\partial^2 u}{\partial x^2} + \dfrac{\partial^2 u}{\partial y^2} = x^2 y^2$.

The complementary function for this equation is the solution to the two-dimensional Laplace equation and [either as a general known result or from substituting the trial form $h(x + \lambda y)$ which leads to $\lambda^2 = -1$ and hence to $\lambda = \pm i$] has the form $f(x + iy) + g(x - iy)$ for arbitrary functions f and g.

It therefore remains only to find a suitable PI. As f and g are not specified, there are infinitely many possibilities and which one we finish up with will depend upon the details of the approach adopted. When a solution has been obtained it should be checked by substitution.

As no PI is obvious by inspection, we make a change of variables with the object of obtaining one by means of an explicit integration. To do this, we use as new variables the arguments of the arbitrary functions appearing in the CF.

Setting $\xi = x + iy$ and $\eta = x - iy$, with $u(x, y) = v(\xi, \eta)$, gives

$$\left(\frac{\partial}{\partial \xi} + \frac{\partial}{\partial \eta}\right)\left(\frac{\partial v}{\partial \xi} + \frac{\partial v}{\partial \eta}\right)$$

$$+ \left(i\frac{\partial}{\partial \xi} - i\frac{\partial}{\partial \eta}\right)\left(i\frac{\partial v}{\partial \xi} - i\frac{\partial v}{\partial \eta}\right) = \left(\frac{\xi + \eta}{2}\right)^2 \left(\frac{\xi - \eta}{2i}\right)^2,$$

$$(1 - 1)\frac{\partial^2 v}{\partial \xi^2} + (2 + 2)\frac{\partial^2 v}{\partial \xi \partial \eta} + (1 - 1)\frac{\partial^2 v}{\partial \eta^2} = -\frac{1}{16}(\xi^2 - \eta^2)^2,$$

$$\frac{\partial^2 v}{\partial \xi \partial \eta} = -\frac{1}{64}(\xi^2 - \eta^2)^2.$$

When we integrate this we can set all constants of integration and all arbitrary functions equal to zero as *any* solution will suffice:

$$\frac{\partial^2 v}{\partial \xi \partial \eta} = -\frac{1}{64}(\xi^4 - 2\xi^2\eta^2 + \eta^4),$$

$$\frac{\partial v}{\partial \eta} = -\frac{1}{64}\left(\frac{\xi^5}{5} - \frac{2\xi^3\eta^2}{3} + \xi\eta^4\right),$$

$$v = -\frac{1}{64}\left(\frac{\xi^5\eta}{5} - \frac{2\xi^3\eta^3}{9} + \frac{\xi\eta^5}{5}\right).$$

Re-expressing this solution as a function of x and y (noting that $\xi\eta = x^2 + y^2$) gives

$$u(x, y) = \frac{1}{(64)(45)}[10\xi^3\eta^3 - 9\xi\eta(\xi^4 + \eta^4)]$$

$$= \frac{1}{(64)(45)}[10(x^2 + y^2)^3 - 18(x^2 + y^2)(x^4 - 6x^2y^2 + y^4)]$$

$$= \frac{x^2 + y^2}{(64)(45)}(10x^4 + 20x^2y^2 + 10y^4 - 18x^4 + 108x^2y^2 - 18y^4)$$

$$= \frac{x^2 + y^2}{(64)(45)}(128x^2y^2 - 8x^4 - 8y^4)$$

$$= \frac{1}{360}(15x^4y^2 - x^6 + 15x^2y^4 - y^6).$$

Check

Applying $\dfrac{\partial^2}{\partial x^2} + \dfrac{\partial^2}{\partial y^2}$ to the final expression yields

$$\frac{1}{360}[15(12)x^2y^2 - 30x^4 + 30y^4 + 30x^4 + 15(12)x^2y^2 - 30y^4] = x^2y^2,$$

as it should.

10.15 The non-relativistic Schrödinger equation,

$$-\frac{\hbar^2}{2m}\nabla^2 u + V(\mathbf{r})u = i\hbar\frac{\partial u}{\partial t},$$

is similar to the diffusion equation in having different orders of derivatives in its various terms; this precludes solutions that are arbitrary functions of particular linear combinations of variables. However, since exponential functions do not change their forms under differentiation, solutions in the form of exponential functions of combinations of the variables may still be possible.

Consider the Schrödinger equation for the case of a constant potential, i.e. for a free particle, and show that it has solutions of the form $A\exp(lx + my + nz + \lambda t)$, where the only requirement is that

$$-\frac{\hbar^2}{2m}\left(l^2 + m^2 + n^2\right) = i\hbar\lambda.$$

In particular, identify the equation and wavefunction obtained by taking λ as $-iE/\hbar$, and l, m and n as $ip_x/\hbar$, $ip_y/\hbar$ and $ip_z/\hbar$, respectively, where E is the energy and $\mathbf{p}$ the momentum of the particle; these identifications are essentially the content of the de Broglie and Einstein relationships.

For a free particle we may omit the potential term $V(\mathbf{r})$ from the Schrödinger equation, which then reads (in Cartesian coordinates)

$$-\frac{\hbar^2}{2m}\left(\frac{\partial^2 u}{\partial x^2} + \frac{\partial^2 u}{\partial y^2} + \frac{\partial^2 u}{\partial z^2}\right) = i\hbar\frac{\partial u}{\partial t}.$$

We try $u(x, y, z, t) = A\exp(lx + my + nz + \lambda t)$, i.e. the product of four exponential functions, and obtain

$$-\frac{\hbar^2}{2m}(l^2 + m^2 + n^2)u = i\hbar\lambda u.$$

This equation is clearly satisfied provided

$$-\frac{\hbar^2}{2m}(l^2 + m^2 + n^2) = i\hbar\lambda.$$

With λ as $-iE/\hbar$, and l, m and n as $ip_x/\hbar$, $ip_y/\hbar$ and $ip_z/\hbar$, respectively, where E is the energy and $\mathbf{p}$ is the momentum of the particle, we have

$$-\frac{\hbar^2}{2m}\left(-\frac{p_x^2}{\hbar^2} - \frac{p_y^2}{\hbar^2} - \frac{p_z^2}{\hbar^2}\right) = E,$$

which can be written more compactly as $E = p^2/2m$, the classical non-relativistic relationship between the (kinetic) energy and momentum of a free particle.

The wavefunction obtained is

$$u(\mathbf{r}, t) = A\exp\left[\frac{i}{\hbar}(p_x x + p_y y + p_z z - Et)\right]$$

$$= A\exp\left[\frac{i}{\hbar}(\mathbf{p}\cdot\mathbf{r} - Et)\right],$$

i.e. a classical plane wave of wave number $\mathbf{k} = \mathbf{p}/\hbar$ and angular frequency $\omega = E/\hbar$ traveling in the direction $\mathbf{p}/p$.

10.17 An incompressible fluid of density ρ and negligible viscosity flows with velocity v along a thin, straight, perfectly light and flexible tube, of cross-section A which is held under tension T. Assume that small transverse displacements u of the tube are governed by

$$\frac{\partial^2 u}{\partial t^2} + 2v\frac{\partial^2 u}{\partial x \partial t} + \left(v^2 - \frac{T}{\rho A}\right)\frac{\partial^2 u}{\partial x^2} = 0.$$

(a) Show that the general solution consists of a superposition of two waveforms traveling with different speeds.
(b) The tube initially has a small transverse displacement $u = a \cos kx$ and is suddenly released from rest. Find its subsequent motion.

(a) This is a second-order equation and will (in general) have two solutions of the form $u(x, t) = f(x + \lambda t)$, where both λ satisfy

$$\lambda^2 + 2v\lambda + \left(v^2 - \frac{T}{\rho A}\right) = 0 \quad \Rightarrow \quad \lambda = -v \pm \sqrt{v^2 - v^2 + \frac{T}{\rho A}} \equiv -v \pm \alpha,$$

and gives (minus) the speed of the corresponding profile. Thus the general displacement consists of a superposition of waveforms traveling with speeds $v \mp \alpha$.

(b) Now $u(x, 0) = a \cos kx$ and $\dot{u}(x, 0) = 0$, where the dot denotes differentiation with respect to time t. Let the general solution be given by

$$u(x, t) = f[x - (v + \alpha)t] + g[x - (v - \alpha)t],$$

with $a \cos kx = f(x) + g(x)$

and $0 = -(v + \alpha)f'(x) - (v - \alpha)g'(x).$

We differentiate the first of these with respect to x and then eliminate the function $f'(x)$:

$$-ka \sin kx = f'(x) + g'(x),$$
$$-ka(v + \alpha) \sin kx = (v + \alpha - v + \alpha)g'(x),$$
$$g'(x) = -\frac{ka(v + \alpha)}{2\alpha} \sin kx,$$
$$\Rightarrow \quad g(x) = \frac{v + \alpha}{2\alpha} a \cos kx + c,$$
$$\Rightarrow \quad f(x) = \frac{\alpha - v}{2\alpha} a \cos kx - c.$$

Now that the forms of the initially arbitrary functions $f(x)$ and $g(x)$ have been determined, it follows that, for a *general* time t,

$$u(x, t) = \frac{\alpha - v}{2\alpha} a \cos[kx - k(v + \alpha)t] + \frac{\alpha + v}{2\alpha} a \cos[kx - k(v - \alpha)t]$$
$$= \frac{a}{2} 2\cos(kx - kvt)\cos k\alpha t + \frac{va}{2\alpha} 2\sin(kx - kvt)\sin(-k\alpha t)$$
$$= a \cos[k(x - vt)]\cos k\alpha t - \frac{va}{\alpha} \sin[k(x - vt)]\sin k\alpha t.$$

10.19 In an electrical cable of resistance R and capacitance C, each per unit length, voltage signals obey the equation $\partial^2 V/\partial x^2 = RC\partial V/\partial t$. This (diffusion-type) equation has solutions of the form

$$f(\zeta) = \frac{2}{\sqrt{\pi}} \int_0^{\zeta} \exp(-v^2)\, dv, \quad \text{where } \zeta = \frac{x(RC)^{1/2}}{2t^{1/2}}.$$

It also has solutions of the form $V = Ax + D$.

(a) Find a combination of these that represents the situation after a steady voltage V_0 is applied at $x = 0$ at time $t = 0$.

(b) Obtain a solution describing the propagation of the voltage signal resulting from the application of the signal $V = V_0$ for $0 < t < T$, $V = 0$ otherwise, to the end $x = 0$ of an infinite cable.

(c) Show that for $t \gg T$ the maximum signal occurs at a value of x proportional to $t^{1/2}$ and has a magnitude proportional to t^{-1}.

(a) Consider the given function

$$f(\zeta) = \frac{2}{\sqrt{\pi}} \int_0^{\zeta} \exp(-v^2)\, dv, \quad \text{where } \zeta = \frac{x(RC)^{1/2}}{2t^{1/2}}.$$

The requirements to be satisfied by the correct combination of this function and $V(x,t) = Ax + D$ are (i) that, at $t = 0$, V is zero for all x, except $x = 0$ where it is V_0, and (ii) that, as $t \to \infty$, V is V_0 for all x.

(i) At $t = 0$, $\zeta = \infty$ and $f(\zeta) = 1$ for all $x \neq 0$.
(ii) As $t \to \infty$, $\zeta \to 0$ and $f(\zeta) \to 0$ for all finite x.

The required combination is therefore $D = V_0$ and $-V_0 f(\zeta)$, i.e.

$$V(x,t) = V_0 \left[1 - \frac{2}{\sqrt{\pi}} \int_0^{\frac{1}{2}x(CR/t)^{1/2}} \exp(-v^2)\, dv \right].$$

(b) The equation is linear and so we may superpose solutions. The response to the input $V = V_0$ for $0 < t < T$ can be considered as that to V_0 applied at $t = 0$ and continued, together with $-V_0$ applied at $t = T$ and continued. The solution is therefore the difference between two solutions of the form found in part (a):

$$V(x,t) = \frac{2V_0}{\sqrt{\pi}} \int_{\frac{1}{2}x(CR/t)^{1/2}}^{\frac{1}{2}x[CR/(t-T)]^{1/2}} \exp\left(-v^2\right)\, dv.$$

(c) To find the maximum signal we set $\partial V/\partial x$ equal to zero. Remembering that we are differentiating with respect to the limits of an integral (whose integrand does not contain x explicitly), we obtain

$$\frac{1}{2}\left(\frac{CR}{t-T}\right)^{1/2} \exp\left[-\frac{x^2 CR}{4(t-T)}\right] - \frac{1}{2}\left(\frac{CR}{t}\right)^{1/2} \exp\left[-\frac{x^2 CR}{4t}\right] = 0.$$

This requires

$$\left(\frac{t-T}{t}\right)^{1/2} = \exp\left[-\frac{x^2 CR}{4(t-T)} + \frac{x^2 CR}{4t}\right]$$

$$= \exp\left[\frac{x^2 CR(-t+t-T)}{4t(t-T)}\right].$$

For $t \gg T$, we expand both sides:

$$1 - \frac{1}{2}\frac{T}{t} + \cdots = 1 - \frac{Tx^2CR}{4t^2} + \cdots,$$

$$\Rightarrow \quad x^2 \approx \frac{2t}{CR} \quad \Rightarrow \quad v = \frac{1}{2}\sqrt{\frac{2t}{CR}}\left(\frac{CR}{t}\right)^{1/2} = \frac{1}{\sqrt{2}}.$$

The corresponding value of V is approximately equal to the value of the integrand, evaluated at this value of v, multiplied by the difference between the two limits of the integral. Thus

$$V_{\text{max}} \approx \frac{2V_0}{\sqrt{\pi}}\exp(-v^2)\frac{x\sqrt{CR}}{2}\left[\frac{1}{(t-T)^{1/2}} - \frac{1}{t^{1/2}}\right]$$

$$\approx \frac{2V_0}{\sqrt{\pi}}e^{-1/2}\frac{x\sqrt{CR}}{2}\frac{1}{2}\frac{T}{t^{3/2}}$$

$$= \frac{V_0 T e^{-1/2}}{\sqrt{2\pi}\, t}.$$

In summary, for $t \gg T$ the maximum signal occurs at a value of x proportional to $t^{1/2}$ and has a magnitude proportional to t^{-1}.

10.21 Consider each of the following situations in a qualitative way and determine the equation type, the nature of the boundary curve and the type of boundary conditions involved:

(a) a conducting bar given an initial temperature distribution and then thermally isolated;
(b) two long conducting concentric cylinders, on each of which the voltage distribution is specified;
(c) two long conducting concentric cylinders, on each of which the charge distribution is specified;
(d) a semi-infinite string, the end of which is made to move in a prescribed way.

We use the notation

$$A\frac{\partial^2 u}{\partial x^2} + B\frac{\partial^2 u}{\partial x \partial y} + C\frac{\partial^2 u}{\partial y^2} + D\frac{\partial u}{\partial x} + E\frac{\partial u}{\partial y} + Fu = R(x, y)$$

to express the most general type of PDE, and the following table

Equation type	Boundary	Conditions
hyperbolic	open	Cauchy
parabolic	open	Dirichlet or Neumann
elliptic	closed	Dirichlet or Neumann

to determine the appropriate boundary type and hence conditions.

(a) The diffusion equation $\kappa\dfrac{\partial^2 T}{\partial x^2} = \dfrac{\partial T}{\partial t}$ has $A = \kappa$, $B = 0$ and $C = 0$; thus $B^2 = 4AC$ and the equation is parabolic. This needs an open boundary. In the present case, the initial heat distribution (at the $t = 0$ boundary) is a Dirichlet condition and the insulation (no temperature gradient at the external surfaces) is a Neumann condition.

(b) The governing equation in two-dimensional Cartesians (not the natural choice for this situation, but this does not matter for the present purpose) is the Laplace equation, $\dfrac{\partial^2 \phi}{\partial x^2} + \dfrac{\partial^2 \phi}{\partial y^2} = 0$, which has $A = 1$, $B = 0$ and $C = 1$ and therefore $B^2 < 4AC$. The equation is therefore elliptic and requires a closed boundary. Since ϕ is specified on the cylinders, the boundary conditions are Dirichlet in this particular situation.

(c) This is the same as part (b) except that the specified charge distribution σ determines $\partial \phi / \partial n$, through $\partial \phi / \partial n = \sigma / \epsilon_0$, and imposes Neumann boundary conditions.

(d) For the wave equation $\dfrac{\partial^2 u}{\partial x^2} - \dfrac{1}{c^2} \dfrac{\partial^2 u}{\partial t^2} = 0$, we have $A = 1$, $B = 0$ and $C = -c^{-2}$, thus making $B^2 > 4AC$ and the equation hyperbolic. We thus require an open boundary and Cauchy conditions, with the displacement of the end of the string having to be specified at all times – this is equivalent to the displacement and the velocity of the end of the string being specified at all times.

10.23 The Klein–Gordon equation (which is satisfied by the quantum-mechanical wavefunction $\Phi(\mathbf{r})$ of a relativistic spinless particle of non-zero mass m) is

$$\nabla^2 \Phi - m^2 \Phi = 0.$$

Show that the solution for the scalar field $\Phi(\mathbf{r})$ in any volume V bounded by a surface S is unique if either Dirichlet or Neumann boundary conditions are specified on S.

Suppose that, for a given set of boundary conditions ($\Phi = f$ or $\partial \Phi / \partial n = g$ on S), there are two solutions to the Klein–Gordon equation, Φ_1 and Φ_2. Then consider $\Phi_3 = \Phi_1 - \Phi_2$, which satisfies

$$\nabla^2 \Phi_3 = \nabla^2 \Phi_1 - \nabla^2 \Phi_2 = m^2 \Phi_1 - m^2 \Phi_2 = m^2 \Phi_3$$

and

$$\text{either } \Phi_3 = f - f = 0, \text{ or } \frac{\partial \Phi_3}{\partial n} = g - g = 0 \text{ on } S.$$

Now apply Green's first theorem with the scalar functions equal to Φ_3 and Φ_3^*:

$$\int^S \Phi_3^* \frac{\partial \Phi_3}{\partial n} \, dS = \int_V [\Phi_3^* \nabla^2 \Phi_3 + (\nabla \Phi_3^*) \cdot (\nabla \Phi_3)] \, dV,$$

$$\Rightarrow \quad 0 = \int_V (m^2 |\Phi_3|^2 + |\nabla \Phi_3|^2) \, dV,$$

whichever set of boundary conditions applies. Since both terms in the integrand on the RHS are non-negative, each must be equal to zero. In particular, $|\Phi_3| = 0$ implies that $\Phi_3 = 0$ everywhere, i.e. $\Phi_1 = \Phi_2$ everywhere; the solution is unique.

11 Solution methods for PDEs

11.1 Solve the following first-order partial differential equations by separating the variables:

$$\text{(a) } \frac{\partial u}{\partial x} - x\frac{\partial u}{\partial y} = 0; \quad \text{(b) } x\frac{\partial u}{\partial x} - 2y\frac{\partial u}{\partial y} = 0.$$

In each case we write $u(x, y) = X(x)Y(y)$, separate the variables into groups that each depend on only one variable, and then assert that each must be equal to a constant, with the several constants satisfying an arithmetic identity.

(a)
$$\frac{\partial u}{\partial x} - x\frac{\partial u}{\partial y} = 0,$$
$$X'Y - xXY' = 0,$$
$$\frac{X'}{xX} = \frac{Y'}{Y} = k \quad \Rightarrow \quad \ln X = \tfrac{1}{2}kx^2 + c_1, \quad \ln Y = ky + c_2,$$
$$\Rightarrow \quad X = Ae^{kx^2/2}, \quad Y = Be^{ky},$$
$$\Rightarrow \quad u(x, y) = Ce^{\lambda(x^2+2y)}, \text{ where } k = 2\lambda.$$

(b)
$$x\frac{\partial u}{\partial x} - 2y\frac{\partial u}{\partial y} = 0,$$
$$xX'Y - 2yXY' = 0,$$
$$\frac{xX'}{X} = \frac{2yY'}{Y} = k \quad \Rightarrow \quad \ln X = k\ln x + c_1,$$
$$\ln Y = \tfrac{1}{2}k\ln y + c_2,$$
$$\Rightarrow \quad X = Ax^k, \quad Y = By^{k/2},$$
$$\Rightarrow \quad u(x, y) = C(x^2 y)^\lambda, \text{ where } k = 2\lambda.$$

11.3 The wave equation describing the transverse vibrations of a stretched membrane under tension T and having a uniform surface density ρ is

$$T\left(\frac{\partial^2 u}{\partial x^2} + \frac{\partial^2 u}{\partial y^2}\right) = \rho\frac{\partial^2 u}{\partial t^2}.$$

149

Find a separable solution appropriate to a membrane stretched on a frame of length a and width b, showing that the natural angular frequencies of such a membrane are given by

$$\omega^2 = \frac{\pi^2 T}{\rho} \left(\frac{n^2}{a^2} + \frac{m^2}{b^2} \right),$$

where n and m are any positive integers.

We seek solutions $u(x, y, t)$ that are periodic in time and have $u(0, y, t) = u(a, y, t) = u(x, 0, t) = u(x, b, t) = 0$. Write $u(x, y, t) = X(x)Y(y)S(t)$ and substitute, obtaining

$$T(X''YS + XY''S) = \rho XYS'',$$

which, when divided through by XYS, gives

$$\frac{X''}{X} + \frac{Y''}{Y} = \frac{\rho}{T} \frac{S''}{S} = -\frac{\omega^2 \rho}{T}.$$

The second equality, obtained by applying the separation of variables principle with separation constant $-\omega^2 \rho / T$, gives $S(t)$ as a sinusoidal function of t of frequency ω, i.e. $A \cos(\omega t) + B \sin(\omega t)$.

We then have, on applying the separation of variables principle a second time, that

$$\frac{X''}{X} = \lambda \text{ and } \frac{Y''}{Y} = \mu, \text{ where } \lambda + \mu = -\frac{\omega^2 \rho}{T}. \qquad (*)$$

These equations must also have sinusoidal solutions. This is because, since $u(0, y, t) = u(a, y, t) = u(x, 0, t) = u(x, b, t) = 0$, each solution has to have zeros at two different values of its argument. We are thus led to

$$X = A \sin(p\,x) \text{ and } Y = B \sin(q\,x), \text{ where } p^2 = -\lambda \text{ and } q^2 = -\mu.$$

Further, since $u(a, y, t) = u(x, b, t) = 0$, we must have $p = n\pi/a$ and $q = m\pi/b$, where n and m are integers. Putting these values back into $(*)$ gives

$$-p^2 - q^2 = -\frac{\omega^2 \rho}{T} \quad \Rightarrow \quad \pi^2 \left(\frac{n^2}{a^2} + \frac{m^2}{b^2} \right) = \frac{\omega^2 \rho}{T}.$$

Hence the quoted result.

11.5 Denoting the three terms of ∇^2 in spherical polars by $\nabla_r^2, \nabla_\theta^2, \nabla_\phi^2$ in an obvious way, evaluate $\nabla_r^2 u$, etc. for the two functions given below and verify that, in each case, although the individual terms are not necessarily zero their sum $\nabla^2 u$ is zero. Identify the corresponding values of ℓ and m.

(a) $u(r, \theta, \phi) = \left(Ar^2 + \dfrac{B}{r^3} \right) \dfrac{3 \cos^2 \theta - 1}{2}$.

(b) $u(r, \theta, \phi) = \left(Ar + \dfrac{B}{r^2} \right) \sin \theta \exp i\phi$.

In both cases we write $u(r, \theta, \phi)$ as $R(r)\Theta(\theta)\Phi(\phi)$ with

$$\nabla_r^2 = \frac{1}{r^2}\frac{\partial}{\partial r}\left(r^2\frac{\partial}{\partial r}\right), \quad \nabla_\theta^2 = \frac{1}{r^2\sin\theta}\frac{\partial}{\partial\theta}\left(\sin\theta\frac{\partial}{\partial\theta}\right), \quad \nabla_\phi^2 = \frac{1}{r^2\sin^2\theta}\frac{\partial^2}{\partial\phi^2}.$$

(a) $u(r, \theta, \phi) = \left(Ar^2 + \dfrac{B}{r^3}\right)\dfrac{3\cos^2\theta - 1}{2}.$

$$\nabla_r^2 u = \frac{1}{r^2}\frac{\partial}{\partial r}\left(2Ar^3 - \frac{3B}{r^2}\right)\Theta = \left(6A + \frac{6B}{r^5}\right)\Theta = \frac{6u}{r^2},$$

$$\nabla_\theta^2 u = \frac{R}{r^2}\frac{1}{\sin\theta}\frac{\partial}{\partial\theta}(-3\sin^2\theta\cos\theta) = \frac{R}{r^2}\left(\frac{-6\sin\theta\cos^2\theta + 3\sin^3\theta}{\sin\theta}\right)$$

$$= \frac{R}{r^2}(-9\cos^2\theta + 3) = -\frac{6u}{r^2},$$

$$\nabla_\phi^2 u = 0.$$

Thus, although $\nabla_r^2 u$ and $\nabla_\theta^2 u$ are not individually zero, their sum is. From $\nabla_r^2 u = \ell(\ell + 1)u = 6u$, we deduce that $\ell = 2$ (or -3) and from $\nabla_\phi^2 u = 0$ that $m = 0$.

(b) $u(r, \theta, \phi) = \left(Ar + \dfrac{B}{r^2}\right)\sin\theta\, e^{i\phi}.$

$$\nabla_r^2 u = \frac{1}{r^2}\frac{\partial}{\partial r}\left(Ar^2 - \frac{2B}{r}\right)\Theta\Phi = \left(\frac{2A}{r} + \frac{2B}{r^4}\right)\Theta\Phi = \frac{2u}{r^2},$$

$$\nabla_\theta^2 u = \frac{R\Phi}{r^2}\frac{1}{\sin\theta}\frac{\partial}{\partial\theta}(\sin\theta\cos\theta) = \frac{R\Phi}{r^2}\left(\frac{-\sin^2\theta + \cos^2\theta}{\sin\theta}\right)$$

$$= -\frac{u}{r^2} + \frac{\cos^2\theta}{\sin^2\theta}\frac{u}{r^2},$$

$$\nabla_\phi^2 u = \frac{R\Theta}{r^2\sin^2\theta}\frac{\partial^2}{\partial\phi^2}(e^{i\phi}) = -\frac{u}{r^2\sin^2\theta}.$$

Hence,

$$\nabla^2 u = \frac{2u}{r^2} - \frac{u}{r^2} + \frac{\cos^2\theta}{\sin^2\theta}\frac{u}{r^2} - \frac{u}{r^2\sin^2\theta} = \frac{u}{r^2}\left(1 + \frac{\cos^2\theta - 1}{\sin^2\theta}\right) = 0.$$

Here each individual term is non-zero, but their sum *is* zero. Further, $\ell(\ell + 1) = 2$ and so $\ell = 1$ (or -2), and from $\nabla_\phi^2 u = -u/(r^2\sin\theta)$ it follows that $m^2 = 1$. In fact, from the normal definition of spherical harmonics, $m = +1$.

11.7 If the stream function $\psi(r, \theta)$ for the flow of a very viscous fluid past a sphere is written as $\psi(r, \theta) = f(r) \sin^2 \theta$, then $f(r)$ satisfies the equation

$$f^{(4)} - \frac{4f''}{r^2} + \frac{8f'}{r^3} - \frac{8f}{r^4} = 0.$$

At the surface of the sphere $r = a$ the velocity field $\mathbf{u} = \mathbf{0}$, whilst far from the sphere $\psi \simeq (Ur^2 \sin^2 \theta)/2$.

Show that $f(r)$ can be expressed as a superposition of powers of r, and determine which powers give acceptable solutions. Hence show that

$$\psi(r, \theta) = \frac{U}{4} \left(2r^2 - 3ar + \frac{a^3}{r} \right) \sin^2 \theta.$$

For solutions of

$$f^{(4)} - \frac{4f''}{r^2} + \frac{8f'}{r^3} - \frac{8f}{r^4} = 0$$

that are powers of r, i.e. have the form Ar^n, n must satisfy the quartic equation

$$n(n-1)(n-2)(n-3) - 4n(n-1) + 8n - 8 = 0,$$
$$(n-1)[n(n-2)(n-3) - 4n + 8] = 0,$$
$$(n-1)(n-2)[n(n-3) - 4] = 0,$$
$$(n-1)(n-2)(n-4)(n+1) = 0.$$

Thus the possible powers are 1, 2, 4 and -1.

Since $\psi \to \frac{1}{2} Ur^2 \sin^2 \theta$ as $r \to \infty$, the solution can contain no higher (positive) power of r than the second. Thus there is no $n = 4$ term and the solution has the form

$$\psi(r, \theta) = \left(\frac{Ur^2}{2} + Ar + \frac{B}{r} \right) \sin^2 \theta.$$

On the surface of the sphere $r = a$ both velocity components, u_r and u_θ, are zero. These components are given in terms of the stream functions, as shown below; note that u_r is found by differentiating with respect to θ and u_θ by differentiating with respect to r.

$$u_r = 0 \quad \Rightarrow \quad \frac{1}{a^2 \sin \theta} \frac{\partial \psi}{\partial \theta} = 0 \quad \Rightarrow \quad \frac{Ua^2}{2} + Aa + \frac{B}{a} = 0,$$

$$u_\theta = 0 \quad \Rightarrow \quad \frac{-1}{a \sin \theta} \frac{\partial \psi}{\partial r} = 0 \quad \Rightarrow \quad Ua + A - \frac{B}{a^2} = 0,$$

$$\Rightarrow \quad A = -\tfrac{3}{4} Ua \text{ and } B = \tfrac{1}{4} Ua^3.$$

The full solution is thus

$$\psi(r, \theta) = \frac{U}{4} \left(2r^2 - 3ar + \frac{a^3}{r} \right) \sin^2 \theta.$$

11.9 A circular disc of radius a is heated in such a way that its perimeter $\rho = a$ has a steady temperature distribution $A + B \cos^2 \phi$, where ρ and ϕ are plane polar coordinates and A and B are constants. Find the temperature $T(\rho, \phi)$ everywhere in the region $\rho < a$.

This is a steady state problem, for which the (heat) diffusion equation becomes the Laplace equation. The most general single-valued solution to the Laplace equation in plane polar coordinates is given by

$$T(\rho, \phi) = C \ln \rho + D + \sum_{n=1}^{\infty} (A_n \cos n\phi + B_n \sin n\phi)(C_n \rho^n + D_n \rho^{-n}).$$

The region $\rho < a$ contains the point $\rho = 0$; since $\ln \rho$ and all ρ^{-n} become infinite at that point, $C = D_n = 0$ for all n.

On $\rho = a$

$$T(a, \phi) = A + B \cos^2 \phi = A + \tfrac{1}{2} B(\cos 2\phi + 1).$$

Equating the coefficients of $\cos n\phi$, including $n = 0$, gives $A + \tfrac{1}{2}B = D$, $A_2 C_2 a^2 = \tfrac{1}{2}B$ and $A_n C_n a^n = 0$ for all $n \neq 2$; further, all $B_n = 0$. The solution everywhere (not just on the perimeter) is therefore

$$T(\rho, \phi) = A + \frac{B}{2} + \frac{B\rho^2}{2a^2} \cos 2\phi.$$

It should be noted that "equating coefficients" to determine unknown constants is justified by the fact that the sinusoidal functions in the sum are mutually orthogonal over the range $0 \leq \phi < 2\pi$.

11.11 The free transverse vibrations of a thick rod satisfy the equation

$$a^4 \frac{\partial^4 u}{\partial x^4} + \frac{\partial^2 u}{\partial t^2} = 0.$$

Obtain a solution in separated-variable form and, for a rod clamped at one end, $x = 0$, and free at the other, $x = L$, show that the angular frequency of vibration ω satisfies

$$\cosh \left(\frac{\omega^{1/2} L}{a} \right) = -\sec \left(\frac{\omega^{1/2} L}{a} \right).$$

[At a clamped end both u and $\partial u / \partial x$ vanish, whilst at a free end, where there is no bending moment, $\partial^2 u / \partial x^2$ and $\partial^3 u / \partial x^3$ are both zero.]

The general solution is written as the product $u(x, t) = X(x)T(t)$, which, on substitution, produces the separated equation

$$a^4 \frac{X^{(4)}}{X} = -\frac{T''}{T} = \omega^2.$$

Here the separation constant has been chosen so as to give oscillatory behavior (in the time variable). The spatial equation then becomes

$$X^{(4)} - \mu^4 X = 0, \quad \text{where } \mu = \omega^{1/2}/a.$$

The required auxiliary equation is $\lambda^4 - \mu^4 = 0$, leading to the general solution

$$X(x) = A \sin \mu x + B \cos \mu x + C \sinh \mu x + D \cosh \mu x.$$

The constants A, B, C and D are to be determined by requiring $X(0) = X'(0) = 0$ and $X''(L) = X'''(L) = 0$.

At the clamped end,

$$X(0) = 0 \quad \Rightarrow \quad D = -B,$$
$$X' \quad = \quad \mu(A \cos \mu x - B \sin \mu x + C \cosh \mu x - B \sinh \mu x),$$
$$X'(0) = 0 \quad \Rightarrow \quad C = -A.$$

At the free end,

$$X'' = \mu^2(-A \sin \mu x - B \cos \mu x - A \sinh \mu x - B \cosh \mu x),$$
$$X''' = \mu^3(-A \cos \mu x + B \sin \mu x - A \cosh \mu x - B \sinh \mu x),$$
$$X''(L) = 0 \quad \Rightarrow \quad A(\sin \mu L + \sinh \mu L) + B(\cos \mu L + \cosh \mu L) = 0,$$
$$X'''(L) = 0 \quad \Rightarrow \quad A(-\cos \mu L - \cosh \mu L) + B(\sin \mu L - \sinh \mu L) = 0.$$

Cross-multiplying then gives

$$-\sin^2 \mu L + \sinh^2 \mu L = \cos^2 \mu L + 2 \cos \mu L \cosh \mu L + \cosh^2 \mu L,$$
$$0 = 1 + 2 \cos \mu L \cosh \mu L + 1,$$
$$-1 = \cos \mu L \cosh \mu L,$$
$$\cosh \left(\frac{\omega^{1/2} L}{a} \right) = -\sec \left(\frac{\omega^{1/2} L}{a} \right).$$

Because sinusoidal and hyperbolic functions can all be written in terms of exponential functions, this problem could also be approached by assuming solutions that are (exponential) functions of linear combinations of x and t (as in Chapter 10). However, in practice, eliminating the t-dependent terms leads to involved algebra.

11.13 A string of length L, fixed at its two ends, is plucked at its mid-point by an amount A and then released. Prove that the subsequent displacement is given by

$$u(x, t) = \sum_{n=0}^{\infty} \frac{8A}{\pi^2(2n + 1)^2} \sin \left[\frac{(2n + 1)\pi x}{L} \right] \cos \left[\frac{(2n + 1)\pi ct}{L} \right],$$

where, in the usual notation, $c^2 = T/\rho$.

Find the total kinetic energy of the string when it passes through its unplucked position, by calculating it in each mode (each n) and summing, using the result

$$\sum_{0}^{\infty} \frac{1}{(2n + 1)^2} = \frac{\pi^2}{8}.$$

Confirm that the total energy is equal to the work done in plucking the string initially.

We start with the wave equation:

$$\frac{\partial^2 u}{\partial x^2} - \frac{1}{c^2}\frac{\partial^2 u}{\partial t^2} = 0$$

and assume a separated-variable solution $u(x, t) = X(x)S(t)$. This leads to

$$\frac{X''}{X} = \frac{1}{c^2}\frac{S''}{S} = -k^2.$$

The solution to the spatial equation is given by

$$X(x) = B\cos kx + C\sin kx.$$

Taking the string as anchored at $x = 0$ and $x = L$, we must have $B = 0$ and k constrained by $\sin kL = 0 \Rightarrow k = n\pi/L$ with n an integer.

The solution to the corresponding temporal equation is

$$S(t) = D\cos kct + E\sin kct.$$

Since there is no initial motion, i.e. $\dot{S}(0) = 0$, it follows that $E = 0$.

For any particular value of k, the constants C and D can be amalgamated. The general solution is given by a superposition of the allowed functions, i.e.

$$u(x, t) = \sum_{n=1}^{\infty} C_n \sin\frac{n\pi x}{L}\cos\frac{n\pi ct}{L}.$$

We now have to determine the C_n by making $u(x, 0)$ match the given initial configuration, which is

$$u(x, 0) = \begin{cases} \dfrac{2Ax}{L} & \text{for } 0 \le x \le \dfrac{L}{2}, \\ \dfrac{2A(L-x)}{L} & \dfrac{L}{2} < x \le L. \end{cases}$$

This is now a Fourier series calculation yielding

$$\frac{C_n L}{2} = \int_0^{L/2}\frac{2Ax}{L}\sin\frac{n\pi x}{L}dx + \int_{L/2}^{L}\frac{2A(L-x)}{L}\sin\frac{n\pi x}{L}dx$$

$$= \frac{2A}{L}J_1 + 2AJ_2 - \frac{2A}{L}J_3,$$

with

$$J_1 = \left[-\frac{xL}{n\pi} \cos \frac{n\pi x}{L} \right]_0^{L/2} + \int_0^{L/2} \frac{L}{n\pi} \cos \frac{n\pi x}{L} \, dx$$

$$= -\frac{L^2}{2\pi n} \cos \frac{n\pi}{2} + \frac{L^2}{n^2\pi^2} \sin \frac{n\pi}{2},$$

$$J_2 = \int_{L/2}^L \sin \frac{n\pi x}{L} \, dx = -\frac{L}{n\pi} \left[\cos \frac{n\pi x}{L} \right]_{L/2}^L$$

$$= -\frac{L}{n\pi} \left[(-1)^n - \cos \frac{n\pi}{2} \right],$$

$$J_3 = \left[-\frac{xL}{n\pi} \cos \frac{n\pi x}{L} \right]_{L/2}^L + \int_{L/2}^L \frac{L}{n\pi} \cos \frac{n\pi x}{L} \, dx$$

$$= \frac{L^2}{2\pi n} \cos \frac{n\pi}{2} - \frac{L^2}{n\pi}(-1)^n - \frac{L^2}{n^2\pi^2} \sin \frac{n\pi}{2}.$$

Thus

$$J_1 - J_3 = -\frac{2L^2}{2\pi n} \cos \frac{n\pi}{2} + \frac{L^2}{n\pi}(-1)^n + \frac{2L^2}{n^2\pi^2} \sin \frac{n\pi}{2}$$

$$= -L J_2 + \frac{2L^2}{n^2\pi^2} \sin \frac{n\pi}{2},$$

and so it follows that

$$\frac{C_n L}{2} = \frac{2A}{L}(J_1 - J_3 + L J_2) = 2A \frac{2L}{n^2\pi^2} \sin \frac{n\pi}{2}.$$

This is zero if n is even and $C_n = 8A(-1)^{(n-1)/2}/(n^2\pi^2)$ if n is odd. Write $n = 2m + 1$, $m = 0, 1, 2, \ldots$, with $C_{2m+1} = \dfrac{8A(-1)^m}{(2m+1)^2\pi^2}$.

The final solution (in which m is replaced by n, to match the question) is thus

$$u(x, t) = \sum_{n=0}^{\infty} \frac{8A(-1)^n}{\pi^2(2n+1)^2} \sin \left[\frac{(2n+1)\pi x}{L} \right] \cos \left[\frac{(2n+1)\pi ct}{L} \right].$$

The velocity profile derived from this is given by

$$\dot{u}(x, t) = \sum_{n=0}^{\infty} \frac{8A(-1)^n}{\pi^2(2n+1)^2} \left(\frac{-(2n+1)\pi c}{L} \right)$$

$$\times \sin \left[\frac{(2n+1)\pi x}{L} \right] \sin \left[\frac{(2n+1)\pi ct}{L} \right],$$

giving the energy in the $(2n + 1)$th mode (evaluated when the time-dependent sine function is maximal) as

$$E_{2n+1} = \int_0^L \tfrac{1}{2}\rho \dot{u}_n^2 \, dx$$

$$= \int_0^L \frac{\rho}{2} \frac{(8A)^2 c^2}{L^2(2n+1)^2\pi^2} \sin^2 \frac{(2n+1)\pi x}{L}$$

$$= \frac{32A^2\rho c^2}{L^2(2n+1)^2\pi^2} \frac{L}{2}.$$

Therefore

$$E = \sum_{n=0}^{\infty} E_{2n+1} = \frac{16A^2\rho c^2}{\pi^2 L} \sum_{n=0}^{\infty} \frac{1}{(2n+1)^2} = \frac{2A^2\rho c^2}{L}.$$

When the mid-point of the string has been displaced sideways by y ($\ll L$), the net (resolved) restoring force is $2T[y/(L/2)] = 4Ty/L$. Thus the total work done to produce a displacement of A is

$$W = \int_0^A \frac{4Ty}{L} \, dy = \frac{2TA^2}{L} = \frac{2\rho c^2 A^2}{L},$$

i.e. the same as the total energy of the subsequent motion.

11.15 Prove that the potential for $\rho < a$ associated with a vertical split cylinder of radius a, the two halves of which ($\cos \phi > 0$ and $\cos \phi < 0$) are maintained at equal and opposite potentials $\pm V$, is given by

$$u(\rho, \phi) = \frac{4V}{\pi} \sum_{n=0}^{\infty} \frac{(-1)^n}{2n+1} \left(\frac{\rho}{a}\right)^{2n+1} \cos(2n+1)\phi.$$

The most general solution of the Laplace equation in cylindrical polar coordinates that is independent of z is

$$T(\rho, \phi) = C \ln \rho + D + \sum_{n=1}^{\infty} (A_n \cos n\phi + B_n \sin n\phi)(C_n \rho^n + D_n \rho^{-n}).$$

The required potential must be single-valued and finite in the space inside the cylinder (which includes $\rho = 0$), and on the cylinder it must take the boundary values $u = V$ for $\cos \phi > 0$ and $u = -V$ for $\cos \phi < 0$, i.e the boundary-value function is a square-wave function with average value zero. Although the function is antisymmetric in $\cos \phi$, it is symmetric in ϕ and so the solution will contain only cosine terms (and no sine terms).

These considerations already determine that $C = D = B_n = D_n = 0$, and so have reduced the solution to the form

$$u(\rho, \phi) = \sum_{n=1}^{\infty} A_n \rho^n \cos n\phi.$$

On $\rho = a$ this must match the stated boundary conditions, and so we are faced with a Fourier cosine series calculation. Multiplying through by $\cos m\phi$ and integrating yields

$$A_m a^m \frac{1}{2} 2\pi = 2 \int_0^{\pi/2} V \cos m\phi \, d\phi + 2 \int_{\pi/2}^\pi (-V) \cos m\phi \, d\phi$$

$$= 2V \left[\frac{\sin m\phi}{m} \right]_0^{\pi/2} - 2V \left[\frac{\sin m\phi}{m} \right]_{\pi/2}^\pi$$

$$= \frac{2V}{m} \left(\sin \frac{m\pi}{2} + \sin \frac{m\pi}{2} \right)$$

$$= (-1)^{(m-1)/2} \frac{4V}{m} \quad \text{for } m \text{ odd}, \ = 0 \text{ for } m \text{ even.}$$

Writing $m = 2n + 1$ gives the solution as

$$u(\rho, \phi) = \frac{4V}{\pi} \sum_{n=0}^\infty \frac{(-1)^n}{2n+1} \left(\frac{\rho}{a} \right)^{2n+1} \cos(2n+1)\phi.$$

11.17 Two identical copper bars are each of length a. Initially, one is at $0\,^\circ$C and the other at $100\,^\circ$C; they are then joined together end to end and thermally isolated. Obtain in the form of a Fourier series an expression $u(x, t)$ for the temperature at any point a distance x from the join at a later time t. Bear in mind the heat flow conditions at the free ends of the bars.

Taking $a = 0.5$ m estimate the time it takes for one of the free ends to attain a temperature of $55\,^\circ$C. The thermal conductivity of copper is $3.8 \times 10^2\,\mathrm{J\,m^{-1}\,K^{-1}\,s^{-1}}$, and its specific heat capacity is $3.4 \times 10^6\,\mathrm{J\,m^{-3}\,K^{-1}}$.

The equation governing the heat flow is

$$k \frac{\partial^2 u}{\partial x^2} = s \frac{\partial u}{\partial t},$$

which is the diffusion equation with diffusion constant $\kappa = k/s = 3.8 \times 10^2/3.4 \times 10^6 = 1.12 \times 10^{-4}\,\mathrm{m^2\,s^{-1}}$.

Making the usual separation of variables substitution shows that the time variation is of the form $T(t) = T(0)e^{-\kappa\lambda^2 t}$ when the spatial solution is a sinusoidal function of λx. The final common temperature is $50\,^\circ$C and we make this explicit by writing the general solution as

$$u(x, t) = 50 + \sum_\lambda (A_\lambda \sin \lambda x + B_\lambda \cos \lambda x) e^{-\kappa\lambda^2 t}.$$

This term having been taken out, the summation must be antisymmetric about $x = 0$ and therefore contain no cosine terms, i.e. $B_\lambda = 0$.

The boundary condition is that there is no heat flow at $x = \pm a$; this means that $\partial u/\partial x = 0$ at these points and requires

$$\lambda A_\lambda \cos \lambda x |_{x=\pm a} = 0 \quad \Rightarrow \quad \lambda a = (n + \tfrac{1}{2})\pi \quad \Rightarrow \quad \lambda = \frac{(2n+1)\pi}{2a},$$

where n is an integer. This corresponds to a fundamental Fourier period of $4a$. The solution thus takes the form

$$u(x, t) = 50 + \sum_{n=0}^{\infty} A_n \sin \frac{(2n+1)\pi x}{2a} \exp\left(-\frac{(2n+1)^2 \pi^2 \kappa t}{4a^2}\right).$$

At $t = 0$, the sum must take the values $+50$ for $0 < x < 2a$ and -50 for $-2a < x < 0$. This is (yet) another square-wave function – one that is antisymmetric about $x = 0$ and has amplitude 50. The calculation will not be repeated here but gives $A_n = 200/[(2n+1)\pi]$, making the complete solution

$$u(x, t) = 50 + \frac{200}{\pi} \sum_{n=0}^{\infty} \frac{1}{2n+1} \sin \frac{(2n+1)\pi x}{2a} \exp\left(-\frac{(2n+1)^2 \pi^2 \kappa t}{4a^2}\right).$$

For a free end, where $x = a$ and $\sin[(2n+1)\pi x/2a] = (-1)^n$, to attain $55\,^\circ$C needs

$$\sum_{n=0}^{\infty} \frac{(-1)^n}{2n+1} \exp\left(-\frac{(2n+1)^2 \pi^2 \, 1.12 \times 10^{-4}}{4 \times 0.25} t\right) = \frac{5\pi}{200} = 0.0785.$$

In principle this is an insoluble equation but, because the RHS $\ll 1$, the $n = 0$ term alone will give a good approximation to t:

$$\exp(-1.105 \times 10^{-3}t) \approx 0.0785 \quad \Rightarrow \quad t \approx 2300 \text{ s}.$$

11.19 For an infinite metal bar that has an initial ($t = 0$) temperature distribution $f(x)$ along its length, the temperature distribution at a general time t can be shown to be given by

$$u(x, t) = \frac{1}{\sqrt{4\pi \kappa t}} \int_{-\infty}^{\infty} \exp\left[-\frac{(x-\xi)^2}{4\kappa t}\right] f(\xi)\, d\xi.$$

Find an explicit expression for $u(x, t)$ given that $f(x) = \exp(-x^2/a^2)$.

The given initial distribution is $f(\xi) = \exp(-\xi^2/a^2)$ and so

$$u(x, t) = \frac{1}{\sqrt{4\pi \kappa t}} \int_{-\infty}^{\infty} \exp\left[-\frac{(x-\xi)^2}{4\kappa t}\right] \exp\left(-\frac{\xi^2}{a^2}\right) d\xi.$$

Now consider the exponent in the integrand, writing $1 + \dfrac{4\kappa t}{a^2}$ as τ^2 for compactness:

$$\text{exponent} = -\frac{\xi^2\tau^2 - 2\xi x + x^2}{4\kappa t}$$

$$= -\frac{(\xi\tau - x\tau^{-1})^2 - x^2\tau^{-2} + x^2}{4\kappa t}$$

$$\equiv -\eta^2 + \frac{x^2\tau^{-2} - x^2}{4\kappa t}, \quad \text{defining } \eta,$$

$$\text{with } d\eta = \frac{\tau \, d\xi}{\sqrt{4\kappa t}}.$$

With a change of variable from ξ to η, the integral becomes

$$u(x, t) = \frac{1}{\sqrt{4\pi\kappa t}} \exp\left(\frac{x^2\tau^{-2} - x^2}{4\kappa t}\right) \int_{-\infty}^{\infty} \exp(-\eta^2) \frac{\sqrt{4\kappa t}}{\tau} d\eta$$

$$= \frac{1}{\sqrt{\pi}} \frac{1}{\tau} \exp\left(x^2 \frac{1 - \tau^2}{4\kappa t \tau^2}\right) \sqrt{\pi}$$

$$= \frac{a}{\sqrt{a^2 + 4\kappa t}} \exp\left(-\frac{x^2}{a^2 + 4\kappa t}\right).$$

In words, although it retains a Gaussian shape, the initial distribution spreads symmetrically about the origin, its variance increasing linearly with time ($a^2 \to a^2 + 4\kappa t$). As is typical with diffusion processes, for large enough times the width varies as $\sqrt{t}$.

11.21 In the region $-\infty < x, y < \infty$ and $-t \leq z \leq t$, a charge-density wave $\rho(\mathbf{r}) = A \cos qx$, in the x-direction, is represented by

$$\rho(\mathbf{r}) = \frac{e^{iqx}}{\sqrt{2\pi}} \int_{-\infty}^{\infty} \tilde{\rho}(\alpha)e^{i\alpha z} \, d\alpha.$$

The resulting potential is represented by

$$V(\mathbf{r}) = \frac{e^{iqx}}{\sqrt{2\pi}} \int_{-\infty}^{\infty} \tilde{V}(\alpha)e^{i\alpha z} \, d\alpha.$$

Determine the relationship between $\tilde{V}(\alpha)$ and $\tilde{\rho}(\alpha)$, and hence show that the potential at the point $(0, 0, 0)$ is given by

$$\frac{A}{\pi \epsilon_0} \int_{-\infty}^{\infty} \frac{\sin kt}{k(k^2 + q^2)} \, dk.$$

Poisson's equation,

$$\nabla^2 V(\mathbf{r}) = -\frac{\rho(\mathbf{r})}{\epsilon_0},$$

provides the link between a charge density and the potential it produces.

Taking $V(\mathbf{r})$ in the form of its Fourier representation gives $\nabla^2 V$ as

$$\frac{\partial^2 V(\mathbf{r})}{\partial x^2} + \frac{\partial^2 V(\mathbf{r})}{\partial y^2} + \frac{\partial^2 V(\mathbf{r})}{\partial z^2} = \frac{e^{iqx}}{\sqrt{2\pi}} \int_{-\infty}^{\infty} (-q^2 - \alpha^2)\tilde{V}(\alpha)e^{i\alpha z} \, d\alpha,$$

with the $-q^2$ arising from the x-differentiation and the $-\alpha^2$ from the z-differentiation; the $\partial^2 V/\partial y^2$ term contributes nothing.

Comparing this with the integral expression for $-\rho(\mathbf{r})/\epsilon_0$ shows that

$$-\tilde{\rho}(\alpha) = \epsilon_0(-q^2 - \alpha^2)\tilde{V}(\alpha).$$

With the charge-density wave confined in the z-direction to $-t \leq z \leq t$, the expression for $\rho(\mathbf{r})$ in Cartesian coordinates is (in terms of Heaviside functions)

$$\rho(\mathbf{r}) = Ae^{iqx}[H(z + t) - H(z - t)].$$

The Fourier transform $\tilde{\rho}(\alpha)$ is therefore given by

$$
\begin{aligned}
\tilde{\rho}(\alpha) &= \frac{1}{\sqrt{2\pi}} \int_{-\infty}^{\infty} A[H(z+t) - H(z-t)] e^{-i\alpha z} \, dz \\
&= \frac{A}{\sqrt{2\pi}} \int_{-t}^{t} e^{-i\alpha z} \, dz \\
&= \frac{A}{\sqrt{2\pi}} \frac{e^{-i\alpha t} - e^{i\alpha t}}{-i\alpha} \\
&= \frac{A}{\sqrt{2\pi}} \frac{2 \sin \alpha t}{\alpha}.
\end{aligned}
$$

Now,
$$
\begin{aligned}
V(x, 0, z) &= \frac{e^{iqx}}{\sqrt{2\pi}} \int_{-\infty}^{\infty} \frac{\tilde{\rho}(\alpha)}{\epsilon_0(q^2 + \alpha^2)} e^{i\alpha z} \, d\alpha \\
&= \frac{e^{iqx}}{\sqrt{2\pi}} \int_{-\infty}^{\infty} \frac{e^{i\alpha z}}{\epsilon_0(q^2 + \alpha^2)} \frac{A}{\sqrt{2\pi}} \frac{2 \sin \alpha t}{\alpha} \, d\alpha,
\end{aligned}
$$
$$
\Rightarrow \quad V(0, 0, 0) = \frac{A}{\pi \epsilon_0} \int_{-\infty}^{\infty} \frac{\sin \alpha t}{\alpha(\alpha^2 + q^2)} \, d\alpha,
$$

as stated in the question.

11.23 Find the Green's function $G(\mathbf{r}, \mathbf{r}_0)$ in the half-space $z > 0$ for the solution of $\nabla^2 \Phi = 0$ with Φ specified in cylindrical polar coordinates (ρ, ϕ, z) on the plane $z = 0$ by

$$
\Phi(\rho, \phi, z) = \begin{cases} 1 & \text{for } \rho \leq 1, \\ 1/\rho & \text{for } \rho > 1. \end{cases}
$$

Determine the variation of $\Phi(0, 0, z)$ along the z-axis.

For the half-space $z > 0$ the bounding surface consists of the plane $z = 0$ and the (hemi-spherical) surface at infinity; the Green's function must take zero value on these surfaces. In order to ensure this when a unit point source is introduced at $\mathbf{r} = \mathbf{y}$, we must place a compensating negative unit source at $\mathbf{y}$'s reflection point in the plane. If, in cylindrical polar coordinates, $\mathbf{y} = (\rho, \phi, z_0)$, then the image charge has to be at $\mathbf{y}' = (\rho, \phi, -z_0)$. The resulting Green's function $G(\mathbf{x}, \mathbf{y})$ is given by

$$
G(\mathbf{x}, \mathbf{y}) = -\frac{1}{4\pi |\mathbf{x} - \mathbf{y}|} + \frac{1}{4\pi |\mathbf{x} - \mathbf{y}'|}.
$$

The solution to the problem with a given potential distribution $f(\rho, \phi)$ on the $z = 0$ part of the bounding surface S is given by

$$
\Phi(\mathbf{y}) = \int_S f(\rho, \phi) \left(-\frac{\partial G}{\partial z} \right) \rho \, d\phi \, d\rho,
$$

the minus sign arising because the outward normal to the region is in the negative z-direction. Calculating these functions explicitly gives

$$G(\mathbf{x}, \mathbf{y}) = -\frac{1}{4\pi[\rho^2 + (z - z_0)^2]^{1/2}} + \frac{1}{4\pi[\rho^2 + (z + z_0)^2]^{1/2}},$$

$$\frac{\partial G}{\partial z} = \frac{z - z_0}{4\pi[\rho^2 + (z - z_0)^2]^{3/2}} - \frac{(z + z_0)}{4\pi[\rho^2 + (z + z_0)^2]^{3/2}},$$

$$-\frac{\partial G}{\partial z}\bigg|_{z=0} = -\frac{-2z_0}{4\pi[\rho^2 + z_0^2]^{3/2}}.$$

Substituting the various factors into the general integral gives

$$\Phi(0, 0, z_0) = \int_0^\infty f(\rho) \frac{2z_0}{4\pi[\rho^2 + z_0^2]^{3/2}} \, 2\pi \, \rho \, d\rho$$

$$= \int_0^1 \frac{z_0 \, \rho}{(\rho^2 + z_0^2)^{3/2}} \, d\rho + \int_1^\infty \frac{z_0}{(\rho^2 + z_0^2)^{3/2}} \, d\rho$$

$$= -z_0 \left[(\rho^2 + z_0^2)^{-1/2} \right]_0^1 + \int_\theta^{\pi/2} \frac{z_0^2 \sec^2 u}{z_0^3 \sec^3 u} \, du,$$

where, in the second integral, we have set $\rho = z_0 \tan u$ with $d\rho = z_0 \sec^2 u \, du$ and $\theta = \tan^{-1}(1/z_0)$. The integral can now be obtained in closed form as

$$\Phi(0, 0, z_0) = -\frac{z_0}{(1 + z_0^2)^{1/2}} + 1 + \frac{1}{z_0} \left[\sin u \right]_\theta^{\pi/2}$$

$$= 1 - \frac{z_0}{(1 + z_0^2)^{1/2}} + \frac{1}{z_0} - \frac{1}{z_0(1 + z_0^2)^{1/2}}.$$

Thus the variation of Φ along the z-axis is given by

$$\Phi(0, 0, z) = \frac{z(1 + z^2)^{1/2} - z^2 + (1 + z^2)^{1/2} - 1}{z(1 + z^2)^{1/2}}.$$

11.25 Find, in the form of an infinite series, the Green's function of the ∇^2 operator for the Dirichlet problem in the region $-\infty < x < \infty$, $-\infty < y < \infty$, $-c \leq z \leq c$.

The fundamental solution in three dimensions of $\nabla^2 \psi = \delta(\mathbf{r})$ is $\psi(\mathbf{r}) = -1/(4\pi r)$.

For the given problem, $G(\mathbf{r}, \mathbf{r}_0)$ has to take the value zero on $z = \pm c$ and $\to 0$ for $|x| \to \infty$ and $|y| \to \infty$. Image charges have to be added in the regions $z > c$ and $z < -c$ to bring this about after a charge q has been placed at $\mathbf{r}_0 = (x_0, y_0, z_0)$ with $-c < z_0 < c$. Clearly all images will be on the line $x = x_0$, $y = y_0$.

Each image placed at $z = \xi$ in the region $z > c$ will require a further image of the same strength but opposite sign at $z = -c - \xi$ (in the region $z < -c$) so as to maintain the plane $z = -c$ as an equipotential. Likewise, each image placed at $z = -\chi$ in the region $z < -c$ will require a further image of the same strength but opposite sign at $z = c + \chi$ (in the region $z > c$) so as to maintain the plane $z = c$ as an equipotential. Thus successive image charges appear as follows:

$$
\begin{array}{ccc}
-q & 2c - z_0 & -2c - z_0 \\
+q & -3c + z_0 & 3c + z_0 \\
-q & 4c - z_0 & -4c - z_0 \\
+q & \text{etc.} & \text{etc.}
\end{array}
$$

The terms in the Green's function that are additional to the fundamental solution,

$$
-\frac{1}{4\pi}[(x - x_0)^2 + (y - y_0)^2 + (z - z_0)^2]^{-1/2},
$$

are therefore

$$
-\frac{(-1)}{4\pi} \sum_{n=2}^{\infty} \left\{ \frac{(-1)^n}{[(x - x_0)^2 + (y - y_0)^2 + (z + (-1)^n z_0 - nc)^2]^{1/2}} \right.
$$
$$
\left. + \frac{(-1)^n}{[(x - x_0)^2 + (y - y_0)^2 + (z + (-1)^n z_0 + nc)^2]^{1/2}} \right\}.
$$

11.27 Determine the Green's function for the Klein–Gordon equation in a half-space as follows.

(a) By applying the divergence theorem to the volume integral

$$
\int_V \left[\phi(\nabla^2 - m^2)\psi - \psi(\nabla^2 - m^2)\phi \right] dV,
$$

obtain a Green's function expression, as the sum of a volume integral and a surface integral, for the function $\phi(\mathbf{r}')$ that satisfies

$$
\nabla^2 \phi - m^2 \phi = \rho
$$

in V and takes the specified form $\phi = f$ on S, the boundary of V. The Green's function, $G(\mathbf{r}, \mathbf{r}')$, to be used satisfies

$$
\nabla^2 G - m^2 G = \delta(\mathbf{r} - \mathbf{r}')
$$

and vanishes when $\mathbf{r}$ is on S.

(b) When V is all space, $G(\mathbf{r}, \mathbf{r}')$ can be written as $G(t) = g(t)/t$, where $t = |\mathbf{r} - \mathbf{r}'|$ and $g(t)$ is bounded as $t \to \infty$. Find the form of $G(t)$.

(c) Find $\phi(\mathbf{r})$ in the half-space $x > 0$ if $\rho(\mathbf{r}) = \delta(\mathbf{r} - \mathbf{r}_1)$ and $\phi = 0$ both on $x = 0$ and as $r \to \infty$.

(a) For general ϕ and ψ we have

$$
\int_V \left[\phi(\nabla^2 - m^2)\psi - \psi(\nabla^2 - m^2)\phi \right] dV = \int_V \left[\phi \nabla^2 \psi - \psi \nabla^2 \phi \right] dV
$$
$$
= \int_V \nabla \cdot (\phi \nabla \psi - \psi \nabla \phi) \, dV
$$
$$
= \int_S (\phi \nabla \psi - \psi \nabla \phi) \cdot \mathbf{n} \, dS.
$$

Now take ϕ as ϕ, with $\nabla^2\phi - m^2\phi = \rho$ and $\phi = f$ on the surface S, and ψ as $G(\mathbf{r}, \mathbf{r}')$ with $\nabla^2 G - m^2 G = \delta(\mathbf{r} - \mathbf{r}')$ and $G(\mathbf{r}, \mathbf{r}') = 0$ on S:

$$\int_V [\phi(\mathbf{r})\delta(\mathbf{r} - \mathbf{r}') - G(\mathbf{r}, \mathbf{r}')\rho(\mathbf{r})]\, dV = \int_S [f(\mathbf{r})\nabla G(\mathbf{r}, \mathbf{r}') - 0] \cdot \mathbf{n}\, dS,$$

which, on rearrangement, gives

$$\phi(\mathbf{r}') = \int_V G(\mathbf{r}, \mathbf{r}')\rho(\mathbf{r})\, dV + \int_S f(\mathbf{r})\nabla G(\mathbf{r}, \mathbf{r}') \cdot \mathbf{n}\, dS.$$

(b) In the following calculation we start by formally integrating the defining Green's equation,

$$\nabla^2 G - m^2 G = \delta(\mathbf{r} - \mathbf{r}'),$$

over a sphere of radius t centered on $\mathbf{r}'$. Having replaced the volume integral of $\nabla^2 G$ with the corresponding surface integral given by the divergence theorem, we move the origin to $\mathbf{r}'$, denote $|\mathbf{r} - \mathbf{r}'|$ by t' and integrate both sides of the equation from $t' = 0$ to $t' = t$:

$$\int_V \nabla^2 G\, dV - \int_V m^2 G\, dV = \int_V \delta(\mathbf{r} - \mathbf{r}')\, dV,$$

$$\int_S \nabla G \cdot \mathbf{n}\, dS - m^2 \int_V G\, dV = 1,$$

$$4\pi t^2 \frac{dG}{dt} - m^2 \int_0^t G(t')4\pi t'^2\, dt' = 1, \qquad (*)$$

$$4\pi t^2 G'' + 8\pi t G' - 4\pi m^2 t^2 G = 0, \text{ from differentiating w.r.t. } t,$$

$$t G'' + 2G' - m^2 t G = 0.$$

With $G(t) = g(t)/t$,

$$G' = -\frac{g}{t^2} + \frac{g'}{t} \quad \text{and} \quad G'' = \frac{2g}{t^3} - \frac{2g'}{t^2} + \frac{g''}{t},$$

and the equation becomes

$$0 = \frac{2g}{t^2} - \frac{2g'}{t} + g'' - \frac{2g}{t^2} + \frac{2g'}{t} - m^2 g,$$

$$0 = g'' - m^2 g,$$

$$\Rightarrow \quad g(t) = Ae^{-mt}, \text{ since } g \text{ is bounded as } t \to \infty.$$

The value of A is determined by resubstituting into $(*)$, which then reads

$$4\pi t^2 \left(-\frac{Ae^{-mt}}{t^2} - \frac{mAe^{-mt}}{t} \right) - m^2 \int_0^t \frac{Ae^{-mt'}}{t'} 4\pi t'^2\, dt' = 1,$$

$$-4\pi Ae^{-mt}(1 + mt) - 4\pi Am^2 \left(-\frac{te^{-mt}}{m} + \frac{1 - e^{-mt}}{m^2} \right) = 1,$$

$$-4\pi A = 1,$$

making the solution

$$G(\mathbf{r}, \mathbf{r}') = -\frac{e^{-mt}}{4\pi t}, \text{ where } t = |\mathbf{r} - \mathbf{r}'|.$$

(c) For the situation in which $\rho(\mathbf{r}) = \delta(\mathbf{r} - \mathbf{r}_1)$, i.e. a unit positive charge at $\mathbf{r}_1 = (x_1, y_1, z_1)$, and $\phi = 0$ on the plane $x = 0$, we must have a unit negative image charge at $\mathbf{r}_2 = (-x_1, y_1, z_1)$. The solution in the region $x > 0$ is then

$$\phi(\mathbf{r}) = -\frac{1}{4\pi} \left(\frac{e^{-m|\mathbf{r}-\mathbf{r}_1|}}{|\mathbf{r} - \mathbf{r}_1|} - \frac{e^{-m|\mathbf{r}-\mathbf{r}_2|}}{|\mathbf{r} - \mathbf{r}_2|} \right).$$

12.1 A surface of revolution, whose equation in cylindrical polar coordinates is $\rho = \rho(z)$, is bounded by the circles $\rho = a$, $z = \pm c$ $(a > c)$. Show that the function that makes the surface integral $I = \int \rho^{-1/2} \, dS$ stationary with respect to small variations is given by $\rho(z) = k + z^2/(4k)$, where $k = [a \pm (a^2 - c^2)^{1/2}]/2$.

The surface element lying between z and $z + dz$ is given by

$$dS = 2\pi\rho \, [(d\rho)^2 + (dz)^2]^{1/2} = 2\pi\rho \, (1 + \rho'^2)^{1/2} \, dz$$

and the integral to be made stationary is

$$I = \int \rho^{-1/2} \, dS = 2\pi \int_{-c}^{c} \rho^{-1/2} \rho \, (1 + \rho'^2)^{1/2} \, dz.$$

The integrand $F(\rho', \rho, z)$ does not in fact contain z explicitly, and so a first integral of the EL equation, symbolically given by $F - \rho' \partial F/\partial \rho' = k$, is

$$\rho^{1/2}(1 + \rho'^2)^{1/2} - \rho' \left[\frac{\rho^{1/2}\rho'}{(1 + \rho'^2)^{1/2}} \right] = A,$$

$$\frac{\rho^{1/2}}{(1 + \rho'^2)^{1/2}} = A.$$

On rearrangement and subsequent integration this gives

$$\frac{d\rho}{dz} = \left(\frac{\rho - A^2}{A^2} \right)^{1/2},$$

$$\int \frac{d\rho}{\sqrt{\rho - A^2}} = \int \frac{dz}{A},$$

$$2\sqrt{\rho - A^2} = \frac{z}{A} + C.$$

Now, $\rho(\pm c) = a$ implies both that $C = 0$ and that $a - A^2 = \dfrac{c^2}{4A^2}$. Thus, writing A^2 as k,

$$4k^2 - 4ka + c^2 = 0 \quad \Rightarrow \quad k = \tfrac{1}{2}[a \pm (a^2 - c^2)^{1/2}].$$

The two stationary functions are therefore

$$\rho = \frac{z^2}{4k} + k,$$

with k as given above. A simple sketch shows that the positive sign in k corresponds to a smaller value of the integral.

12.3 The refractive index n of a medium is a function only of the distance r from a fixed point O. Prove that the equation of a light ray, assumed to lie in a plane through O, traveling in the medium satisfies (in plane polar coordinates)

$$\frac{1}{r^2}\left(\frac{dr}{d\phi}\right)^2 = \frac{r^2\,n^2(r)}{a^2\,n^2(a)} - 1,$$

where a is the distance of the ray from O at the point at which $dr/d\phi = 0$.

If $n = [1 + (\alpha^2/r^2)]^{1/2}$ and the ray starts and ends far from O, find its deviation (the angle through which the ray is turned), if its minimum distance from O is a.

An element of path length is $ds = [(dr)^2 + (r\,d\phi)^2]^{1/2}$ and the time taken for the light to traverse it is $n(r)\,ds/c$, where c is the speed of light *in vacuo*. Fermat's principle then implies that the light follows the curve that minimizes

$$T = \int \frac{n(r)\,ds}{c} = \int \frac{n(r'^2 + r^2)^{1/2}}{c}\,d\phi,$$

where $r' = dr/d\phi$. Since the integrand does not contain ϕ explicitly, the EL equation integrates to (see Problem 12.1)

$$n(r'^2 + r^2)^{1/2} - r'\frac{nr'}{(r'^2 + r^2)^{1/2}} = A,$$

$$\frac{nr^2}{(r'^2 + r^2)^{1/2}} = A.$$

Since $r' = 0$ when $r = a$, $A = n(a)a^2/a$, and the equation is as follows:

$$a^2 n^2(a)(r'^2 + r^2) = n^2(r)r^4,$$

$$r'^2 = \frac{n^2(r)r^4}{n^2(a)a^2} - r^2,$$

$$\Rightarrow \quad \frac{1}{r^2}\left(\frac{dr}{d\phi}\right)^2 = \frac{n^2(r)r^2}{n^2(a)a^2} - 1.$$

If $n(r) = [1 + (\alpha/r)^2]^{1/2}$, the minimizing curve satisfies

$$\left(\frac{dr}{d\phi}\right)^2 = \frac{r^2(r^2 + \alpha^2)}{a^2 + \alpha^2} - r^2$$

$$= \frac{r^2(r^2 - a^2)}{a^2 + \alpha^2},$$

$$\Rightarrow \quad \frac{d\phi}{(a^2 + \alpha^2)^{1/2}} = \pm\frac{dr}{r\sqrt{r^2 - a^2}}.$$

By symmetry,

$$\frac{\Delta\phi}{(a^2+\alpha^2)^{1/2}} \equiv \frac{\phi_{\text{final}}-\phi_{\text{initial}}}{(a^2+\alpha^2)^{1/2}}$$

$$= 2\int_a^\infty \frac{dr}{r\sqrt{r^2-a^2}}, \qquad \text{set } r = a\cosh\psi$$

$$= 2\int_0^\infty \frac{a\sinh\psi}{a^2\cosh\psi\,\sinh\psi}\,d\psi$$

$$= \frac{2}{a}\int_0^\infty \operatorname{sech}\psi\,d\psi, \qquad \text{set } e^\psi = z$$

$$= \frac{2}{a}\int_1^\infty \frac{z^{-1}\,dz}{\frac{1}{2}(z+z^{-1})}$$

$$= \frac{2}{a}\int_1^\infty \frac{2\,dz}{z^2+1}$$

$$= \frac{4}{a}\left[\tan^{-1}z\right]_1^\infty$$

$$= \frac{4}{a}\left(\frac{\pi}{2}-\frac{\pi}{4}\right) = \frac{\pi}{a}.$$

If the refractive index were everywhere unity ($\alpha=0$), $\Delta\phi$ would be π (no deviation). Thus the deviation is given by

$$\frac{\pi}{a}(a^2+\alpha^2)^{1/2}-\pi.$$

12.5 Prove the following results about general systems.

(a) For a system described in terms of coordinates q_i and t, show that if t does not appear explicitly in the expressions for x, y and z ($x = x(q_i, t)$, etc.) then the kinetic energy T is a homogeneous quadratic function of the $\dot{q}_i$ (it may also involve the q_i). Deduce that $\sum_i \dot{q}_i(\partial T/\partial \dot{q}_i) = 2T$.

(b) Assuming that the forces acting on the system are derivable from a potential V, show, by expressing dT/dt in terms of q_i and $\dot{q}_i$, that $d(T+V)/dt = 0$.

To save space we will use the summation convention for summing over the index of the q_i.

(a) The space variables x, y and z are not explicit functions of t and the kinetic energy, T, is given by

$$T = \tfrac{1}{2}(\alpha_x\dot{x}^2 + \alpha_y\dot{y}^2 + \alpha_z\dot{z}^2)$$

$$= \frac{1}{2}\left[\alpha_x\left(\frac{\partial x}{\partial q_i}\dot{q}_i\right)^2 + \alpha_y\left(\frac{\partial y}{\partial q_j}\dot{q}_j\right)^2 + \alpha_z\left(\frac{\partial z}{\partial q_k}\dot{q}_k\right)^2\right]$$

$$= A_{mn}\dot{q}_m\dot{q}_n,$$

with

$$A_{mn} = \frac{1}{2}\left(\alpha_x\frac{\partial x}{\partial q_m}\frac{\partial x}{\partial q_n} + \alpha_y\frac{\partial y}{\partial q_m}\frac{\partial y}{\partial q_n} + \alpha_z\frac{\partial z}{\partial q_m}\frac{\partial z}{\partial q_n}\right) = A_{nm}.$$

Hence T is a homogeneous quadratic function of the $\dot{q}_i$ (though the A_{mn} may involve the q_i). Further,

$$\frac{\partial T}{\partial \dot{q}_i} = A_{in}\dot{q}_n + A_{mi}\dot{q}_m = 2A_{mi}\dot{q}_m$$

$$\text{and} \quad \dot{q}_i \frac{\partial T}{\partial \dot{q}_i} = 2\dot{q}_i A_{mi}\dot{q}_m = 2T.$$

(b) The Lagrangian is $L = T - V$, with $T = T(q_i, \dot{q}_i)$ and $V = V(q_i)$. Thus

$$\frac{dT}{dt} = \frac{\partial T}{\partial q_i}\dot{q}_i + \frac{dT}{d\dot{q}_i}\ddot{q}_i \quad \text{and} \quad \frac{dV}{dt} = \frac{\partial V}{\partial q_i}\dot{q}_i. \qquad (*)$$

Hamilton's principle requires that

$$\frac{d}{dt}\left(\frac{\partial L}{\partial \dot{q}_i}\right) = \frac{\partial L}{\partial q_i},$$

$$\Rightarrow \quad \frac{d}{dt}\left(\frac{\partial T}{\partial \dot{q}_i}\right) = \frac{\partial T}{\partial q_i} - \frac{\partial V}{\partial q_i}. \qquad (**)$$

But, from part (a),

$$2T = \dot{q}_i \frac{\partial T}{\partial \dot{q}_i},$$

$$\frac{d}{dt}(2T) = \ddot{q}_i \frac{\partial T}{\partial \dot{q}_i} + \dot{q}_i \frac{d}{dt}\left(\frac{\partial T}{\partial \dot{q}_i}\right)$$

$$= \ddot{q}_i \frac{\partial T}{\partial \dot{q}_i} + \dot{q}_i \frac{\partial T}{\partial q_i} - \dot{q}_i \frac{\partial V}{\partial q_i}, \quad \text{using } (**),$$

$$= \frac{dT}{dt} - \frac{dV}{dt}, \qquad\qquad \text{using } (*).$$

This can be rearranged as

$$\frac{d}{dt}(T + V) = 0.$$

12.7 In cylindrical polar coordinates, the curve $(\rho(\theta), \theta, \alpha\rho(\theta))$ lies on the surface of the cone $z = \alpha\rho$. Show that geodesics (curves of minimum length joining two points) on the cone satisfy

$$\rho^4 = c^2[\beta^2\rho'^2 + \rho^2],$$

where c is an arbitrary constant, but β has to have a particular value. Determine the form of $\rho(\theta)$ and hence find the equation of the shortest path on the cone between the points $(R, -\theta_0, \alpha R)$ and $(R, \theta_0, \alpha R)$.

[You will find it useful to determine the form of the derivative of $\cos^{-1}(u^{-1})$.]

In cylindrical polar coordinates the element of length is given by

$$(ds)^2 = (d\rho)^2 + (\rho\, d\theta)^2 + (dz)^2,$$

and the total length of a curve between two points parameterized by θ_0 and θ_1 is

$$
\begin{aligned}
s &= \int_{\theta_0}^{\theta_1} \sqrt{\left(\frac{d\rho}{d\theta}\right)^2 + \rho^2 + \left(\frac{dz}{d\theta}\right)^2}\, d\theta \\
&= \int_{\theta_0}^{\theta_1} \sqrt{\rho^2 + (1+\alpha^2)\left(\frac{d\rho}{d\theta}\right)^2}\, d\theta, \quad \text{since } z = \alpha\rho.
\end{aligned}
$$

Since the independent variable θ does not occur explicitly in the integrand, a first integral of the EL equation is

$$
\sqrt{\rho^2 + (1+\alpha^2)\rho'^2} - \rho'\frac{(1+\alpha^2)\rho'}{\sqrt{\rho^2 + (1+\alpha^2)\rho'^2}} = c.
$$

After being multiplied through by the square root, this can be arranged as follows:

$$
\rho^2 + (1+\alpha^2)\rho'^2 - (1+\alpha^2)\rho'^2 = c\sqrt{\rho^2 + (1+\alpha^2)\rho'^2},
$$

$$
\rho^4 = c^2[\rho^2 + (1+\alpha^2)\rho'^2].
$$

This is the given equation of the geodesic, in which c is arbitrary but β^2 must have the value $1 + \alpha^2$.

Guided by the hint, we first determine the derivative of $y(u) = \cos^{-1}(u^{-1})$:

$$
\frac{dy}{du} = \frac{-1}{\sqrt{1-u^{-2}}}\frac{-1}{u^2} = \frac{1}{u\sqrt{u^2-1}}.
$$

Now, returning to the geodesic, rewrite it as

$$
\rho^4 - c^2\rho^2 = c^2\beta^2\rho'^2,
$$

$$
\rho(\rho^2 - c^2)^{1/2} = c\beta\frac{d\rho}{d\theta}.
$$

Setting $\rho = cu$,

$$
uc^2(u^2-1)^{1/2} = c^2\beta\frac{du}{d\theta},
$$

$$
d\theta = \frac{\beta\,du}{u(u^2-1)^{1/2}},
$$

which integrates to

$$
\theta = \beta\cos^{-1}\left(\frac{1}{u}\right) + k,
$$

using the result from the hint.

Since the geodesic must pass through both $(R, -\theta_0, \alpha R)$ and $(R, \theta_0, \alpha R)$, we must have $k = 0$ and

$$
\cos\frac{\theta_0}{\beta} = \frac{c}{R}.
$$

Further, at a general point on the geodesic,

$$\cos \frac{\theta}{\beta} = \frac{c}{\rho}.$$

Eliminating c then shows that the geodesic on the cone that joins the two given points is

$$\rho(\theta) = \frac{R \cos(\theta_0/\beta)}{\cos(\theta/\beta)}.$$

12.9 You are provided with a line of length $\pi a/2$ and negligible mass and some lead shot of total mass M. Use a variational method to determine how the lead shot must be distributed along the line if the loaded line is to hang in a circular arc of radius a when its ends are attached to two points at the same height. Measure the distance s along the line from its center.

We first note that the total mass of shot available is merely a scaling factor and not a constraint on the minimization process.

The length of string is sufficient to form one-quarter of a complete circle of radius a, and so the ends of the string must be fixed to points that are $2a \sin(\pi/4) = \sqrt{2}a$ apart.

We take the distribution of shot as $\rho = \rho(s)$ and have to minimize the integral $\int gy(s)\rho(s)\,ds$, but subject to the requirement $\int dx = a/\sqrt{2}$. Expressed as an integral over s, this requirement can be written

$$\frac{a}{\sqrt{2}} = \int_{s=0}^{s=\pi a/4} dx = \int_0^{\pi a/4} (1 - y'^2)^{1/2}\,ds,$$

where the derivative y' of y is with respect to s (not x).

We therefore consider the minimization of $\int F(y, y', s)\,ds$, where

$$F(y, y', s) = gy\rho + \lambda\sqrt{1 - y'^2}.$$

The EL equation takes the form

$$\frac{d}{ds}\left(\frac{\partial F}{\partial y'}\right) = \frac{\partial F}{\partial y},$$

$$\lambda \frac{d}{ds}\left(\frac{-y'}{\sqrt{1 - y'^2}}\right) = g\rho(s),$$

$$\frac{-\lambda y'}{\sqrt{1 - y'^2}} = \int_0^s g\rho(s')\,ds' \equiv gP(s),$$

since $y'(0) = 0$ by symmetry.

Now we require $P(s)$ to be such that the solution to this equation takes the form of an arc of a circle, $y(s) = y_0 - a\cos(s/a)$. If this is so, then $y'(s) = \sin(s/a)$ and

$$\frac{-\lambda \sin(s/a)}{\cos(s/a)} = gP(s).$$

When $s = \pi a/4$, $P(s)$ must have the value $M/2$, implying that $\lambda = -Mg/2$ and that, consequently,

$$P(s) = \frac{M}{2} \tan\left(\frac{s}{a}\right).$$

The required distribution $\rho(s)$ is recovered by differentiating this to obtain

$$\rho(s) = \frac{dP}{ds} = \frac{M}{2a} \sec^2\left(\frac{s}{a}\right).$$

12.11 A general result is that light travels through a variable medium by a path that minimizes the travel time (this is an alternative formulation of Fermat's principle). With respect to a particular cylindrical polar coordinate system (ρ, ϕ, z), the speed of light $v(\rho, \phi)$ is independent of z. If the path of the light is parameterized as $\rho = \rho(z)$, $\phi = \phi(z)$, show that

$$v^2(\rho'^2 + \rho^2\phi'^2 + 1)$$

is constant along the path.

   For the particular case when $v = v(\rho) = b(a^2 + \rho^2)^{1/2}$, show that the two Euler–Lagrange equations have a common solution in which the light travels along a helical path given by $\phi = Az + B$, $\rho = C$, provided that A has a particular value.

In cylindrical polar coordinates with $\rho = \rho(z)$ and $\phi = \phi(z)$,

$$ds = \left[1 + \left(\frac{d\rho}{dz}\right)^2 + \rho^2\left(\frac{d\phi}{dz}\right)^2\right]^{1/2} dz.$$

The total travel time of the light is therefore given by

$$\tau = \int \frac{(1 + \rho'^2 + \rho^2\phi'^2)^{1/2}}{v(\rho, \phi)} \, dz.$$

Since z does not appear explicitly in the integrand, we have from the general first integral of the EL equations for more than one dependent variable that

$$\frac{(1 + \rho'^2 + \rho^2\phi'^2)^{1/2}}{v(\rho, \phi)} - \frac{1}{v}\frac{\rho'^2}{(1 + \rho'^2 + \rho^2\phi'^2)^{1/2}} - \frac{1}{v}\frac{\rho^2\phi'^2}{(1 + \rho'^2 + \rho^2\phi'^2)^{1/2}} = k.$$

Rearranging this gives

$$1 + \rho'^2 + \rho^2\phi'^2 - \rho'^2 - \rho^2\phi'^2 = kv(1 + \rho'^2 + \rho^2\phi'^2)^{1/2},$$

$$1 = kv(1 + \rho'^2 + \rho^2\phi'^2)^{1/2},$$

$$\Rightarrow \quad v^2(1 + \rho'^2 + \rho^2\phi'^2) = c, \quad \text{along the path.}$$

Denoting $(1 + \rho'^2 + \rho^2\phi'^2)$ by $(**)$ for brevity, the EL equations for ρ and ϕ are, respectively,

$$\frac{\rho\phi'^2}{v(**)^{1/2}} - \frac{(**)^{1/2}}{v^2}\frac{\partial v}{\partial \rho} = \frac{d}{dz}\left[\frac{\rho'}{v(**)^{1/2}}\right], \quad (1)$$

and

$$-\frac{(**)^{1/2}}{v^2}\frac{\partial v}{\partial \phi} = \frac{d}{dz}\left[\frac{\rho^2\phi'}{v(**)^{1/2}}\right]. \quad (2)$$

Now, if $v = b(a^2 + \rho^2)^{1/2}$, the only dependence on z in a possible solution $\phi = Az + B$ with $\rho = C$ is through the first of these equations. To see this we note that the square brackets on the RHSs of the two EL equations do not contain any undifferentiated ϕ-terms and so the derivatives (with respect to z) of both are zero. Since $\partial v/\partial \phi$ is also zero, equation (2) is identically satisfied as $0 = 0$. This leaves only (1), which reads

$$\frac{CA^2}{b(a^2 + C^2)^{1/2}(1 + 0 + C^2 A^2)^{1/2}} - \frac{(1 + 0 + C^2 A^2)^{1/2} bC}{b^2(a^2 + C^2)(a^2 + C^2)^{1/2}} = 0.$$

This is satisfied provided

$$A^2(a^2 + C^2) = 1 + C^2 A^2,$$
$$\text{i.e.} \quad A = a^{-1}.$$

Thus, a solution in the form of a helix is possible provided that the helix has a particular pitch, $2\pi a$.

12.13 The Schwarzchild metric for the static field of a non-rotating spherically symmetric black hole of mass M is given by

$$(ds)^2 = c^2 \left(1 - \frac{2GM}{c^2 r}\right)(dt)^2 - \frac{(dr)^2}{1 - 2GM/(c^2 r)} - r^2 (d\theta)^2 - r^2 \sin^2 \theta \, (d\phi)^2.$$

Considering only motion confined to the plane $\theta = \pi/2$, and assuming that the path of a small test particle is such as to make $\int ds$ stationary, find two first integrals of the equations of motion. From their Newtonian limits, in which GM/r, $\dot{r}^2$ and $r^2\dot{\phi}^2$ are all $\ll c^2$, identify the constants of integration.

For motion confined to the plane $\theta = \pi/2$, $d\theta = 0$ and the corresponding term in the metric can be ignored. With this simplification, we can write

$$ds = \left\{ c^2 \left(1 - \frac{2GM}{c^2 r}\right) - \frac{\dot{r}^2}{1 - (2GM)/(c^2 r)} - r^2\dot{\phi}^2 \right\}^{1/2} dt.$$

Writing the terms in braces as $\{**\}$, the EL equation for ϕ reads

$$\frac{d}{dt}\left(\frac{-r^2\dot{\phi}}{\{**\}^{1/2}}\right) - 0 = 0,$$

$$\Rightarrow \quad \frac{r^2\dot{\phi}}{\{**\}^{1/2}} = A.$$

In the Newtonian limit $\{**\} \to c^2$ and the equation becomes $r^2\dot{\phi} = Ac$. Thus, Ac is a measure of the angular momentum of the particle about the origin.

The EL equation for r is more complicated but, because ds does not contain t explicitly, we can use the general result for the first integral of the EL equations when there is more than one dependent variable: $F - \sum_i \dot{q}_i \dfrac{\partial F}{\partial \dot{q}_i} = k$. This gives us a second equation as follows:

$$F - \dot{r}\frac{\partial F}{\partial \dot{r}} - \dot{\phi}\frac{\partial F}{\partial \dot{\phi}} = B,$$

$$\{**\}^{1/2} + \frac{\dot{r}}{\{**\}^{1/2}} \frac{\dot{r}}{[1 - (2GM)/(c^2 r)]} + \frac{\dot{\phi}}{\{**\}^{1/2}} r^2 \dot{\phi} = B.$$

Multiplying through by $\{**\}^{1/2}$ and canceling the terms in $\dot{r}^2$ and $\dot{\phi}^2$ now gives

$$c^2 - \frac{2GM}{r} = B\left\{c^2 - \frac{2GM}{r} - \frac{\dot{r}^2}{[1 - (2GM)/(c^2 r)]} - r^2\dot{\phi}^2\right\}^{1/2}.$$

In the Newtonian limits, in which GM/r, $\dot{r}^2$ and $r^2\dot{\phi}^2$ are all $\ll c^2$, the equation can be rearranged and the braces expanded to first order in small quantities to give

$$B = \left(c^2 - \frac{2GM}{r}\right)\left\{c^2 - \frac{2GM}{r} - \frac{\dot{r}^2}{[1 - (2GM)/(c^2 r)]} - r^2\dot{\phi}^2\right\}^{-1/2},$$

$$cB = c^2 - \frac{2GM}{r} + \frac{c^2 GM}{c^2 r} + \frac{c^2 \dot{r}^2}{2c^2} + \frac{c^2 r^2 \dot{\phi}^2}{2c^2} + \cdots,$$

$$= c^2 - \frac{GM}{r} + \frac{1}{2}(\dot{r}^2 + r^2\dot{\phi}^2) + \cdots,$$

which can be read as "total energy = rest mass energy + gravitational energy + radial and azimuthal kinetic energy". Thus Bc is a measure of the total energy of the test particle.

12.15 Determine the minimum value that the integral

$$J = \int_0^1 [x^4(y'')^2 + 4x^2(y')^2]\,dx$$

can have, given that y is not singular at $x = 0$ and that $y(1) = y'(1) = 1$. Assume that the Euler–Lagrange equation gives the lower limit.

We first set $y'(x) = u(x)$ with $u(1) = y'(1) = 1$. The integral then becomes

$$J = \int_0^1 [x^4(u')^2 + 4x^2 u^2]\,dx. \qquad (*)$$

This will be stationary if (using the EL equation)

$$\frac{d}{dx}(2x^4 u') - 8x^2 u = 0,$$

$$8x^3 u' + 2x^4 u'' - 8x^2 u = 0,$$

$$x^2 u'' + 4xu' - 4u = 0.$$

As this is a homogeneous equation, we try $u(x) = Ax^n$, obtaining

$$n(n-1) + 4n - 4 = 0 \quad \Rightarrow \quad n = -4, \text{ or } n = 1.$$

The form of $y'(x)$ is thus

$$y'(x) = u(x) = \frac{A}{x^4} + Bx \quad \text{with} \quad A + B = 1.$$

Further,

$$y(x) = -\frac{A}{3x^3} + \frac{Bx^2}{2} + C.$$

Since y is not singular at $x = 0$ and $y(1) = 1$, we have that $A = 0$, $B = 1$ and $C = \frac{1}{2}$, yielding $y(x) = \frac{1}{2}(1 + x^2)$. The minimal value of J is thus

$$J_{\text{min}} = \int_0^1 [x^4(1)^2 + 4x^2(x)^2] \, dx = \int_0^1 5x^4 \, dx = [x^5]_0^1 = 1.$$

12.17 Find an appropriate but simple trial function and use it to estimate the lowest eigenvalue λ_0 of Stokes' equation

$$\frac{d^2y}{dx^2} + \lambda xy = 0, \qquad y(0) = y(\pi) = 0.$$

Explain why your estimate must be strictly greater than λ_0.

Stokes' equation is an SL equation with $p = 1$, $q = 0$ and $\rho = x$. For the given boundary conditions the obvious trial function is $y(x) = \sin x$. The lowest eigenvalue $\lambda_0 \leq I/J$, where

$$I = \int_0^\pi py'^2 \, dx = \int_0^\pi \cos^2 x \, dx = \frac{\pi}{2}$$

and

$$J = \int_0^\pi \rho y^2 \, dx = \int_0^\pi x \sin^2 x \, dx$$

$$= \int_0^\pi \tfrac{1}{2}x(1 - \cos 2x) \, dx$$

$$= \left[\frac{x^2}{4}\right]_0^\pi - \left[\frac{x}{2}\frac{\sin 2x}{2}\right]_0^\pi + \frac{1}{2}\int_0^\pi \frac{\sin 2x}{2} \, dx$$

$$= \frac{\pi^2}{4}.$$

Thus $\lambda_0 \leq (\frac{1}{2}\pi)/(\frac{1}{4}\pi^2) = 2/\pi$.

However, if we substitute the trial function directly into the equation we obtain

$$-\sin x + \frac{2}{\pi} x \sin x = 0, \qquad \bullet$$

which is clearly not satisfied. Thus the trial function is not an eigenfunction, and the actual lowest eigenvalue must be strictly less than the estimate of $2/\pi$.

12.19 A drum skin is stretched across a fixed circular rim of radius a. Small transverse vibrations of the skin have an amplitude $z(\rho, \phi, t)$ that satisfies

$$\nabla^2 z = \frac{1}{c^2} \frac{\partial^2 z}{\partial t^2}$$

in plane polar coordinates. For a normal mode independent of azimuth, in which case $z = Z(\rho)\cos\omega t$, find the differential equation satisfied by $Z(\rho)$. By using a trial function of the form $a^\nu - \rho^\nu$, with adjustable parameter ν, obtain an estimate for the lowest normal mode frequency.

[The exact answer is $(5.78)^{1/2} c/a$.]

In cylindrical polar coordinates, (ρ, ϕ), the wave equation

$$\nabla^2 z = \frac{1}{c^2} \frac{\partial^2 z}{\partial t^2}$$

has azimuth-independent solutions (i.e. independent of ϕ) of the form $z(\rho, t) = Z(\rho)\cos\omega t$, and reduces to

$$\frac{1}{\rho}\frac{d}{d\rho}\left(\rho\frac{dZ}{d\rho}\right)\cos\omega t = -\frac{Z\omega^2}{c^2}\cos\omega t,$$

$$\frac{d}{d\rho}\left(\rho\frac{dZ}{d\rho}\right) + \frac{\omega^2}{c^2}\rho Z = 0.$$

The boundary conditions require that $Z(a) = 0$ and, so that there is no physical discontinuity in the slope of the drum skin at the origin, $Z'(0) = 0$.

This is an SL equation with $p = \rho$, $q = 0$ and weight function $w = \rho$. A suitable trial function is $Z(\rho) = a^\nu - \rho^\nu$, which automatically satisfies $Z(a) = 0$ and, provided $\nu > 1$, has $Z'(0) = -\nu\rho^{\nu-1}|_{\rho=0} = 0$.

We recall that the lowest eigenfrequency satisfies the general formula

$$\frac{\omega^2}{c^2} \leq \frac{\displaystyle\int_0^a [(pZ')^2 - qZ^2]\,d\rho}{\displaystyle\int_0^a wZ^2\,d\rho}.$$

In this case

$$\frac{\omega^2}{c^2} \leq \frac{\displaystyle\int_0^a \rho\, \nu^2 \rho^{2\nu-2}\,d\rho}{\displaystyle\int_0^a \rho(a^\nu - \rho^\nu)^2\,d\rho}$$

$$= \frac{\displaystyle\int_0^a \nu^2 \rho^{2\nu-1}\,d\rho}{\displaystyle\int_0^a (\rho a^{2\nu} - 2\rho^{\nu+1}a^\nu + \rho^{2\nu+1})\,d\rho}$$

$$= \frac{(v^2 a^{2v})/2v}{\dfrac{a^{2v+2}}{2} - \dfrac{2a^{2v+2}}{v+2} + \dfrac{a^{2v+2}}{2v+2}}$$

$$= \frac{1}{a^2} \frac{v(v+2)(2v+2)}{(v+2)(2v+2) - 4(2v+2) + 2(v+2)}$$

$$= \frac{(v+2)(v+1)}{va^2}.$$

Since v is an adjustable parameter and we know that, however we choose it, the resulting estimate can *never* be less than the lowest true eigenvalue, we choose the value that minimizes the above estimate. Differentiating the estimate with respect to v gives

$$v(2v+3) - (v^2 + 3v + 2) = 0 \quad \Rightarrow \quad v^2 - 2 = 0 \quad \Rightarrow \quad v = \sqrt{2}.$$

Thus the least upper bound to be found with this parameterization is

$$\omega^2 \le \frac{c^2}{a^2} \frac{(\sqrt{2}+2)(\sqrt{2}+1)}{\sqrt{2}} = \frac{c^2}{2a^2}(\sqrt{2}+2)^2 \quad \Rightarrow \quad \omega = (5.83)^{1/2} \frac{c}{a}.$$

As noted, the actual lowest eigenfrequency is very little below this.

12.21 For the boundary conditions given below, obtain a functional $\Lambda(y)$ whose stationary values give the eigenvalues of the equation

$$(1+x)\frac{d^2 y}{dx^2} + (2+x)\frac{dy}{dx} + \lambda y = 0, \qquad y(0) = 0, \ y'(2) = 0.$$

Derive an approximation to the lowest eigenvalue λ_0 using the trial function $y(x) = xe^{-x/2}$. For what value(s) of γ would

$$y(x) = xe^{-x/2} + \beta \sin \gamma x$$

be a suitable trial function for attempting to obtain an improved estimate of λ_0?

Since the derivative of $1 + x$ is not equal to $2 + x$, the given equation is not in self-adjoint form and an integrating factor for the standard form equation,

$$\frac{d^2 y}{dx^2} + \frac{2+x}{1+x}\frac{dy}{dx} + \frac{\lambda y}{1+x} = 0,$$

is needed. This will be

$$\exp\left\{ \int^x \frac{2+u}{1+u} \, du \right\} = \exp\left\{ \int^x \left(1 + \frac{1}{1+u} \right) du \right\} = e^x(1+x).$$

Thus, after multiplying through by this IF, the equation takes the SL form

$$[(1+x)e^x y']' + \lambda e^x y = 0,$$

with $p(x) = (1+x)e^x$, $q(x) = 0$ and $\rho(x) = e^x$.

The required functional is therefore

$$\Lambda(y) = \frac{\int_0^2 [(1+x)e^x y'^2 + 0]\,dx}{\int_0^2 y^2 e^x\,dx},$$

provided that, for the eigenfunctions y_i of the equation, $\left[y_i p(x) y'_j(x)\right]_0^2 = 0$; this condition is automatically satisfied with the given boundary conditions.

For the trial function $y(x) = xe^{-x/2}$, clearly $y(0) = 0$ and, less obviously, $y'(x) = (1 - \frac{1}{2}x)e^{-x/2}$, making $y'(2) = 0$. The functional takes the following form:

$$\Lambda = \frac{\int_0^2 (1+x)e^x \left(1 - \frac{1}{2}x\right)^2 e^{-x}\,dx}{\int_0^2 x^2 e^{-x} e^x\,dx}$$

$$= \frac{\int_0^2 (1+x)\left(1 - \frac{1}{2}x\right)^2 dx}{\int_0^2 x^2\,dx}$$

$$= \frac{\int_0^2 \left(1 - x^2 + \frac{1}{4}x^2 + \frac{1}{4}x^3\right) dx}{8/3}$$

$$= \frac{3}{8}\left(2 - \frac{3}{4}\frac{8}{3} + \frac{16}{16}\right) = \frac{3}{8}.$$

Thus the lowest eigenvalue is $\leq \frac{3}{8}$.

We already know that $xe^{-x/2}$ is a suitable trial function and thus $y_2(x) = \sin\gamma x$ can be considered on its own. It satisfies $y_2(0) = 0$, but must also satisfy $y'_2(2) = \gamma\cos(2\gamma) = 0$. This requires that $\gamma = \frac{1}{2}(n + \frac{1}{2})\pi$ for some integer n; trial functions with γ of this form can be used to try to obtain a better bound on λ_0 by choosing the best value for n and adjusting the parameter β.

12.23 The unnormalized ground-state (i.e. the lowest-energy) wavefunction of the simple harmonic oscillator of classical frequency ω is $\exp(-\alpha x^2)$, where $\alpha = m\omega/2\hbar$. Take as a trial function the orthogonal wavefunction $x^{2n+1}\exp(-\alpha x^2)$, using the integer n as a variable parameter, and apply either Sturm–Liouville theory or the Rayleigh–Ritz principle to show that the energy of the second lowest state of a quantum harmonic oscillator is $\leq 3\hbar\omega/2$.

We first note that, for n a non-negative integer,

$$\int_{-\infty}^{\infty} x^{2n+1} e^{-\alpha x^2} e^{-\alpha x^2}\,dx = 0$$

on symmetry grounds and so confirm that the ground-state wavefunction, $\exp(-\alpha x^2)$, and the trial function, $\psi_{2n+1} = x^{2n+1}\exp(-\alpha x^2)$, are orthogonal with respect to a unit weight function.

The Hamiltonian for the quantum harmonic oscillator in one dimension is given by

$$H = -\frac{\hbar^2}{2m}\frac{d^2}{dx^2} + \frac{k}{2}x^2.$$

Calculus of variations

This means that to prepare the elements required for a Rayleigh–Ritz analysis we will need to find the second derivative of the trial function and evaluate integrals with integrands of the form $x^n \exp(-2\alpha x^2)$. To this end, define

$$I_n = \int_{-\infty}^{\infty} x^n e^{-2\alpha x^2} \, dx, \quad \text{with recurrence relation } I_n = \frac{n-1}{4\alpha} I_{n-2}.$$

Using Leibnitz' formula shows that

$$\frac{d^2 \psi_{2n+1}}{dx^2} = \left[2n(2n+1)x^{2n-1} + 2(2n+1)(-2\alpha)x^{2n+1} \right.$$
$$\left. + (4\alpha^2 x^2 - 2\alpha)x^{2n+1} \right] e^{-\alpha x^2}$$
$$= \left[2n(2n+1)x^{2n-1} - 2(4n+3)\alpha x^{2n+1} + 4\alpha^2 x^{2n+3} \right] e^{-\alpha x^2}.$$

Hence, we find that $\langle H \rangle$ is given by

$$-\frac{\hbar^2}{2m} \int_{-\infty}^{\infty} x^{2n+1} e^{-\alpha x^2} \frac{d^2 \psi_{2n+1}}{dx^2} \, dx + \frac{k}{2} \int_{-\infty}^{\infty} x^2 x^{4n+2} e^{-2\alpha x^2} \, dx$$

$$= -\frac{\hbar^2}{2m} \left[2n(2n+1)I_{4n} - 2(4n+3)\alpha I_{4n+2} + 4\alpha^2 I_{4n+4} \right] + \frac{k}{2} I_{4n+4}$$

$$= I_{4n+2} \left\{ -\frac{\hbar^2}{2m} \left[\frac{2n(2n+1)4\alpha}{4n+1} - 2(4n+3)\alpha + \frac{4\alpha^2(4n+3)}{4\alpha} \right] + \frac{k(4n+3)}{8\alpha} \right\},$$

where we have used the recurrence relation to express all integrals in terms of I_{4n+2}. This has been done because the denominator of the Rayleigh–Ritz quotient is this (same) normalization integral, namely

$$\int_{-\infty}^{\infty} \psi^*_{2n+1} \psi_{2n+1} \, dx = I_{4n+2}.$$

Thus, the estimate $E_{2n+1} = \langle H \rangle / I_{4n+2}$ is given by

$$E_{2n+1} = -\frac{\hbar^2 \alpha}{2m} \left(\frac{16n^2 + 8n - 16n^2 - 16n - 3}{4n+1} \right) + \frac{k(4n+3)}{8\alpha}$$

$$= \frac{\hbar^2 \alpha}{2m} \frac{8n+3}{4n+1} + \frac{k(4n+3)}{8\alpha}.$$

Using $\omega^2 = \dfrac{k}{m}$ and $\alpha = \dfrac{m\omega}{2\hbar}$ then yields

$$E_{2n+1} = \frac{\hbar\omega}{4} \left(\frac{8n+3}{4n+1} + 4n + 3 \right) = \frac{\hbar\omega}{2} \frac{8n^2 + 12n + 3}{4n+1}.$$

For non-negative integers n (the orthogonality requirement is not satisfied for non-integer values), this has a minimum value of $\frac{3}{2}\hbar\omega$ when $n = 0$. Thus the second lowest energy level is less than or equal to this value. In fact, it is equal to this value, as can be shown by substituting ψ_1 into $H\psi = E\psi$.

12.25 The upper and lower surfaces of a film of liquid, which has surface energy per unit area (surface tension) γ and density ρ, have equations $z = p(x)$ and $z = q(x)$, respectively. The film has a given volume V (per unit depth in the y-direction) and lies in the region $-L < x < L$, with $p(0) = q(0) = p(L) = q(L) = 0$. The total energy (per unit depth) of the film consists of its surface energy and its gravitational energy, and is expressed by

$$E = \tfrac{1}{2}\rho g \int_{-L}^{L} (p^2 - q^2)\,dx + \gamma \int_{-L}^{L} \left[(1 + p'^2)^{1/2} + (1 + q'^2)^{1/2}\right] dx.$$

(a) Express V in terms of p and q.

(b) Show that, if the total energy is minimized, p and q must satisfy

$$\frac{p'^2}{(1 + p'^2)^{1/2}} - \frac{q'^2}{(1 + q'^2)^{1/2}} = \text{constant}.$$

(c) As an approximate solution, consider the equations

$$p = a(L - |x|), \qquad q = b(L - |x|),$$

where a and b are sufficiently small that a^3 and b^3 can be neglected compared with unity. Find the values of a and b that minimize E.

(a) The total volume constraint is given simply by

$$V = \int_{-L}^{L} [p(x) - q(x)]\,dx.$$

(b) To take account of the constraint, consider the minimization of $E - \lambda V$, where λ is an undetermined Lagrange multiplier. The integrand does not contain x explicitly and so we have two first integrals of the EL equations, one for $p(x)$ and the other for $q(x)$. They are

$$\frac{1}{2}\rho g(p^2 - q^2) + \gamma(1 + p'^2)^{1/2} + \gamma(1 + q'^2)^{1/2} - \lambda(p - q) - p'\frac{\gamma p'}{(1 + p'^2)^{1/2}} = k_1$$

and

$$\frac{1}{2}\rho g(p^2 - q^2) + \gamma(1 + p'^2)^{1/2} + \gamma(1 + q'^2)^{1/2} - \lambda(p - q) - q'\frac{\gamma q'}{(1 + q'^2)^{1/2}} = k_2.$$

Subtracting these two equations gives

$$\frac{p'^2}{(1 + p'^2)^{1/2}} - \frac{q'^2}{(1 + q'^2)^{1/2}} = \text{constant}.$$

(c) If

$$p = a(L - |x|), \qquad q = b(L - |x|),$$

the derivatives of p and q only take the values $\pm a$ and $\pm b$, respectively, and the volume constraint becomes

$$V = \int_{-L}^{L} (a - b)(L - |x|)\,dx = (a - b)L^2 \quad \Rightarrow \quad b = a - \frac{V}{L^2}.$$

The total energy can now be expressed entirely in terms of a and the given parameters, as follows:

$$
\begin{aligned}
E &= \frac{1}{2}\rho g \int_{-L}^{L} (a^2 - b^2)(L - |x|)^2 \, dx + 2\gamma L(1 + a^2)^{1/2} + 2\gamma L(1 + b^2)^{1/2} \\
&= \frac{1}{2}\rho g (a^2 - b^2)\frac{2L^3}{3} + 2\gamma L(1 + \tfrac{1}{2}a^2 + 1 + \tfrac{1}{2}b^2) + O(a^4) + O(b^4) \\
&\approx \frac{\rho g L^3}{3}\left[a^2 - \left(a - \tfrac{V}{L^2}\right)^2\right] + 2\gamma L\left[2 + \tfrac{1}{2}a^2 + \tfrac{1}{2}\left(a - \tfrac{V}{L^2}\right)^2\right] \\
&= \frac{\rho g L^3}{3}\left(\frac{2aV}{L^2} - \frac{V^2}{L^4}\right) + 2\gamma L\left(2 + a^2 - \frac{aV}{L^2} + \frac{V^2}{2L^4}\right).
\end{aligned}
$$

This is minimized with respect to a when

$$
\frac{2\rho g L^3 V}{3L^2} + 4\gamma L a - \frac{2\gamma L V}{L^2} = 0,
$$

$$
\Rightarrow \quad a = \frac{V}{2L^2} - \frac{\rho g V}{6\gamma},
$$

$$
\Rightarrow \quad b = -\frac{V}{2L^2} - \frac{\rho g V}{6\gamma}.
$$

As might be expected, $|b| > |a|$ and there is more of the liquid below the $z = 0$ plane than there is above it.

13 Integral equations

13.1 Solve the integral equation

$$\int_0^\infty \cos(xv) y(v)\, dv = \exp(-x^2/2)$$

for the function $y = y(x)$ for $x > 0$. Note that for $x < 0$, $y(x)$ can be chosen as is most convenient.

Since $\cos uv$ is an even function of v, we will make $y(-v) = y(v)$ so that the complete integrand is also an even function of v. The integral I on the LHS can then be written as

$$I = \frac{1}{2} \int_{-\infty}^\infty \cos(xv) y(v)\, dv = \frac{1}{2} \mathrm{Re} \int_{-\infty}^\infty e^{ixv} y(v)\, dv = \frac{1}{2} \int_{-\infty}^\infty e^{ixv} y(v)\, dv,$$

the last step following because $y(v)$ is symmetric in v. The integral is now $\sqrt{2\pi} \times$ a Fourier transform, and it follows from the inversion theorem for Fourier transforms applied to

$$\frac{1}{2} \int_{-\infty}^\infty e^{ixv} y(v)\, dv = \exp(-x^2/2)$$

that

$$y(x) = \frac{2}{2\pi} \int_{-\infty}^\infty e^{-u^2/2} e^{-iux}\, du$$

$$= \frac{1}{\pi} \int_{-\infty}^\infty e^{-(u+ix)^2/2} e^{-x^2/2}\, dx$$

$$= \frac{1}{\pi} \sqrt{2\pi}\, e^{-x^2/2}$$

$$= \sqrt{\frac{2}{\pi}}\, e^{-x^2/2}.$$

Although, as noted in the question, $y(x)$ is arbitrary for $x < 0$, because its form in this range does not affect the value of the integral, for $x > 0$ it *must* have the form given. This is tricky to prove formally, but any second solution $w(x)$ has to satisfy

$$\int_0^\infty \cos(xv)[y(v) - w(v)]\, dv = 0$$

for all $x > 0$. Intuitively, this implies that $y(x)$ and $w(x)$ are identical functions.

13.3 Convert

$$f(x) = \exp x + \int_0^x (x - y) f(y)\, dy$$

into a differential equation, and hence show that its solution is

$$(\alpha + \beta x)\exp x + \gamma \, \exp(-x),$$

where α, β, γ are constants that should be determined.

We differentiate the integral equation twice and obtain

$$f'(x) = e^x + (x - x)f(x) + \int_0^x f(y)\, dy,$$

$$f''(x) = e^x + f(x).$$

Expressed in the usual differential equation form, this last equation is

$$f''(x) - f(x) = e^x, \text{ for which the CF is } f(x) = Ae^x + Be^{-x}.$$

Since the complementary function contains the RHS of the equation, we try as a PI $f(x) = Cxe^x$:

$$Cxe^x + 2Ce^x - Cxe^x = e^x \quad \Rightarrow \quad \beta = C = \tfrac{1}{2}.$$

The general solution is therefore $f(x) = Ae^x + Be^{-x} + \tfrac{1}{2}xe^x$.

The boundary conditions needed to evaluate A and B are constructed by considering the integral equation and its derivative(s) at $x = 0$, because with $x = 0$ the integral on the RHS contributes nothing. We have

$$f(0) = e^0 + 0 = 1 \quad \Rightarrow \quad A + B = 1$$

and $\qquad f'(0) = e^0 + 0 = 1 \quad \Rightarrow \quad A - B + \tfrac{1}{2} = 1.$

Solving these yields $\alpha = A = \tfrac{3}{4}$ and $\gamma = B = \tfrac{1}{4}$ and makes the complete solution

$$f(x) = \tfrac{3}{4}e^x + \tfrac{1}{4}e^{-x} + \tfrac{1}{2}xe^x.$$

13.5 Solve for $\phi(x)$ the integral equation

$$\phi(x) = f(x) + \lambda \int_0^1 \left[\left(\frac{x}{y}\right)^n + \left(\frac{y}{x}\right)^n \right] \phi(y)\, dy,$$

where $f(x)$ is bounded for $0 < x < 1$ and $-\tfrac{1}{2} < n < \tfrac{1}{2}$, expressing your answer in terms of the quantities $F_m = \int_0^1 f(y)y^m\, dy$.

(a) Give the explicit solution when $\lambda = 1$.
(b) For what values of λ are there no solutions unless $F_{\pm n}$ are in a particular ratio? What is this ratio?

This equation has a symmetric degenerate kernel, and so we set

$$\phi(x) = f(x) + a_1 x^n + a_2 x^{-n},$$

giving

$$\frac{\phi(x) - f(x)}{\lambda} = \int_0^1 \left(\frac{x^n}{y^n} + \frac{y^n}{x^n}\right) [f(y) + a_1 y^n + a_2 y^{-n}] \, dy$$

$$= x^n \int_0^1 \frac{f(y)}{y^n} \, dy + x^{-n} \int_0^1 y^n f(y) \, dy + a_1 x^n$$

$$+ a_2 x^{-n} + a_1 x^{-n} \int_0^1 y^{2n} \, dy + a_2 x^n \int_0^1 y^{-2n} \, dy$$

$$= x^n \left(F_{-n} + a_1 + \frac{a_2}{1 - 2n}\right) + x^{-n} \left(F_n + a_2 + \frac{a_1}{2n + 1}\right).$$

This is consistent with the assumed form of $\phi(x)$, provided

$$a_1 = \lambda \left(F_{-n} + a_1 + \frac{a_2}{1 - 2n}\right) \quad \text{and} \quad a_2 = \lambda \left(F_n + a_2 + \frac{a_1}{2n + 1}\right).$$

These two simultaneous linear equations can now be solved for a_1 and a_2.
(a) For $\lambda = 1$, the equations simplify and decouple to yield

$$a_2 = -(1 - 2n)F_{-n} \quad \text{and} \quad a_1 = -(1 + 2n)F_n,$$

respectively, giving as the explicit solution

$$\phi(x) = f(x) - (1 + 2n)F_n x^n - (1 - 2n)F_{-n} x^{-n}.$$

(b) For a general value of λ,

$$(1 - \lambda)a_1 - \frac{\lambda}{1 - 2n} a_2 = \lambda F_{-n},$$

$$-\frac{\lambda}{1 + 2n} a_1 + (1 - \lambda)a_2 = \lambda F_n.$$

The case $\lambda = 0$ is trivial, with $\phi(x) = f(x)$, and so suppose that $\lambda \neq 0$. Then, after being divided through by λ, the equations can be written in the matrix and vector form $\mathbf{Aa} = \mathbf{F}$:

$$\begin{pmatrix} \frac{1}{\lambda} - 1 & -\frac{1}{1 - 2n} \\ -\frac{1}{1 + 2n} & \frac{1}{\lambda} - 1 \end{pmatrix} \begin{pmatrix} a_1 \\ a_2 \end{pmatrix} = \begin{pmatrix} F_{-n} \\ F_n \end{pmatrix}.$$

In general, this matrix equation will have no solution if $|\mathbf{A}| = 0$. This will be the case if

$$\left(\frac{1}{\lambda} - 1\right)^2 - \frac{1}{1 - 4n^2} = 0,$$

which, on rearrangement, shows that λ would have to be given by

$$\frac{1}{\lambda} = 1 \pm \frac{1}{\sqrt{1 - 4n^2}}.$$

We note that this value for λ is real because n lies in the range $-\frac{1}{2} < n < \frac{1}{2}$. In fact $-\infty < \lambda < \frac{1}{2}$. Even for these two values of λ, however, if *either* $F_n = F_{-n} = 0$ *or the* matrix equation

$$\left(\begin{array}{cc} \pm\dfrac{1}{\sqrt{1-4n^2}} & -\dfrac{1}{1-2n} \\ -\dfrac{1}{1+2n} & \pm\dfrac{1}{\sqrt{1-4n^2}} \end{array} \right) \left(\begin{array}{c} a_1 \\ a_2 \end{array} \right) = \left(\begin{array}{c} F_{-n} \\ F_n \end{array} \right)$$

is equivalent to two linear equations that are multiples of each other, there will still be a solution. In this latter case, we must have

$$\frac{F_n}{F_{-n}} = \mp\sqrt{\frac{1-2n}{1+2n}}.$$

Again we note that, because of the range in which n lies, this ratio is real; this condition can, however, require any value in the range $-\infty$ to ∞ for F_n/F_{-n}.

13.7 The kernel of the integral equation

$$\psi(x) = \lambda \int_a^b K(x,y)\psi(y)\,dy$$

has the form

$$K(x,y) = \sum_{n=0}^{\infty} h_n(x)g_n(y),$$

where the $h_n(x)$ form a complete orthonormal set of functions over the interval $[a,b]$.

(a) Show that the eigenvalues λ_i are given by

$$|\mathbf{M} - \lambda^{-1}\mathbf{I}| = 0,$$

where $\mathbf{M}$ is the matrix with elements

$$M_{kj} = \int_a^b g_k(u)h_j(u)\,du.$$

If the corresponding solutions are $\psi^{(i)}(x) = \sum_{n=0}^{\infty} a_n^{(i)}h_n(x)$, find an expression for $a_n^{(i)}$.

(b) Obtain the eigenvalues and eigenfunctions over the interval $[0, 2\pi]$ if

$$K(x,y) = \sum_{n=1}^{\infty} \frac{1}{n}\cos nx \cos ny.$$

(a) We write the ith eigenfunction as

$$\psi^{(i)}(x) = \sum_{n=0}^{\infty} a_n^{(i)}h_n(x).$$

From the orthonormality of the $h_n(x)$, it follows immediately that

$$a_m^{(i)} = \int_a^b h_m(x)\psi^{(i)}(x)\,dx.$$

However, the coefficients $a_m^{(i)}$ have to be found as the components of the eigenvectors $\mathbf{a}^{(i)}$ defined below, since the $\psi^{(i)}$ are not initially known.

Substituting this assumed form of solution, we obtain

$$\sum_{m=0}^{\infty} a_m^{(i)} h_m(x) = \lambda_i \int_a^b \sum_{n=0}^{\infty} h_n(x) g_n(y) \sum_{l=0}^{\infty} a_l^{(i)} h_l(y)\, dy$$

$$= \lambda_i \sum_{n,l} a_l^{(i)} M_{nl} h_n(x).$$

Since the $\{h_n\}$ are an orthonormal set, it follows that

$$a_m^{(i)} = \lambda_i \sum_{n,l} a_l^{(i)} M_{nl} \delta_{mn} = \lambda_i \sum_{l=0}^{\infty} M_{ml} a_l^{(i)},$$

i.e. $(\mathbf{M} - \lambda_i^{-1}\mathbf{I})\mathbf{a}^{(i)} = 0.$

Thus, the allowed values of λ_i are given by $|\mathbf{M} - \lambda^{-1}\mathbf{I}| = 0$, and the expansion coefficients $a_m^{(i)}$ by the components of the corresponding eigenvectors.

(b) To make the set $\{h_n(x) = \cos nx\}$ into a complete orthonormal set we need to add the set of functions $\{\eta_\nu(x) = \sin \nu x\}$ and then normalize all the functions by multiplying them by $1/\sqrt{\pi}$. For this particular kernel the general functions $g_n(x)$ are given by $g_n(x) = n^{-1}\sqrt{\pi}\cos nx$.

The matrix elements are then

$$M_{kj} = \int_0^{2\pi} \frac{1}{\sqrt{\pi}} \cos ju \, \frac{\sqrt{\pi}}{k} \cos ku \, du = \frac{\pi}{k}\delta_{kj},$$

$$M_{k\nu} = \int_0^{2\pi} \frac{1}{\sqrt{\pi}} \sin \nu u \, \frac{\sqrt{\pi}}{k} \cos ku \, du = 0.$$

Thus the matrix $\mathbf{M}$ is diagonal and particularly simple. The eigenvalue equation reads

$$\sum_{j=0}^{\infty} \left(\frac{\pi}{k}\delta_{kj} - \lambda_i^{-1}\delta_{kj} \right) a_j^{(i)} = 0,$$

giving the immediate result that $\lambda_k = k/\pi$ with $a_k^{(k)} = 1$ and all other $a_j^{(k)} = a_\nu^{(k)} = 0$. The eigenfunction corresponding to eigenvalue k/π is therefore

$$\psi^{(k)}(x) = h_k(x) = \frac{1}{\sqrt{\pi}} \cos kx.$$

13.9 For $f(t) = \exp(-t^2/2)$, use the relationships of the Fourier transforms of $f'(t)$ and $tf(t)$ to that of $f(t)$ itself to find a simple differential equation satisfied by $\tilde{f}(\omega)$, the Fourier transform of $f(t)$, and hence determine $\tilde{f}(\omega)$ to within a constant. Use this result to solve for $h(t)$ the integral equation

$$\int_{-\infty}^{\infty} e^{-t(t-2x)/2} h(t)\, dt = e^{3x^2/8}.$$

As a standard result,

$$\mathcal{F}\left[f'(t) \right] = i\omega \tilde{f}(\omega),$$

though we will not need this relationship in the following solution.

From its definition,

$$\mathcal{F}[tf(t)] = \frac{1}{\sqrt{2\pi}} \int_{-\infty}^{\infty} tf(t)\, e^{-i\omega t}\, dt$$

$$= \frac{1}{-i} \frac{d}{d\omega} \left(\frac{1}{\sqrt{2\pi}} \int_{-\infty}^{\infty} f(t)\, e^{-i\omega t}\, dt \right) = i \frac{d\tilde{f}}{d\omega}.$$

Now, for the particular given function,

$$\tilde{f}(\omega) = \frac{1}{\sqrt{2\pi}} \int_{-\infty}^{\infty} e^{-t^2/2}\, e^{-i\omega t}\, dt$$

$$= \frac{1}{\sqrt{2\pi}} \left[\frac{e^{-t^2/2}\, e^{-i\omega t}}{-i\omega} \right]_{-\infty}^{\infty} + \frac{1}{\sqrt{2\pi}} \int_{-\infty}^{\infty} \frac{t e^{-t^2/2}\, e^{-i\omega t}}{-i\omega}\, dt$$

$$= 0 - \frac{1}{i\omega}\, i \frac{d\tilde{f}}{d\omega}.$$

Hence,

$$\frac{d\tilde{f}}{d\omega} = -\omega \tilde{f} \quad \Rightarrow \quad \ln \tilde{f} = -\tfrac{1}{2}\omega^2 + k \quad \Rightarrow \quad \tilde{f} = Ae^{-\omega^2/2}, \qquad (*)$$

giving $\tilde{f}(\omega)$ to within a multiplicative constant. Now, we are also given

$$\int_{-\infty}^{\infty} e^{-t(t-2x)/2}\, h(t)\, dt = e^{3x^2/8},$$

$$\Rightarrow \quad \int_{-\infty}^{\infty} e^{-(t-x)^2/2}\, e^{x^2/2}\, h(t)\, dt = e^{3x^2/8},$$

$$\Rightarrow \quad \int_{-\infty}^{\infty} e^{-(x-t)^2/2}\, h(t)\, dt = e^{-x^2/8}. \qquad (**)$$

The LHS of $(**)$ is a convolution integral, and so applying the convolution theorem for Fourier transforms and result $(*)$, used twice, yields

$$\sqrt{2\pi}\, Ae^{-\omega^2/2} \tilde{h}(\omega) = \mathcal{F}\left[e^{-(x/2)^2/2} \right] = Ae^{-(2\omega)^2/2},$$

$$\Rightarrow \quad \sqrt{2\pi}\, \tilde{h}(\omega) = e^{-3\omega^2/2} = e^{-(\sqrt{3}\omega)^2/2},$$

$$\Rightarrow \quad h(t) = \frac{1}{\sqrt{2\pi}\, A}\, e^{-(t/\sqrt{3})^2/2} = \frac{1}{\sqrt{2\pi}\, A}\, e^{-t^2/6}.$$

We now substitute in $(**)$ to determine A:

$$\int_{-\infty}^{\infty} e^{-(x-t)^2/2}\, \frac{1}{\sqrt{2\pi}\, A}\, e^{-t^2/6}\, dt = e^{-x^2/8},$$

$$\frac{1}{\sqrt{2\pi}\, A} \int_{-\infty}^{\infty} e^{-2t^2/3}\, e^{xt}\, e^{-x^2/2}\, e^{x^2/8}\, dt = 1,$$

$$\frac{1}{\sqrt{2\pi}\, A} \int_{-\infty}^{\infty} \exp\left[-\frac{2}{3}\left(t - \frac{3x}{4} \right)^2 \right]\, dt = 1.$$

From the normalization of the Gaussian integral, this implies that

$$\frac{1}{\sqrt{2\pi A}} = \frac{2}{\sqrt{2\pi}\sqrt{3}},$$

which in turn means $A = \sqrt{3}/2$, giving finally that

$$h(t) = \sqrt{\frac{2}{3\pi}}\, e^{-t^2/6}.$$

13.11 At an international "peace" conference a large number of delegates are seated around a circular table with each delegation sitting near its allies and diametrically opposite the delegation most bitterly opposed to it. The position of a delegate is denoted by θ, with $0 \le \theta \le 2\pi$. The fury $f(\theta)$ felt by the delegate at θ is the sum of his own natural hostility $h(\theta)$ and the influences on him of each of the other delegates; a delegate at position ϕ contributes an amount $K(\theta - \phi)f(\phi)$. Thus

$$f(\theta) = h(\theta) + \int_0^{2\pi} K(\theta - \phi)f(\phi)\,d\phi.$$

Show that if $K(\psi)$ takes the form $K(\psi) = k_0 + k_1 \cos\psi$ then

$$f(\theta) = h(\theta) + p + q\cos\theta + r\sin\theta$$

and evaluate p, q and r. A positive value for k_1 implies that delegates tend to placate their opponents but upset their allies, whilst negative values imply that they calm their allies but infuriate their opponents. A walkout will occur if $f(\theta)$ exceeds a certain threshold value for some θ. Is this more likely to happen for positive or for negative values of k_1?

Given that $K(\psi) = k_0 + k_1 \cos\psi$, we try a solution $f(\theta) = h(\theta) + p + q\cos\theta + r\sin\theta$, reducing the equation to

$$\begin{aligned}
p + &q\cos\theta + r\sin\theta \\
&= \int_0^{2\pi} [k_0 + k_1(\cos\theta\cos\phi + \sin\theta\sin\phi)] \\
&\quad \times [h(\phi) + p + q\cos\phi + r\sin\phi]\,d\phi \\
&= k_0(H + 2\pi p) + k_1(H_c\cos\theta + H_s\sin\theta + \pi q\cos\theta + \pi r\sin\theta),
\end{aligned}$$

where $H = \int_0^{2\pi} h(z)\,dz$, $H_c = \int_0^{2\pi} h(z)\cos z\,dz$ and $H_s = \int_0^{2\pi} h(z)\sin z\,dz$.

Thus, on equating the constant terms and the coefficients of $\cos\theta$ and $\sin\theta$, we have

$$p = k_0 H + 2\pi k_0 p \quad \Rightarrow \quad p = \frac{k_0 H}{1 - 2\pi k_0},$$

$$q = k_1 H_c + k_1\pi q \quad \Rightarrow \quad q = \frac{k_1 H_c}{1 - k_1\pi},$$

$$r = k_1 H_s + k_1\pi r \quad \Rightarrow \quad r = \frac{k_1 H_s}{1 - k_1\pi}.$$

And so the full solution for $f(\theta)$ is given by

$$f(\theta) = h(\theta) + \frac{k_0 H}{1 - 2\pi k_0} + \frac{k_1 H_c}{1 - k_1 \pi} \cos \theta + \frac{k_1 H_s}{1 - k_1 \pi} \sin \theta$$

$$= h(\theta) + \frac{k_0 H}{1 - 2\pi k_0} + \frac{k_1}{1 - k_1 \pi} \left(H_c^2 + H_s^2 \right)^{1/2} \cos(\theta - \alpha),$$

where $\tan \alpha = H_s / H_c$.

Clearly, the maximum value of $f(\theta)$ will depend upon $h(\theta)$ and its various integrals, but it is most likely to exceed any particular value if k_1 is positive and $\approx \pi^{-1}$. Stick with your friends!

13.13 The operator $\mathcal{M}$ is defined by

$$\mathcal{M}f(x) \equiv \int_{-\infty}^{\infty} K(x, y) f(y) \, dy,$$

where $K(x, y) = 1$ inside the square $|x| < a$, $|y| < a$ and $K(x, y) = 0$ elsewhere. Consider the possible eigenvalues of $\mathcal{M}$ and the eigenfunctions that correspond to them; show that the only possible eigenvalues are 0 and $2a$ and determine the corresponding eigenfunctions. Hence find the general solution of

$$f(x) = g(x) + \lambda \int_{-\infty}^{\infty} K(x, y) f(y) \, dy.$$

From the given properties of $K(x, y)$ we can assert the following.

(i) No matter what the form of $f(x)$, $\mathcal{M}f(x) = 0$ if $|x| > a$.

(ii) All functions for which both $\int_{-a}^{a} f(y) \, dy = 0$ and $f(x) = 0$ for $|x| > a$ are eigenfunctions corresponding to eigenvalue 0.

(iii) For any function $f(x)$, the integral $\int_{-a}^{a} f(y) \, dy$ is equal to a constant whose value is independent of x; thus $f(x)$ can only be an eigenfunction if it is equal to a constant, μ, for $-a \le x \le a$ and is zero otherwise. For this case $\int_{-a}^{a} f(y) \, dy = 2a\mu$ and the eigenvalue is $2a$.

Point (iii) gives the only possible non-zero eigenvalue, whilst point (ii) shows that eigenfunctions corresponding to zero eigenvalues do exist.

Denote by $S(x, a)$ the function that has unit value for $|x| \le a$ and zero value otherwise; $K(x, y)$ could be expressed as $K(x, y) = S(x, a)S(y, a)$. Substitute the trial solution $f(x) = g(x) + kS(x, a)$ into

$$f(x) = g(x) + \lambda \int_{-\infty}^{\infty} K(x, y) f(y) \, dy.$$

This gives

$$g(x) + kS(x, a) = g(x) + \lambda \int_{-\infty}^{\infty} K(x, y)[\, g(y) + kS(y, a)\,] \, dy,$$

$$kS(x, a) = \lambda S(x, a) \int_{-a}^{a} g(y) \, dy + \lambda k \, 2a \, S(x, a).$$

Here, having replaced $K(x, y)$ by $S(x, a)S(y, a)$, we use the factor $S(y, a)$ to reduce the limits of the y-integration from $\pm\infty$ to $\pm a$. As this result is to hold for all x we must have

$$k = \frac{\lambda G}{1 - 2a\lambda}, \quad \text{where } G = \int_{-a}^{a} g(y)\, dy.$$

The general solution is thus

$$f(x) = \begin{cases} g(x) + \dfrac{\lambda G}{1 - 2a\lambda} & \text{for } |x| \leq a, \\ g(x) & \text{for } |x| > a. \end{cases}$$

13.15 Use Fredholm theory to show that, for the kernel

$$K(x, z) = (x + z)\exp(x - z)$$

over the interval $[0, 1]$, the resolvent kernel is

$$R(x, z; \lambda) = \frac{\exp(x - z)\left[(x + z) - \lambda\left(\frac{1}{2}x + \frac{1}{2}z - xz - \frac{1}{3}\right)\right]}{1 - \lambda - \frac{1}{12}\lambda^2},$$

and hence solve

$$y(x) = x^2 + 2\int_0^1 (x + z)\exp(x - z)\, y(z)\, dz,$$

expressing your answer in terms of I_n, where $I_n = \int_0^1 u^n \exp(-u)\, du$.

We calculate successive values of d_n and $D_n(x, z)$ using the Fredholm recurrence relations:

$$d_n = \int_a^b D_{n-1}(x, x)\, dx,$$

$$D_n(x, z) = K(x, z)d_n - n\int_a^b K(x, z_1)D_{n-1}(z_1, z)\, dz_1,$$

starting from $d_0 = 1$ and $D_0(x, z) = (x + z)e^{x-z}$. In the first iteration we obtain

$$d_1 = \int_0^1 (u + u)e^{u-u}\, du = 1,$$

$$D_1(x, z) = (x + z)e^{x-z}(1) - 1\int_0^1 (x + u)e^{x-u}(u + z)e^{u-z}\, du$$

$$= (x + z)e^{x-z} - e^{x-z}\int_0^1 [xz + (x + z)u + u^2]\, du$$

$$= e^{x-z}\left[\frac{1}{2}(x + z) - xz - \frac{1}{3}\right].$$

Performing the second iteration gives

$$d_2 = \int_0^1 e^{u-u}\left(u - u^2 - \tfrac{1}{3}\right) du = \tfrac{1}{2} - \tfrac{1}{3} - \tfrac{1}{3} = -\tfrac{1}{6},$$

$$D_2(x, z) = (x + z)e^{x-z}\left(-\tfrac{1}{6}\right)$$

$$-2\int_0^1 (x + u)e^{x-u}e^{u-z}\left[\tfrac{1}{2}(u + z) - uz - \tfrac{1}{3}\right] du$$

$$= e^{x-z}\left\{-\tfrac{1}{6}(x + z) - 2\left[x\left(\tfrac{1}{4} + \tfrac{z}{2} - \tfrac{z}{2} - \tfrac{1}{3}\right) + \left(\tfrac{1}{6} + \tfrac{z}{4} - \tfrac{z}{3} - \tfrac{1}{6}\right)\right]\right\}$$

$$= e^{x-z}\left\{-\tfrac{1}{6}(x + z) - 2\left[-\tfrac{x}{12} - \tfrac{z}{12}\right]\right\} = 0.$$

Since $D_2(x, z) = 0$, $d_3 = 0$, $D_3(x, z) = 0$, etc. Consequently both $D(x, z; \lambda)$ and $d(\lambda)$ are finite, rather than infinite, series:

$$D(x, z; \lambda) = (x + z)e^{x-z} - \lambda\left[\tfrac{1}{2}(x + z) - xz - \tfrac{1}{3}\right]e^{x-z},$$

$$d(\lambda) = 1 - \lambda + \left(-\tfrac{1}{6}\right)\frac{\lambda^2}{2!} = 1 - \lambda - \tfrac{1}{12}\lambda^2.$$

The resolvent kernel $R(x, z; \lambda)$, given by the ratio $D(x, z; \lambda)/d(\lambda)$, is therefore as stated in the question.

For the particular integral equation, $\lambda = 2$ and $f(x) = x^2$. It follows that

$$d(\lambda) = 1 - 2 - \tfrac{4}{12} = -\tfrac{4}{3} \quad \text{and} \quad D(x, z : \lambda) = \left(2xz + \tfrac{2}{3}\right)e^{x-z}.$$

The solution is therefore given by

$$y(x) = f(x) + \lambda \int_0^1 R(x, z; \lambda)f(z)\, dz$$

$$= x^2 + 2\int_0^1 \frac{\left(2xz + \tfrac{2}{3}\right)z^2 e^{x-z}}{-\tfrac{4}{3}}\, dz$$

$$= x^2 - \int_0^1 (3xz^3 + z^2)e^{x-z}\, dz$$

$$= x^2 - (3x I_3 + I_2)e^x.$$

Complex variables

14.1 Find an analytic function of $z = x + iy$ whose imaginary part is

$$(y \cos y + x \sin y) \exp x.$$

If the required function is $f(z) = u + iv$, with $v = (y \cos y + x \sin y) \exp x$, then, from the Cauchy–Riemann equations,

$$\frac{\partial v}{\partial x} = e^x (y \cos y + x \sin y + \sin y) = -\frac{\partial u}{\partial y}.$$

Integrating with respect to y gives

$$u = -e^x \int (y \cos y + x \sin y + \sin y) \, dy + f(x)$$

$$= -e^x \left(y \sin y - \int \sin y \, dy - x \cos y - \cos y \right) + f(x)$$

$$= -e^x (y \sin y + \cos y - x \cos y - \cos y) + f(x)$$

$$= e^x (x \cos y - y \sin y) + f(x).$$

We determine $f(x)$ by applying the second Cauchy–Riemann equation, which equates $\partial u / \partial x$ with $\partial v / \partial y$:

$$\frac{\partial u}{\partial x} = e^x (x \cos y - y \sin y + \cos y) + f'(x),$$

$$\frac{\partial v}{\partial y} = e^x (\cos y - y \sin y + x \cos y).$$

By comparison, $\quad f'(x) = 0 \quad \Rightarrow \quad f(x) = k,$

where k is a real constant that can be taken as zero. Hence, the analytic function is given by

$$f(z) = u + iv = e^x (x \cos y - y \sin y + iy \cos y + ix \sin y)$$

$$= e^x [(\cos y + i \sin y)(x + iy)]$$

$$= e^x e^{iy} (x + iy)$$

$$= ze^z.$$

The final line confirms explicitly that this is a function of z alone (as opposed to a function of both z and z^*).

14.3 Find the radii of convergence of the following Taylor series:

$$\text{(a)} \sum_{n=2}^{\infty} \frac{z^n}{\ln n}, \quad \text{(b)} \sum_{n=1}^{\infty} \frac{n!z^n}{n^n},$$

$$\text{(c)} \sum_{n=1}^{\infty} z^n n^{\ln n}, \quad \text{(d)} \sum_{n=1}^{\infty} \left(\frac{n+p}{n}\right)^{n^2} z^n, \quad \text{with } p \text{ real.}$$

In each case we consider the series as $\sum_n a_n z^n$ and apply the formula

$$\frac{1}{R} = \lim_{n \to \infty} |a_n|^{1/n}$$

derived from considering the Cauchy root test for absolute convergence.

(a) $\quad \dfrac{1}{R} = \lim_{n \to \infty} \left(\dfrac{1}{\ln n}\right)^{1/n} = 1$, since $-n^{-1} \ln \ln n \to 0$ as $n \to \infty$.

Thus $R = 1$. For interest, we also note that at the point $z = 1$ the series is

$$\sum_{n=2}^{\infty} \frac{1}{\ln n} > \sum_{n=2}^{\infty} \frac{1}{n},$$

which diverges. This shows that the given series diverges at this point on its circle of convergence.

(b) $\quad \dfrac{1}{R} = \lim_{n \to \infty} \left(\dfrac{n!}{n^n}\right)^{1/n}$.

Since the nth root of $n!$ tends to n as $n \to \infty$, the limit of this ratio is that of n/n, namely unity. Thus $R = 1$ and the series converges inside the unit circle.

(c) $\quad \dfrac{1}{R} = \lim_{n \to \infty} \left(n^{\ln n}\right)^{1/n} = \lim_{n \to \infty} n^{(\ln n)/n}$

$$= \lim_{n \to \infty} \exp\left[\frac{\ln n}{n} \ln n\right] = \exp(0) = 1.$$

Thus $R = 1$ and the series converges inside the unit circle. It is obvious that the series diverges at the point $z = 1$.

(d) $\quad \dfrac{1}{R} = \lim_{n \to \infty} \left[\left(\dfrac{n+p}{n}\right)^{n^2}\right]^{1/n} = \lim_{n \to \infty} \left(\dfrac{n+p}{n}\right)^{n}$

$$= \lim_{n \to \infty} \left(1 + \frac{p}{n}\right)^{n} = e^{p}.$$

Thus $R = e^{-p}$ and the series converges inside a circle of this radius centered on the origin $z = 0$.

14.5 Determine the types of singularities (if any) possessed by the following functions at $z = 0$ and $z = \infty$:

$$(a)\ (z-2)^{-1}, \quad (b)\ (1+z^3)/z^2, \quad (c)\ \sinh(1/z),$$
$$(d)\ e^z/z^3, \quad (e)\ z^{1/2}/(1+z^2)^{1/2}.$$

(a) Although $(z-2)^{-1}$ has a simple pole at $z = 2$, at both $z = 0$ and $z = \infty$ it is well behaved and analytic.

b) Near $z = 0$, $f(z) = (1+z^3)/z^2$ behaves like $1/z^2$ and so has a double pole there. It is clear that as $z \to \infty$ $f(z)$ behaves as z and so has a simple pole there; this can be made more formal by setting $z = 1/\xi$ to obtain $g(\xi) = \xi^2 + \xi^{-1}$ and considering $\xi \to 0$. This leads to the same conclusion.

(c) As $z \to \infty$, $f(z) = \sinh(1/z)$ behaves like $\sinh \xi$ as $\xi \to 0$, i.e. analytically. However, the definition of the sinh function involves an infinite series – in this case an infinite series of inverse powers of z. Thus, no finite n for which

$$\lim_{z \to 0} [\, z^n f(z) \,] \text{ is finite}$$

can be found, and $f(z)$ has an essential singularity at $z = 0$.

(d) Near $z = 0$, $f(z) = e^z/z^3$ behaves as $1/z^3$ and has a pole of order 3 at the origin. At $z = \infty$ it has an obvious essential singularity; formally, the series expansion of $e^{1/\xi}$ about $\xi = 0$ contains arbitrarily high inverse powers of ξ.

(e) Near $z = 0$, $f(z) = z^{1/2}/(1+z^2)^{1/2}$ behaves as $z^{1/2}$ and therefore has a branch point there. To investigate its behavior as $z \to \infty$, we set $z = 1/\xi$ and obtain

$$f(z) = g(\xi) = \left(\frac{\xi^{-1}}{1+\xi^{-2}}\right)^{1/2} = \left(\frac{\xi}{\xi^2+1}\right)^{1/2} \sim \xi^{1/2} \text{ as } \xi \to 0.$$

Hence $f(z)$ also has a branch point at $z = \infty$.

14.7 Find the real and imaginary parts of the functions (i) z^2, (ii) e^z, and (iii) $\cosh \pi z$. By considering the values taken by these parts on the boundaries of the region $x \geq 0$, $y \leq 1$, determine the solution of Laplace's equation in that region that satisfies the boundary conditions

$$\phi(x, 0) = 0, \qquad \phi(0, y) = 0,$$
$$\phi(x, 1) = x, \qquad \phi(1, y) = y + \sin \pi y.$$

Writing $f_k(z) = u_k(x, y) + i v_k(x, y)$, we have

$$
\begin{aligned}
\text{(i)} \quad f_1(z) &= z^2 = (x+iy)^2 \\
&\Rightarrow u_1 = x^2 - y^2 \text{ and } v_1 = 2xy, \\
\text{(ii)} \quad f_2(z) &= e^z = e^{x+iy} = e^x(\cos y + i \sin y) \\
&\Rightarrow u_2 = e^x \cos y \text{ and } v_2 = e^x \sin y, \\
\text{(iii)} \quad f_3(z) &= \cosh \pi z = \cosh \pi x \cos \pi y + i \sinh \pi x \sin \pi y \\
&\Rightarrow u_3 = \cosh \pi x \cos \pi y \text{ and } v_3 = \sinh \pi x \sin \pi y.
\end{aligned}
$$

All of these u and v are necessarily solutions of Laplace's equation (this follows from the Cauchy–Riemann equations), and, since Laplace's equation is linear, we can form any linear combination of them and it will also be a solution. We need to choose the combination that matches the given boundary conditions.

Since the third and fourth conditions involve x and $\sin \pi y$, and these appear only in v_1 and v_3, respectively, let us try a linear combination of them:

$$\phi(x, y) = A(2xy) + B(\sinh \pi x \sin \pi y).$$

The requirement $\phi(x, 0) = 0$ is clearly satisfied, as is $\phi(0, y) = 0$. The condition $\phi(x, 1) = x$ becomes $2Ax + 0 = x$, requiring $A = \frac{1}{2}$, and the remaining condition, $\phi(1, y) = y + \sin \pi y$, takes the form $y + B \sinh \pi \sin \pi y = y + \sin \pi y$, thus determining B as $1/\sinh \pi$.

With ϕ a solution of Laplace's equation and all of the boundary conditions satisfied, the uniqueness theorem guarantees that

$$\phi(x, y) = xy + \frac{\sinh \pi x \sin \pi y}{\sinh \pi}$$

is the correct solution.

14.9 The *fundamental theorem of algebra* states that, for a complex polynomial $p_n(z)$ of degree n, the equation $p_n(z) = 0$ has precisely n complex roots. By applying Liouville's theorem, which reads

If $f(z)$ is analytic and bounded for all z then f is a constant,

to $f(z) = 1/p_n(z)$, prove that $p_n(z) = 0$ has at least one complex root. Factor out that root to obtain $p_{n-1}(z)$ and, by repeating the process, prove the fundamental theorem.

We prove this result by the method of contradiction. Suppose $p_n(z) = 0$ has no roots in the complex plane, then $f_n(z) = 1/p_n(z)$ is bounded for all z and, by Liouville's theorem, is therefore a constant. It follows that $p_n(z)$ is also a constant and that $n = 0$. However, if $n > 0$ we have a contradiction and it was wrong to suppose that $p_n(z) = 0$ has no roots; it must have at least one. Let one of them be $z = z_1$; i.e. $p_n(z)$, being a polynomial, can be written $p_n(z) = (z - z_1)p_{n-1}(z)$.

Now, by considering $f_{n-1}(z) = 1/p_{n-1}(z)$ in just the same way, we can conclude that either $n - 1 = 0$ or a further reduction is possible. It is clear that n such reductions are needed to make f_0 a constant, thus establishing that $p_n(z) = 0$ has precisely n (complex) roots.

14.11 The function

$$f(z) = (1 - z^2)^{1/2}$$

of the complex variable z is defined to be real and positive on the real axis for $-1 < x < 1$. Using cuts running along the real axis for $1 < x < +\infty$ and $-\infty < x < -1$, show how $f(z)$ is made single-valued and evaluate it on the upper and lower sides of both cuts.

Use these results and a suitable contour in the complex z-plane to evaluate the integral

$$I = \int_1^\infty \frac{dx}{x(x^2 - 1)^{1/2}}.$$

Confirm your answer by making the substitution $x = \sec \theta$.

As usual when dealing with branch cuts aimed at making a multi-valued function into a single-valued one, we introduce polar coordinates centered on the branch points. For $f(z)$ the branch points are at $z = \pm 1$, and so we define r_1 as the distance of z from the point 1 and θ_1 as the angle the line joining 1 to z makes with the part of the x-axis for which $1 < x < +\infty$, with $0 \le \theta_1 \le 2\pi$. Similarly, r_2 and θ_2 are centered on the point -1, but θ_2 lies in the range $-\pi \le \theta_2 \le \pi$.

With these definitions,

$$f(z) = (1 - z^2)^{1/2} = (1 - z)^{1/2}(1 + z)^{1/2}$$
$$= \left[(-r_1 e^{i\theta_1})(r_2 e^{i\theta_2}) \right]^{1/2}$$
$$= (r_1 r_2)^{1/2} e^{i(\theta_1 + \theta_2 - \pi)/2}.$$

In the final line the choice between $\exp(+i\pi)$ and $\exp(-i\pi)$ for dealing with the minus sign appearing before r_1 in the second line was resolved by the requirement that $f(z)$ is real and positive when $-1 < x < 1$ with $y = 0$. For these values of z, $r_1 = 1 - x$, $r_2 = 1 + x$, $\theta_1 = \pi$ and $\theta_2 = 0$. Thus,

$$f(z) = \left[(1 - x)(1 + x) \right]^{1/2} e^{(\pi + 0 - \pi)/2} = (1 - x^2)^{1/2} e^{i0} = +(1 - x^2)^{1/2},$$

as required.

Now applying the same prescription to points lying just above and just below each of the cuts, we have

$$x > 1, \, y = 0_+ \qquad r_1 = x - 1 \quad r_2 = x + 1 \quad \theta_1 = 0 \quad \theta_2 = 0$$
$$\Rightarrow \quad f(z) = (x^2 - 1)^{1/2} e^{i(0 + 0 - \pi)/2} = -i(x^2 - 1)^{1/2},$$

$$x > 1, \, y = 0_- \qquad r_1 = x - 1 \quad r_2 = x + 1 \quad \theta_1 = 2\pi \quad \theta_2 = 0$$
$$\Rightarrow \quad f(z) = (x^2 - 1)^{1/2} e^{i(2\pi + 0 - \pi)/2} = i(x^2 - 1)^{1/2},$$

$$x < -1, \, y = 0_+ \qquad r_1 = 1 - x \quad r_2 = -x - 1 \quad \theta_1 = \pi \quad \theta_2 = \pi$$
$$\Rightarrow \quad f(z) = (x^2 - 1)^{1/2} e^{i(\pi + \pi - \pi)/2} = i(x^2 - 1)^{1/2},$$

$$x < -1, \, y = 0_- \qquad r_1 = 1 - x \quad r_2 = -x - 1 \quad \theta_1 = \pi \quad \theta_2 = -\pi$$
$$\Rightarrow \quad f(z) = (x^2 - 1)^{1/2} e^{i(\pi - \pi - \pi)/2} = -i(x^2 - 1)^{1/2}.$$

To use these results to evaluate the given integral I, consider the contour integral

$$J = \int_C \frac{dz}{z(1 - z^2)^{1/2}} = \int_c \frac{dz}{z f(z)}.$$

Here C is a large circle (consisting of arcs Γ_1 and Γ_2 in the upper and lower half-planes, respectively) of radius R centered on the origin but indented along the positive and negative x-axes by the cuts considered earlier. At the ends of the cuts are two small circles γ_1 and γ_2 that enclose the branch points $z = 1$ and $z = -1$, respectively. Thus the complete closed contour, starting from γ_1 and moving along the positive real axis, consists of, in order, circle γ_1, cut C_1, arc Γ_1, cut C_2, circle γ_2, cut C_3, arc Γ_2 and cut C_4, leading back to γ_1.

On the arcs Γ_1 and Γ_2 the integrand is $O(R^{-2})$ and the contributions to the contour integral $\to 0$ as $R \to \infty$. For the small circle γ_1, where we can set $z = 1 + \rho e^{i\phi}$ with $dz = i\rho e^{i\phi}\,d\phi$, we have

$$\int_{\gamma_1} \frac{dz}{z(1+z)^{1/2}(1-z)^{1/2}} = \int_0^{2\pi} \frac{i\rho e^{i\phi}}{(1+\rho e^{i\phi})(2+\rho e^{i\phi})^{1/2}(-\rho e^{i\phi})^{1/2}}\,d\phi,$$

and this $\to 0$ as $\rho \to 0$. Similarly, the small circle γ_2 contributes nothing to the contour integral. This leaves only the contributions from the four arms of the branch cuts. To relate these to I we use our previous results about the value of $f(z)$ on the various arms:

$$\text{on } C_1, z = x \text{ and } \int_{C_1} = \int_1^\infty \frac{dx}{x[-i(x^2-1)^{1/2}]} = iI;$$

$$\text{on } C_2, z = -x \text{ and } \int_{C_2} = \int_\infty^1 \frac{-dx}{-x[i(x^2-1)^{1/2}]} = iI;$$

$$\text{on } C_3, z = -x \text{ and } \int_{C_3} = \int_1^\infty \frac{-dx}{-x[-i(x^2-1)^{1/2}]} = iI;$$

$$\text{on } C_4, z = x \text{ and } \int_{C_1} = \int_\infty^1 \frac{dx}{x[i(x^2-1)^{1/2}]} = iI.$$

So the full contour integral around C has the value $4iI$. But, this must be the same as $2\pi i$ times the residue of $z^{-1}(1-z^2)^{-1/2}$ at $z = 0$, which is the only pole of the integrand inside the contour. The residue is clearly unity, and so we deduce that $I = \pi/2$.

This particular integral can be evaluated much more simply using elementary methods. Setting $x = \sec\theta$ with $dx = \sec\theta \tan\theta\,d\theta$ gives

$$I = \int_1^\infty \frac{dx}{x(x^2-1)^{1/2}}$$
$$= \int_0^{\pi/2} \frac{\sec\theta \tan\theta\,d\theta}{\sec\theta\,(\sec^2\theta-1)^{1/2}} = \int_0^{\pi/2} d\theta = \frac{\pi}{2},$$

and so verifies the result obtained by contour integration.

14.13 The following is an alternative (and roundabout!) way of evaluating the Gaussian integral.

(a) Prove that the integral of $[\exp(i\pi z^2)]\operatorname{cosec}\pi z$ around the parallelogram with corners $\pm 1/2 \pm R\exp(i\pi/4)$ has the value $2i$.

(b) Show that the parts of the contour parallel to the real axis give no contribution when $R \to \infty$.

(c) Evaluate the integrals along the other two sides by putting $z' = r\exp(i\pi/4)$ and working in terms of $z' + \frac{1}{2}$ and $z' - \frac{1}{2}$. Hence by letting $R \to \infty$ show that

$$\int_{-\infty}^\infty e^{-\pi r^2}\,dr = 1.$$

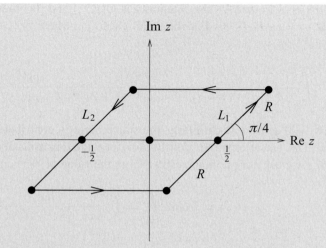

Figure 14.1 The parallelogram contour used in Problem 14.13.

The integral is

$$\int_C e^{i\pi z^2} \operatorname{cosec} \pi z \, dz = \int_C \frac{e^{i\pi z^2}}{\sin \pi z} \, dz$$

and the suggested contour C is shown in Figure 14.1.

(a) The integrand has (simple) poles only on the real axis at $z = n$, where n is an integer. The only such pole enclosed by C is at $z = 0$. The residue there is

$$a_{-1} = \lim_{z \to 0} \frac{z e^{i\pi z^2}}{\sin \pi z} = \frac{1}{\pi}.$$

The value of the integral around C is therefore $2\pi i \times (\pi^{-1}) = 2i$.

(b) On the parts of C parallel to the real axis, $z = \pm R e^{i\pi/4} + x'$, where $-\frac{1}{2} \le x' \le \frac{1}{2}$. The integrand is thus given by

$$\begin{aligned}
f(z) &= \frac{1}{\sin \pi z} \exp\left[i\pi \left(\pm R e^{i\pi/4} + x' \right)^2 \right] \\
&= \frac{1}{\sin \pi z} \exp\left[i\pi \left(R^2 e^{i\pi/2} \pm 2Rx' e^{i\pi/4} + x'^2 \right) \right] \\
&= \frac{1}{\sin \pi z} \exp\left[-\pi R^2 \pm \frac{2\pi i R x'}{\sqrt{2}} (1 + i) + i\pi x'^2 \right] \\
&= O\left(\exp[-\pi R^2 \mp \sqrt{2}\pi R x'] \right) \\
&\to 0 \text{ as } R \to \infty.
\end{aligned}$$

Since the integration range is finite ($-\frac{1}{2} \le x' \le \frac{1}{2}$), the integrals $\to 0$ as $R \to \infty$.

(c) On the first of the other two sides, let us set $z = \frac{1}{2} + re^{i\pi/4}$ with $-R \le r \le R$. The corresponding integral I_1 is

$$I_1 = \int_{L_1} e^{i\pi z^2} \operatorname{cosec} \pi z \, dz$$

$$= \int_{-R}^{R} \frac{\exp\left[i\pi\left(\frac{1}{2} + re^{i\pi/4}\right)^2\right]}{\sin\left[\pi\left(\frac{1}{2} + re^{i\pi/4}\right)\right]} e^{i\pi/4} \, dr$$

$$= \int_{-R}^{R} \frac{e^{i\pi/4} \exp(i\pi re^{i\pi/4}) \exp(i\pi r^2 i) e^{i\pi/4}}{\cos(\pi re^{i\pi/4})} \, dr$$

$$= \int_{-R}^{R} \frac{i \exp(i\pi re^{i\pi/4}) e^{-\pi r^2}}{\cos(\pi re^{i\pi/4})} \, dr.$$

Similarly (remembering the sense of integration), the remaining side contributes

$$I_2 = -\int_{-R}^{R} \frac{i \exp(-i\pi re^{i\pi/4}) e^{-\pi r^2}}{-\cos(\pi re^{i\pi/4})} \, dr.$$

Adding together all four contributions gives

$$0 + 0 + \int_{-R}^{R} \frac{i[\exp(i\pi re^{i\pi/4}) + \exp(-i\pi re^{i\pi/4})] e^{-\pi r^2}}{\cos(\pi re^{i\pi/4})} \, dr,$$

which simplifies to

$$\int_{-R}^{R} 2i e^{-\pi r^2} \, dr.$$

From part (a), this must be equal to $2i$ as $R \to \infty$, and so $\int_{-\infty}^{\infty} e^{-\pi r^2} \, dr = 1$.

15 Applications of complex variables

Many of the problems in this chapter involve contour integration and the choice of a suitable contour. In order to save the space taken by drawing several broadly similar contours that differ only in notation, the positions of poles, the values of lengths or angles, or other minor details, we make reference to Figure 15.1 which shows a number of typical contour types.

15.1 In the method of complex impedances for a.c. circuits, an inductance L is represented by a complex impedance $Z_L = i\omega L$ and a capacitance C by $Z_C = 1/(i\omega C)$. Kirchhoff's circuit laws,

$$\sum_i I_i = 0 \text{ at a node and } \sum_i Z_i I_i = \sum_j V_j \text{ around any closed loop,}$$

are then applied as if the circuit were a d.c. one.

Apply this method to the a.c. bridge connected as in Figure 15.2 to show that if the resistance R is chosen as $R = (L/C)^{1/2}$ then the amplitude of the current I_R through it is independent of the angular frequency ω of the applied a.c. voltage $V_0 e^{i\omega t}$.

Determine how the phase of I_R, relative to that of the voltage source, varies with the angular frequency ω.

Omitting the common factor $e^{i\omega t}$ from all currents and voltages, let the current drawn from the voltage source be (the complex quantity) I and the current flowing from A to D be I_1. Then the currents in the remaining branches are $AE: I - I_1$, $DB: I_1 - I_R$ and $EB: I - I_1 + I_R$.

Applying $\sum_i Z_i I_i = \sum_j V_j$ to three separate loops yields

loop $ADBA$

$$i\omega L\, I_1 + \frac{1}{i\omega C}(I_1 - I_R) = V_0,$$

loop $ADEA$

$$i\omega L\, I_1 + R\, I_R - \frac{1}{i\omega C}(I - I_1) = 0,$$

loop $DBED$

$$\frac{1}{i\omega C}(I_1 - I_R) - i\omega L\,(I - I_1 + I_R) - R\, I_R = 0.$$

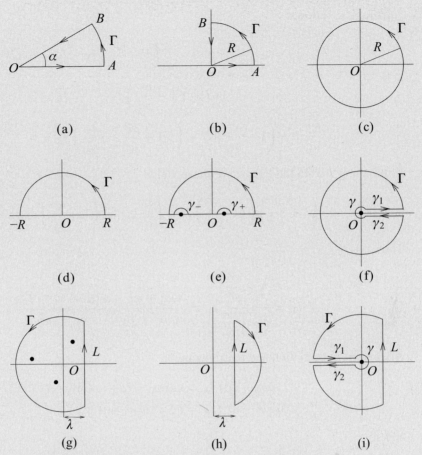

Figure 15.1 Typical contours for use in contour integration.

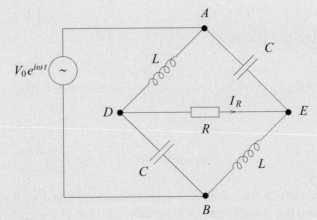

Figure 15.2 The inductor–capacitor–resistor network for Problem 15.1.

Now, denoting $(LC)^{-1}$ by ω_0^2 and choosing R as $(L/C)^{1/2} = (\omega_0 C)^{-1}$, we can write these equations as follows:

$$\left(1 - \frac{\omega^2}{\omega_0^2}\right) I_1 - I_R = i\omega C V_0,$$

$$-I + \left(1 - \frac{\omega^2}{\omega_0^2}\right) I_1 + i\frac{\omega}{\omega_0} I_R = 0,$$

$$\frac{\omega^2}{\omega_0^2} I + \left(1 - \frac{\omega^2}{\omega_0^2}\right) I_1 + \left(-1 + \frac{\omega^2}{\omega_0^2} - i\frac{\omega}{\omega_0}\right) I_R = 0.$$

Eliminating I from the last two of these yields

$$\left(1 + \frac{\omega^2}{\omega_0^2}\right)\left(1 - \frac{\omega^2}{\omega_0^2}\right) I_1 - \left(\frac{i\omega}{\omega_0} + 1\right)\left(1 - \frac{\omega^2}{\omega_0^2}\right) I_R = 0.$$

Thus,

$$I_R = \frac{1 + \dfrac{\omega^2}{\omega_0^2}}{1 + i\dfrac{\omega}{\omega_0}} I_1 = \frac{\omega_0^2 + \omega^2}{\omega_0(\omega_0 + i\omega)} \frac{\omega_0^2(i\omega C V_0 + I_R)}{\omega_0^2 - \omega^2}.$$

After some cancellation and rearrangement,

$$\left(\omega_0^2 - \omega^2\right) I_R = \omega_0(\omega_0 - i\omega)(i\omega C V_0 + I_R),$$

$$(i\omega\omega_0 - \omega^2) I_R = \omega_0\omega(i\omega_0 + \omega)C V_0,$$

and so

$$I_R = \omega_0 C V_0 \frac{i\omega_0 + \omega}{i\omega_0 - \omega} = \omega_0 C V_0 \frac{(i\omega_0 + \omega)(-i\omega_0 - \omega)}{(i\omega_0 - \omega)(-i\omega_0 - \omega)}$$

$$= \omega_0 C V_0 \frac{\omega_0^2 - \omega^2 - 2i\omega\omega_0}{\omega_0^2 + \omega^2}.$$

From this we can read off

$$|I_R| = \omega_0 C V_0 \frac{\left[\left(\omega^2 - \omega_0^2\right)^2 + 4\omega^2\omega_0^2\right]^{1/2}}{\omega_0^2 + \omega^2} = \omega_0 C V_0, \text{ i.e. independent of } \omega,$$

and

$$\phi = \text{phase of } I_R = \tan^{-1} \frac{-2\omega\omega_0}{\omega_0^2 - \omega^2}.$$

Thus I_R (which was arbitrarily and notionally defined as flowing from D to E in the equivalent d.c. circuit) has an imaginary part that is always negative but a real part that changes sign as ω passes through ω_0. Its phase ϕ, relative to that of the voltage source, therefore varies from 0 when ω is small to $-\pi$ when ω is large.

15.3   For the function

$$f(z) = \ln\left(\frac{z+c}{z-c}\right),$$

where c is real, show that the real part u of f is constant on a circle of radius $c \operatorname{cosech} u$ centered on the point $z = c \coth u$. Use this result to show that the electrical capacitance per unit length of two parallel cylinders of radii a, placed with their axes $2d$ apart, is proportional to $[\cosh^{-1}(d/a)]^{-1}$.

From

$$f(z) = \ln\left(\frac{z+c}{z-c}\right) = \ln\left|\frac{z+c}{z-c}\right| + i \arg\left(\frac{z+c}{z-c}\right),$$

we have that

$$u = \ln\left|\frac{z+c}{z-c}\right| = \frac{1}{2}\ln\frac{(x+c)^2+y^2}{(x-c)^2+y^2} \quad \Rightarrow \quad e^{2u} = \frac{(x+c)^2+y^2}{(x-c)^2+y^2}.$$

The curve upon which $u(x,y)$ is constant is therefore given by

$$(x^2 - 2cx + c^2 + y^2)e^{2u} = x^2 + 2xc + c^2 + y^2.$$

This can be rewritten as

$$x^2(e^{2u} - 1) - 2xc(e^{2u} + 1) + y^2(e^{2u} - 1) + c^2(e^{2u} - 1) = 0,$$

$$x^2 - 2xc\frac{e^{2u} + 1}{e^{2u} - 1} + y^2 + c^2 = 0,$$

$$x^2 - 2xc \coth u + y^2 + c^2 = 0,$$

which, in conic-section form, becomes

$$(x - c \coth u)^2 + y^2 = c^2 \coth^2 u - c^2 = c^2 \operatorname{cosech}^2 u.$$

This is a circle with center $(c \coth u, 0)$ and radius $|c \operatorname{cosech} u|$.

Now consider two such circles with the same value of $|c \operatorname{cosech} u|$, equal to a, but different values of u satisfying $c \coth u_1 = -d$ and $c \coth u_2 = +d$. These two equations imply that $u_1 = -u_2$, corresponding physically to equal but opposite charges $-Q$ and $+Q$ placed on identical cylindrical conductors that coincide with the circles; the conductors are raised to potentials u_1 and u_2.

We have already established that we need $c \coth u_2 = d$ and $c \operatorname{cosech} u_2 = a$. Dividing these two equations gives $\cosh u_2 = d/a$.

The capacitance (per unit length) of the arrangement is given by the magnitude of the charge on one conductor divided by the potential difference between the conductors that results from the presence of that charge, i.e.

$$C = \frac{Q}{u_2 - u_1} \propto \frac{1}{2u_2} = \frac{1}{2\cosh^{-1}(d/a)},$$

as stated in the question.

15.5 By considering in turn the transformations

$$z = \tfrac{1}{2}c(w + w^{-1}) \quad \text{and} \quad w = \exp\zeta,$$

where $z = x + iy$, $w = r\exp i\theta$, $\zeta = \xi + i\eta$ and c is a real positive constant, show that $z = c\cosh\zeta$ maps the strip $\xi \geq 0$, $0 \leq \eta \leq 2\pi$, onto the whole z-plane. Which curves in the z-plane correspond to the lines $\xi = $ constant and $\eta = $ constant? Identify those corresponding to $\xi = 0$, $\eta = 0$ and $\eta = 2\pi$.

The electric potential ϕ of a charged conducting strip $-c \leq x \leq c$, $y = 0$, satisfies

$$\phi \sim -k\ln(x^2 + y^2)^{1/2} \text{ for large values of } (x^2 + y^2)^{1/2},$$

with ϕ constant on the strip. Show that $\phi = \operatorname{Re}\left[-k\cosh^{-1}(z/c)\right]$ and that the magnitude of the electric field near the strip is $k(c^2 - x^2)^{-1/2}$.

We first note that the combined transformation is given by

$$z = \frac{c}{2}(e^\zeta + e^{-\zeta}) = c\cosh\zeta \quad \Rightarrow \quad \zeta = \cosh^{-1}\frac{z}{c}.$$

The successive connections linking the strip in the ζ-plane and its image in the z-plane are

$$z = c\cosh\zeta = c\cosh(\xi + i\eta)$$
$$= c\cosh\xi\cos\eta + ic\sinh\xi\sin\eta, \text{ with } \xi > 0, 0 \leq \eta \leq 2\pi,$$
$$re^{i\theta} = w = e^\zeta = e^\xi e^{i\eta}, \text{ with the strip as } 1 < r < \infty, 0 \leq \theta \leq 2\pi,$$
$$x + iy = z = \frac{c}{2}(w + w^{-1})$$
$$= \frac{c}{2}[r(\cos\theta + i\sin\theta) + r^{-1}(\cos\theta - i\sin\theta)]$$
$$= \frac{c}{2}\left(r + \frac{1}{r}\right)\cos\theta + i\frac{c}{2}\left(r - \frac{1}{r}\right)\sin\theta.$$

This last expression for z and the previous specification of the strip in terms of r and θ show that both x and y can take all values, i.e. that the original strip in the ζ-plane is mapped onto the whole of the z-plane. From the two expressions for z we also see that $x = c\cosh\xi\cos\eta$ and $y = c\sinh\xi\sin\eta$.

For ξ constant, the contour in the xy-plane, obtained by eliminating η, is

$$\frac{x^2}{c^2\cosh^2\xi} + \frac{y^2}{c^2\sinh^2\xi} = 1, \quad \text{i.e. an ellipse.}$$

The eccentricity of the ellipse is given by

$$e = \left(\frac{c^2\cosh^2\xi - c^2\sinh^2\xi}{c^2\cosh^2\xi}\right)^{1/2} = \frac{1}{\cosh\xi}.$$

The foci of the ellipse are at $\pm e \times$ the major semi-axis, i.e. $\pm 1/\cosh\xi \times c\cosh\xi = \pm c$. This is independent of ξ and so all the ellipses are confocal.

Similarly, for η constant, the contour is

$$\frac{x^2}{c^2 \cos^2 \eta} - \frac{y^2}{c^2 \sin^2 \eta} = 1.$$

This is one of a set of confocal hyperbolae.

(i) $\xi = 0 \quad \Rightarrow \quad y = 0, \ x = c \cos \eta$.

This is the finite line (degenerate ellipse) on the x-axis, $-c \le x \le c$.

(ii) $\eta = 0 \quad \Rightarrow \quad y = 0, \ x = c \cosh \xi$.

This is a part of the x-axis not covered in (i), $c < x < \infty$. The other part, $-\infty < x < -c$, corresponds to $\eta = \pi$.

(iii) This is the same as (the first case) in (ii).

Now, in the ζ-plane, consider the real part of the function $F(\zeta) = -k\zeta$, with k real. On $\xi = 0$ [case (i) above] it reduces to Re $\{-ik\eta\}$, which is zero for all η, i.e. a constant. This implies that the real part of the transformed function will be a constant (actually zero) on $-c \le x \le c$ in the z-plane.

Further,

$$(x^2 + y^2)^{1/2} = (c^2 \cosh^2 \xi \cos^2 \eta + c^2 \sinh^2 \xi \sin^2 \eta)^{1/2}$$
$$\approx \tfrac{1}{2} c e^{\xi} \text{ for large } \xi,$$
$$\Rightarrow \quad \xi \approx \ln(x^2 + y^2)^{1/2} + \text{fixed constant.}$$

Hence,

$$\text{Re } \{-k\zeta\} = -k\xi \approx -k \ln(x^2 + y^2)^{1/2} \text{ for large } (x^2 + y^2)^{1/2}.$$

Thus, the transformation

$$F(\zeta) = -k\zeta \quad \rightarrow \quad f(z) = -k \cosh^{-1} \frac{z}{c}$$

produces a function in the z-plane that satisfies the stated boundary conditions (as well as satisfying Laplace's equation). It is therefore the required solution.

The electric field near the conducting strip, where $y = 0$ and $z^2 = x^2$, can have no component in the x-direction (except at the points $x = \pm c$), but its magnitude is still given by

$$E = |f'(z)| = \left| -\frac{k}{\sqrt{z^2 - c^2}} \right| = \frac{k}{(c^2 - x^2)^{1/2}}.$$

15.7 Prove that if $f(z)$ has a simple zero at z_0 then $1/f(z)$ has residue $1/f'(z_0)$ there. Hence evaluate

$$\int_{-\pi}^{\pi} \frac{\sin \theta}{a - \sin \theta} \, d\theta,$$

where a is real and > 1.

If $f(z)$ is analytic and has a simple zero at $z = z_0$ then it can be written as

$$f(z) = \sum_{n=1}^{\infty} a_n (z - z_0)^n, \qquad \text{with } a_1 \ne 0.$$

Using a binomial expansion,

$$\frac{1}{f(z)} = \frac{1}{a_1(z - z_0)\left(1 + \sum_{n=2}^{\infty} \frac{a_n}{a_1}(z - z_0)^{n-1}\right)}$$

$$= \frac{1}{a_1(z - z_0)}(1 + b_1(z - z_0) + b_2(z - z_0)^2 + \cdots),$$

for some coefficients, b_i. The residue at $z = z_0$ is clearly a_1^{-1}.

But, from differentiating the Taylor expansion,

$$f'(z) = \sum_{n=1}^{\infty} n a_n (z - z_0)^{n-1},$$

$$\Rightarrow \quad f'(z_0) = a_1 + 0 + 0 + \cdots = a_1,$$

i.e. the residue $= \dfrac{1}{a_1}$ can also be expressed as $\dfrac{1}{f'(z_0)}$.

Denote the required integral by I and consider the contour integral

$$J = \int_C \frac{dz}{a - \dfrac{1}{2i}(z - z^{-1})} = \int_C \frac{2iz \, dz}{2aiz - z^2 + 1},$$

where C is the unit circle, i.e. contour (c) of Figure 15.1 with $R = 1$. The denominator has simple zeros at $z = ai \pm \sqrt{-a^2 + 1} = i(a \pm \sqrt{a^2 - 1})$. Since a is strictly greater than 1, $\alpha = i(a - \sqrt{a^2 - 1})$ lies strictly inside the unit circle, whilst $\beta = i(a + \sqrt{a^2 - 1})$ lies strictly outside it (and need not be considered further).

Extending the previous result to the case of $h(z) = g(z)/f(z)$, where $g(z)$ is analytic at z_0, the residue of $h(z)$ at $z = z_0$ can be seen to be $g(z_0)/f'(z_0)$. Applying this, we find that the residue of the integrand at $z = \alpha$ is given by

$$\left. \frac{2iz}{2ai - 2z} \right|_{z=\alpha} = \frac{i\alpha}{ai - ai + i\sqrt{a^2 - 1}} = \frac{\alpha}{\sqrt{a^2 - 1}}.$$

Now on the unit circle, $z = e^{i\theta}$ with $dz = i e^{i\theta} \, d\theta$, and J can be written as

$$J = \int_{-\pi}^{\pi} \frac{i e^{i\theta} \, d\theta}{a - \dfrac{1}{2i}(e^{i\theta} - e^{-i\theta})} = \int_{-\pi}^{\pi} \frac{i(\cos\theta + i\sin\theta) \, d\theta}{a - \sin\theta}.$$

Hence,

$$I = -\text{Re } J = -\text{Re } 2\pi i \frac{i(a - \sqrt{a^2 - 1})}{\sqrt{a^2 - 1}}$$

$$= 2\pi \left(\frac{a}{\sqrt{a^2 - 1}} - 1 \right).$$

Although it is not asked for, we can also deduce from the fact that the residue at $z = \alpha$ is purely imaginary that

$$\int_{-\pi}^{\pi} \frac{\cos\theta}{a - \sin\theta}\, d\theta = 0,$$

a result that can also be obtained by more elementary means, when it is noted that the numerator of the integrand is the derivative of the denominator.

15.9 Prove that

$$\int_0^\infty \frac{\cos mx}{4x^4 + 5x^2 + 1}\, dx = \frac{\pi}{6}\left(4e^{-m/2} - e^{-m}\right) \qquad \text{for } m > 0.$$

Since, when z is on the real axis, the integrand is equal to

$$\operatorname{Re} \frac{e^{imz}}{(z^2 + 1)(4z^2 + 1)} = \operatorname{Re} \frac{e^{imz}}{(z + i)(z - i)(2z + i)(2z - i)},$$

we consider the integral of $f(z) = \dfrac{e^{imz}}{(z + i)(z - i)(2z + i)(2z - i)}$ around contour (d) in Figure 15.1.

As $|f(z)| \sim |z|^{-4}$ as $z \to \infty$ and $m > 0$, all the conditions for Jordan's lemma to hold are satisfied and the integral around the large semi-circle contributes nothing. For this integrand there are two poles inside the contour, at $z = i$ and at $z = \frac{1}{2}i$. The respective residues are

$$\frac{e^{-m}}{2i\, 3i\, i} = \frac{ie^{-m}}{6} \quad \text{and} \quad \frac{e^{-m/2}}{\frac{3i}{2}\left(-\frac{i}{2}\right)2i} = \frac{-2ie^{-m/2}}{3}.$$

The residue theorem therefore reads

$$\int_{-\infty}^{\infty} \frac{e^{imx}}{4x^4 + 5x^2 + 1}\, dx + 0 = 2\pi i\left(\frac{ie^{-m}}{6} - \frac{2ie^{-m/2}}{3}\right),$$

and the stated result follows from equating real parts and changing the lower integration limit, recognizing that the integrand is symmetric about $x = 0$ and so the integral from 0 to ∞ is equal to half of that from $-\infty$ to ∞.

15.11 Using a suitable cut plane, prove that if α is real and $0 < \alpha < 1$ then

$$\int_0^\infty \frac{x^{-\alpha}}{1 + x}\, dx$$

has the value $\pi \operatorname{cosec} \pi\alpha$.

As α is not an integer, the complex form of the integrand $f(z) = \dfrac{z^{-\alpha}}{1 + z}$ is not single-valued. We therefore need to perform the contour integration in a cut plane; contour (f) of Figure 15.1 is a suitable contour. We will be making use of the fact that, because the integrand takes different values on γ_1 and γ_2, the contributions coming from these two parts of the complete contour, although related, do *not* cancel. The contributions from γ and Γ are both zero because:

(i) around γ, $|zf(z)| \sim \dfrac{z\, z^{-\alpha}}{1} = z^{1-\alpha} \to 0$ as $|z| \to 0$, since $\alpha < 1$;

(ii) around Γ, $|zf(z)| \sim \dfrac{z\, z^{-\alpha}}{z} = z^{-\alpha} \to 0$ as $|z| \to \infty$, since $\alpha > 0$.

Therefore, the only contributions come from the cut; on γ_1, $z = xe^{0i}$, whilst on γ_2, $z = xe^{2\pi i}$. The only pole inside the contour is a simple one at $z = -1 = e^{i\pi}$, where the residue is $e^{-i\pi\alpha}$. The residue theorem now reads

$$0 + \int_0^\infty \frac{x^{-\alpha}}{1+x}\, dx + 0 - \int_0^\infty \frac{x^{-\alpha} e^{-2\pi i \alpha}}{1 + xe^{2\pi i}}\, dx = 2\pi i\, e^{-i\pi\alpha},$$

$$\Rightarrow \quad (1 - e^{-2\pi i \alpha}) \int_0^\infty \frac{x^{-\alpha}}{1+x}\, dx = 2\pi i\, e^{-i\pi\alpha}.$$

This can be rearranged to read

$$\int_0^\infty \frac{x^{-\alpha}}{1+x}\, dx = \frac{2\pi i\, e^{-i\pi\alpha}}{(1 - e^{-2\pi i \alpha})} = \frac{2\pi i}{e^{i\pi\alpha} - e^{-i\pi\alpha}} = \frac{\pi}{\sin \pi\alpha},$$

thus establishing the stated result.

15.13 By integrating a suitable function around a large semi-circle in the upper half plane and a small semi-circle centered on the origin, determine the value of

$$I = \int_0^\infty \frac{(\ln x)^2}{1 + x^2}\, dx$$

and deduce, as a by-product of your calculation, that

$$\int_0^\infty \frac{\ln x}{1 + x^2}\, dx = 0.$$

The suggested contour is that shown in Figure 15.1(e), but with only one indentation γ on the real axis (at $z = 0$) and with $R = \infty$. The appropriate complex function is

$$f(z) = \frac{(\ln z)^2}{1 + z^2}.$$

The only pole inside the contour is at $z = i$, and the residue there is given by

$$\frac{(\ln i)^2}{i + i} = \frac{(\ln 1 + i(\pi/2))^2}{2i} = -\frac{\pi^2}{8i}.$$

To evaluate the integral around γ, we set $z = \rho\, e^{i\theta}$ with $\ln z = \ln \rho + i\theta$ and $dz = i\rho\, e^{i\theta}\, d\theta$; the integral becomes

$$\int_\pi^0 \frac{\ln^2 \rho + 2i\theta \ln \rho - \theta^2}{1 + \rho^2 e^{2i\theta}}\, i\rho\, e^{i\theta}\, d\theta, \text{ which} \to 0 \text{ as } \rho \to 0.$$

Thus γ contributes nothing. Even more obviously, on Γ, $|zf(z)| \sim z^{-1}$ and tends to zero as $|z| \to \infty$, showing that Γ also contributes nothing.

On γ_+, $z = xe^{i0}$ and the contribution is equal to I.

On γ_-, $z = xe^{i\pi}$ and the contribution is (remembering that the contour actually runs from $x = \infty$ to $x = 0$) given by

$$I_- = -\int_0^\infty \frac{(\ln x + i\pi)^2}{1 + x^2} e^{i\pi} \, dx$$

$$= I + 2i\pi \int_0^\infty \frac{\ln x}{1 + x^2} \, dx - \pi^2 \int_0^\infty \frac{1}{1 + x^2} \, dx.$$

The residue theorem for the complete closed contour thus reads

$$0 + I + 0 + I + 2i\pi \int_0^\infty \frac{\ln x}{1 + x^2} \, dx - \pi^2 \left[\tan^{-1} x \right]_0^\infty = 2\pi i \left(\frac{-\pi^2}{8i} \right).$$

Equating the real parts $\Rightarrow$ $2I - \frac{1}{2}\pi^3 = -\frac{1}{4}\pi^3$ $\Rightarrow$ $I = \frac{1}{8}\pi^3$.

Equating the imaginary parts gives the stated by-product.

15.15 Prove that

$$\sum_{-\infty}^\infty \frac{1}{n^2 + \frac{3}{4}n + \frac{1}{8}} = 4\pi.$$

Carry out the summation numerically, say between -4 and 4, and note how much of the sum comes from values near the poles of the contour integration.

In order to evaluate this sum, we must first find a function of z that takes the value of the corresponding term in the sum whenever z is an integer. Clearly this is

$$\frac{1}{z^2 + \frac{3}{4}z + \frac{1}{8}}.$$

Further, to make use of the properties of contour integrals, we need to multiply this function by one that has simple poles at the same points, each with unit residue. An appropriate choice of integrand is therefore

$$f(z) = \frac{\pi \cot \pi z}{z^2 + \frac{3}{4}z + \frac{1}{8}} = \frac{\pi \cot \pi z}{\left(z + \frac{1}{2} \right)\left(z + \frac{1}{4} \right)}.$$

The contour to be used must enclose all integer values of z, both positive and negative and, in practical terms, must give zero contribution for $|z| \to \infty$, except possibly on the real axis. A large circle C, centered on the origin (see contour (c) in Figure 15.1) suggests itself.

As $|zf(z)| \to 0$ on C, the contour integral has value zero. This implies that the residues at the enclosed poles add up to zero. The residues are

$$\frac{\pi \cot\left(-\frac{1}{2}\pi\right)}{-\frac{1}{2}+\frac{1}{4}} = 0 \qquad \text{at } z = -\frac{1}{2},$$

$$\frac{\pi \cot\left(-\frac{1}{4}\pi\right)}{-\frac{1}{4}+\frac{1}{2}} = -4\pi \quad \text{at } z = -\frac{1}{4},$$

$$\sum_{n=-\infty}^{\infty} \frac{1}{\left(n+\frac{1}{2}\right)\left(n+\frac{1}{4}\right)} \qquad \text{at } z = n, \ -\infty < n < \infty.$$

The quoted result follows immediately.

For the rough numerical summation we tabulate n, $D(n) = n^2 + \frac{3}{4}n + \frac{1}{8}$ and the nth term of the series, $1/D(n)$:

n	$D(n)$	$1/D(n)$
-4	13.125	0.076
-3	6.875	0.146
-2	2.625	0.381
-1	0.375	2.667
0	0.125	8.000
1	1.875	0.533
2	5.625	0.178
3	11.375	0.088
4	19.125	0.052

The total of these nine terms is 12.121; this is to be compared with the total for the entire infinite series (of positive terms), which is $4\pi = 12.566$. It will be seen that the sum is dominated by the terms for $n = 0$ and $n = -1$. These two values bracket the positions on the real axis of the poles at $z = -\frac{1}{2}$ and $z = -\frac{1}{4}$.

15.17 By considering the integral of

$$\left(\frac{\sin \alpha z}{\alpha z}\right)^2 \frac{\pi}{\sin \pi z}, \qquad \alpha < \frac{\pi}{2},$$

around a circle of large radius, prove that

$$\sum_{m=1}^{\infty} (-1)^{m-1} \frac{\sin^2 m\alpha}{(m\alpha)^2} = \frac{1}{2}.$$

Denote the given function by $f(z)$ and consider its integral around contour (c) in Figure 15.1.

As $|z| \to \infty$, $\sin \alpha z \sim e^{\alpha |z|}$, and so $f(z) \sim |z|^{-2} e^{2\alpha |z|} e^{-\pi |z|} = z^{-2} e^{(2\alpha - \pi)|z|}$, and, since $\alpha < \frac{1}{2}\pi$, $|z f(z)\, dz| \to 0$ as $|z| \to \infty$ and the integral around the contour has value zero for $R = \infty$.

The function $f(z)$ has simple poles at $z = n$, where n is an integer, $-\infty < n < \infty$. The pole at $z = 0$ is only a first-order pole as the term in parentheses $\to 1$ as $z \to 0$ and has no singularity there. It follows that the sum of the residues of $f(z)$ at all of its poles is zero. For $n \neq 0$, that residue is

$$
\pi \left(\frac{\sin n\alpha}{n\alpha} \right)^2 \left(\frac{d(\sin \pi z)}{dz} \bigg|_{z=n} \right)^{-1} = \left(\frac{\sin n\alpha}{n\alpha} \right)^2 \frac{1}{\cos \pi n}
$$

$$
= (-1)^n \left(\frac{\sin n\alpha}{n\alpha} \right)^2.
$$

For $n = 0$ the residue is 1.

Since the general residue is an even function of n, the sum for $-\infty < n \leq -1$ is equal to that for $1 \leq n < \infty$, and the zero sum of the residues can be written

$$
1 + 2 \sum_{n=1}^{\infty} (-1)^n \left(\frac{\sin n\alpha}{n\alpha} \right)^2 = 0,
$$

leading immediately to the stated result.

15.19 Find the function $f(t)$ whose Laplace transform is

$$
\bar{f}(s) = \frac{e^{-s} - 1 + s}{s^2}.
$$

Consider first the Taylor series expansion of $\bar{f}(s)$ about $s = 0$:

$$
\bar{f}(s) = \frac{e^{-s} - 1 + s}{s^2} = \frac{\left(1 - s + \frac{1}{2}s^2 + \cdots \right) - 1 + s}{s^2} \sim \frac{1}{2} + O(s).
$$

Thus $\bar{f}$ has no pole at $s = 0$, and λ in the Bromwich integral can be as small as we wish (but > 0). When the integration line is made part of a closed contour C, the inversion integral becomes

$$
f(t) = \int_C \frac{e^{-s} e^{st} - e^{st} + s e^{st}}{s^2}\, ds.
$$

For $t < 0$, all the terms $\to 0$ as $\operatorname{Re} s \to \infty$, and so we close the contour in the right half-plane, as in contour (h) of Figure 15.1. On Γ, s times the integrand $\to 0$, and, as the contour encloses no poles, it follows that the integral along L is zero. Thus $f(t) = 0$ for $t < 0$.

For $t > 1$, all terms $\to 0$ as Re $s \to -\infty$, and so we close the contour in the left half-plane, as in contour (g) of Figure 15.1. On Γ, s times the integrand again $\to 0$, and, as this contour also encloses no poles, it again follows that the integral along L is zero. Thus $f(t) = 0$ for $t > 1$, as well as for $t < 0$.

For $0 < t < 1$, we need to separate the Bromwich integral into two parts (guided by the different ways in which the parts behave as $|s| \to \infty$):

$$f(t) = \int_L \frac{e^{-s}e^{st}}{s^2}\, ds + \int_L \frac{(s-1)e^{st}}{s^2}\, ds \equiv I_1 + I_2.$$

For I_1 the exponent is $s(t-1)$; $t-1$ is negative and so, as in the case $t < 0$, we close the contour in the right half-plane [contour (h)]. No poles are included in this contour, and we conclude that $I_1 = 0$.

For I_2 the exponent is st, indicating that (g) is the appropriate contour. However, $(s-1)/s^2$ does have a pole at $s = 0$ and that is inside the contour. The integral around Γ contributes nothing (that is why it was chosen), and the integral along L must be equal to the residue of $(s-1)e^{st}/s^2$ at $s = 0$. Now,

$$\frac{(s-1)e^{st}}{s^2} = \left(\frac{1}{s} - \frac{1}{s^2}\right)\left(1 + st + \frac{s^2 t^2}{2!} + \cdots\right) = -\frac{1}{s^2} + \frac{1}{s}(1-t) + \cdots.$$

The residue, and hence the value of I_2, is therefore $1 - t$. Since I_1 has been shown to have value 0, $1 - t$ is also the expression for $f(t)$ for $0 < t < 1$.

15.21 Use contour (i) in Figure 15.1 to show that the function with Laplace transform $s^{-1/2}$ is $(\pi x)^{-1/2}$.

[For an integrand of the form $r^{-1/2}\exp(-rx)$, change variable to $t = r^{1/2}$.]

With the suggested contour no poles of $s^{-1/2}e^{sx}$ are enclosed and so the integral of $(2\pi i)^{-1}s^{-1/2}e^{sx}$ around the closed curve must have the value zero. It is also clear that the integral along Γ will be zero since Re $s < 0$ on Γ.

For the small circle γ enclosing the origin, set $s = \rho\, e^{i\theta}$, with $ds = i\rho\, e^{i\theta}\, d\theta$, and consider

$$\lim_{\rho \to 0} \int_0^{2\pi} \rho^{-1/2}e^{-i\theta/2}\exp(x\rho\, e^{i\theta})i\rho\, e^{i\theta}\, d\theta.$$

This $\to 0$ as $\rho \to 0$ (as $\rho^{1/2}$).

On the upper cut, γ_1, $s = re^{i\pi}$ and the contribution to the integral is

$$\frac{1}{2\pi i}\int_\infty^0 \frac{e^{-i\pi/2}}{r^{1/2}}\exp(rxe^{i\pi})e^{i\pi}\, dr,$$

whilst, on the lower cut, γ_2, $s = re^{-i\pi}$, and its contribution to the integral is

$$\frac{1}{2\pi i}\int_0^\infty \frac{e^{i\pi/2}}{r^{1/2}}\exp(rxe^{-i\pi})e^{-i\pi}\, dr.$$

Combining the two (and making both integrals run over the same range) gives

$$-\frac{1}{2\pi i}\int_0^\infty \frac{2i}{r^{1/2}}e^{-rx}\,dr = -\frac{1}{\pi}\int_0^\infty \frac{1}{t}e^{-t^2x}2t\,dt, \text{ after setting } r = t^2,$$

$$= -\frac{2}{\pi}\frac{\sqrt{\pi}}{2\sqrt{x}}.$$

Since this must add to the Bromwich integral along L to make zero, it follows that the function with Laplace transform $s^{-1/2}$ is $(\pi x)^{-1/2}$.

16 Probability

16.1 By shading or numbering Venn diagrams, determine which of the following are valid relationships between events. For those that are, prove the relationship using de Morgan's laws.

(a) $\overline{(\bar{X} \cup Y)} = X \cap \bar{Y}$.
(b) $\bar{X} \cup \bar{Y} = \overline{(X \cup Y)}$.
(c) $(X \cup Y) \cap Z = (X \cup Z) \cap Y$.
(d) $X \cup \overline{(Y \cap Z)} = (X \cup \bar{Y}) \cap \bar{Z}$.
(e) $X \cup \overline{(Y \cap Z)} = (X \cup \bar{Y}) \cup \bar{Z}$.

For each part of this question we refer to the corresponding part of Figure 16.1.

(a) This relationship is correct as both expressions define the shaded region that is both inside X and outside Y.

(b) This relationship is *not* valid. The LHS specifies the whole sample space *apart from* the region marked with the heavy shading. The RHS defines the region that is lightly shaded. The unmarked regions of X and Y are included in the former but not in the latter.

(c) This relationship is *not* valid. The LHS specifies the sum of the regions marked 2, 3 and 4 in the figure, whilst the RHS defines the sum of the regions marked 1, 3 and 4.

(d) This relationship is *not* valid. On the LHS, $\overline{Y \cap Z}$ is the whole sample space apart from regions 3 and 4. So $X \cup \overline{(Y \cap Z)}$ consists of all regions except for region 3. On the RHS, $X \cup \bar{Y}$ contains all regions except 3 and 7. The events $\bar{Z}$ contain regions 1, 6, 7 and 8 and so $(X \cup \bar{Y}) \cap \bar{Z}$ consists of regions 1, 6 and 8. Thus regions 2, 4, 5 and 7 are in one specification but not in the other.

(e) This relationship is valid. The LHS is as found in (d), namely all regions except for region 3. The RHS consists of the union (as opposed to the intersection) of the two subregions found in (d) and thus contains those regions found in either or both of $X \cup \bar{Y}$ (1, 2, 4, 5, 6 and 8) and $\bar{Z}$ (1, 6, 7 and 8). This covers all regions except region 3 – in agreement with those found for the LHS.

For the two valid relationships, their proofs using de Morgan's laws are:

(a) $$\overline{(\bar{X} \cup Y)} = \bar{\bar{X}} \cap \bar{Y} = X \cap \bar{Y},$$

(e) $$X \cup \overline{(Y \cap Z)} = X \cup (\bar{Y} \cup \bar{Z}) = (X \cup \bar{Y}) \cup \bar{Z}.$$

Here we have also used the result that the complement of the complement of a set is the set itself.

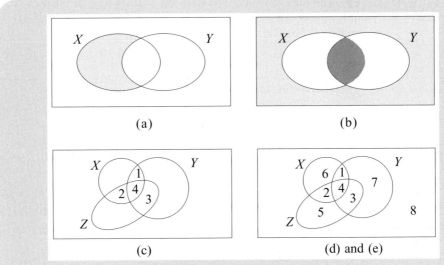

Figure 16.1 The Venn diagrams used in Problem 16.1.

16.3 *A* and *B* each have two unbiased four-faced dice, the four faces being numbered 1, 2, 3 and 4. Without looking, *B* tries to guess the sum x of the numbers on the bottom faces of *A*'s two dice after they have been thrown onto a table. If the guess is correct *B* receives x^2 euros, but if not he loses x euros.

Determine *B*'s expected gain per throw of *A*'s dice when he adopts each of the following strategies:

(a) he selects x at random in the range $2 \leq x \leq 8$;
(b) he throws his own two dice and guesses x to be whatever they indicate;
(c) he takes your advice and always chooses the same value for x. Which number would you advise?

We first calculate the probabilities $p(x)$ and the corresponding gains $g(x) = p(x)x^2 - [1 - p(x)]x$ for each value of the total x. Expressing both in units of $1/16$, they are as follows:

x	2	3	4	5	6	7	8
$p(x)$	1	2	3	4	3	2	1
$g(x)$	-26	-24	-4	40	30	0	-56

(a) If *B*'s guess is random in the range $2 \leq x \leq 8$ then his expected return is

$$\frac{1}{16}\frac{1}{7}(-26 - 24 - 4 + 40 + 30 + 0 - 56) = -\frac{40}{112} = -0.36 \text{ euros.}$$

(b) If he picks by throwing his own dice then his distribution of guesses is the same as that of $p(x)$ and his expected return is

$$\frac{1}{16}\frac{1}{16}[1(-26) + 2(-24) + 3(-4) + 4(40) + 3(30) + 2(0) + 1(-56)]$$

$$= \frac{108}{256} = 0.42 \text{ euros.}$$

(c) If B chooses y, then his expected return is

$$h(y) = p(y)y^2 - \sum_{x \neq y} p(x)x.$$

An additional line in the table (in the same units) would read $h(x)$, -74, -56, -20, 40, 46, 32, -8. You should not advise B, but take his place, guess "6" each time, and expect an average profit of $46/16 = 2.87$ euros per throw.

16.5 Two duelists, A and B, take alternate shots at each other, and the duel is over when a shot (fatal or otherwise!) hits its target. Each shot fired by A has a probability α of hitting B, and each shot fired by B has a probability β of hitting A. Calculate the probabilities P_1 and P_2, defined as follows, that A will win such a duel: P_1, A fires the first shot; P_2, B fires the first shot.

If they agree to fire simultaneously, rather than alternately, what is the probability P_3 that A will win, i.e. hit B without being hit himself?

Each shot has only two possible outcomes, a hit or a miss. P_1 is the probability that A will win when it is his turn to fire the next shot, and he is still able to do so (event W). There are three possible outcomes of the first two shots: C_1, A hits with his shot; C_2, A misses but B hits; C_3, both miss. Thus

$$P_1 = \sum_i \Pr(C_i)\Pr(W|C_i)$$

$$= [\alpha \times 1] + [(1-\alpha)\beta \times 0] + [(1-\alpha)(1-\beta) \times P_1]$$

$$\Rightarrow \quad P_1 = \frac{\alpha}{\alpha + \beta - \alpha\beta}.$$

When B fires first but misses, the situation is the one just considered. But if B hits with his first shot then clearly A's chances of winning are zero. Since these are the only two possible outcomes of B's first shot, we can write

$$P_2 = [\beta \times 0] + [(1-\beta) \times P_1] \quad \Rightarrow \quad P_2 = \frac{(1-\beta)\alpha}{\alpha + \beta - \alpha\beta}.$$

When both fire at the same time there are four possible outcomes D_i to the first round: D_1, A hits and B misses; D_2, B hits but A misses; D_3, they both hit; D_4, they both miss. If getting hit, even if you manage to hit your opponent, does not count as a win, then

$$P_3 = \sum_i \Pr(D_i)\Pr(W|D_i)$$

$$= [\alpha(1-\beta) \times 1] + [(1-\alpha)\beta \times 0] + [\alpha\beta \times 0] + [(1-\alpha)(1-\beta) \times P_3].$$

This can be rearranged as

$$P_3 = \frac{\alpha(1-\beta)}{\alpha + \beta - \alpha\beta} = P_2.$$

Thus the result is the same as if B had fired first. However, we also note that if all that matters to A is that B is hit, whether or not he is hit himself, then the third bracket takes the value $\alpha\beta \times 1$ and P_3 takes the same value as P_1.

16.7 An electronics assembly firm buys its microchips from three different suppliers; half of them are bought from firm X, whilst firms Y and Z supply 30% and 20%, respectively. The suppliers use different quality-control procedures and the percentages of defective chips are 2%, 4% and 4% for X, Y and Z, respectively. The probabilities that a defective chip will fail two or more assembly-line tests are 40%, 60% and 80%, respectively, whilst all defective chips have a 10% chance of escaping detection. An assembler finds a chip that fails only one test. What is the probability that it came from supplier X?

Since the number of tests failed by a defective chip are mutually exclusive outcomes (0, 1 or ≥ 2), a chip supplied by X has a probability of failing just one test given by $0.02(1 - 0.1 - 0.4) = 0.010$. The corresponding probabilities for chips supplied by Y and Z are $0.04(1 - 0.1 - 0.6) = 0.012$ and $0.04(1 - 0.1 - 0.8) = 0.004$, respectively.

Using "1" to denote failing a single test, Bayes' theorem gives the probability that the chip was supplied by X as

$$Pr(X|1) = \frac{Pr(1|X)\,Pr(X)}{Pr(1|X)\,Pr(X) + Pr(1|Y)\,Pr(Y) + Pr(1|Z)\,Pr(Z)}$$

$$= \frac{0.010 \times 0.5}{0.010 \times 0.5 + 0.012 \times 0.3 + 0.004 \times 0.2} = \frac{50}{94}.$$

16.9 A boy is selected at random from amongst the children belonging to families with n children. It is known that he has at least two sisters. Show that the probability that he has $k-1$ brothers is

$$\frac{(n-1)!}{(2^{n-1} - n)(k-1)!(n-k)!},$$

for $1 \leq k \leq n-2$ and zero for other values of k. Assume that boys and girls are equally likely.

The boy has $n-1$ siblings. Let A_j be the event that $j-1$ of them are brothers, i.e. his family contains j boys and $n-j$ girls. The probability of event A_j is

$$Pr(A_j) = \frac{{}^{n-1}C_{j-1}\left(\frac{1}{2}\right)^{n-1}}{\sum_{j=1}^{n} {}^{n-1}C_{j-1}\left(\frac{1}{2}\right)^{n-1}} = \frac{(n-1)!}{2^{n-1}(j-1)!(n-j)!}.$$

If B is the event that the boy has at least two sisters, then

$$Pr(B|A_j) = \begin{cases} 1 & 1 \leq j \leq n-2, \\ 0 & n-1 \leq j \leq n. \end{cases}$$

Now we apply Bayes' theorem to give the probability that he has $k - 1$ brothers:

$$\Pr(A_k|B) = \frac{1 \Pr(A_k)}{\sum_{j=1}^{n-2} 1 \Pr(A_j)},$$

for $1 \leq k \leq n - 2$. The denominator of this expression is the sum $1 = (\frac{1}{2} + \frac{1}{2})^{n-1} = \sum_{j=1}^{n} \, ^{n-1}C_{j-1} \left(\frac{1}{2}\right)^{n-1}$, but omitting the $j = n - 1$ and the $j = n$ terms, and so is equal to

$$1 - \frac{(n-1)!}{2^{n-1}(n-2)! \, 1!} - \frac{(n-1)!}{2^{n-1}(n-1)! \, 0!} = \frac{1}{2^{n-1}} \left[2^{n-1} - (n-1) - 1 \right].$$

Thus,

$$\Pr(A_k|B) = \frac{(n-1)!}{2^{n-1}(k-1)!(n-k)!} \frac{2^{n-1}}{2^{n-1} - n} = \frac{(n-1)!}{(2^{n-1} - n)(k-1)!(n-k)!},$$

as given in the question.

16.11 A set of $2N + 1$ rods consists of one of each integer length $1, 2, \ldots, 2N, 2N + 1$. Three, of lengths a, b and c, are selected, of which a is the longest. By considering the possible values of b and c, determine the number of ways in which a non-degenerate triangle (i.e. one of non-zero area) can be formed (i) if a is even, and (ii) if a is odd. Combine these results appropriately to determine the total number of non-degenerate triangles that can be formed using three of the $2N + 1$ rods, and hence show that the probability that such a triangle can be formed from a random selection (without replacement) of three rods is

$$\frac{(N-1)(4N+1)}{2(4N^2 - 1)}.$$

Rod a is the longest of the three rods. As no two are the same length, let $a > b > c$. To form a non-degenerate triangle we require that $b + c > a$, and, in consequence, $4 \leq a \leq 2N + 1$.

(i) With a even. Consider each b ($< a$) in turn and determine how many values of c allow a triangle to be made:

b	Values of c	Number of c values
$a - 1$	$2, 3, \cdots, a - 2$	$a - 3$
$a - 2$	$3, 4, \cdots, a - 3$	$a - 5$
$\cdots$	$\cdots$	$\cdots$
$\frac{1}{2}a + 1$	$\frac{1}{2}a$	1

Thus, there are $1 + 3 + 5 + \cdots + (a - 3)$ possible triangles when a is even.

(ii) A table for odd a is similar, except that the last line will read $b = \frac{1}{2}(a + 3), c = \frac{1}{2}(a - 1)$ or $\frac{1}{2}(a + 1)$, and the number of c values $= 2$. Thus there are $2 + 4 + 6 + \cdots + (a - 3)$ possible triangles when a is odd.

To find the total number $n(N)$ of possible triangles, we group together the cases $a = 2m$ and $a = 2m + 1$, where $m = 1, 2, \ldots, N$. Then,

$$n(N) = \sum_{m=2}^{N} [1 + 3 + \cdots + (2m - 3)] + [2 + 4 + \cdots + (2m + 1 - 3)]$$

$$= \sum_{m=2}^{N} \sum_{k=1}^{2m-2} k = \sum_{m=2}^{N} \tfrac{1}{2}(2m - 2)(2m - 1) = \sum_{m=2}^{N} 2m^2 - 3m + 1$$

$$= 2\left[\tfrac{1}{6}N(N + 1)(2N + 1) - 1\right] - 3\left[\tfrac{1}{2}N(N + 1) - 1\right] + N - 1$$

$$= \frac{N}{6}[2(N + 1)(2N + 1) - 9(N + 1) + 6]$$

$$= \frac{N}{6}(4N^2 - 3N - 1) = \frac{N}{6}(4N + 1)(N - 1).$$

The number of ways that three rods can be drawn at random (without replacement) is $(2N + 1)(2N)(2N - 1)/3!$ and so the probability that they can form a triangle is

$$\frac{N(4N + 1)(N - 1)}{6} \frac{3!}{(2N + 1)(2N)(2N - 1)} = \frac{(N - 1)(4N + 1)}{2(4N^2 - 1)},$$

as stated in the question.

16.13 The duration (in minutes) of a telephone call made from a public call-box is a random variable T. The probability density function of T is

$$f(t) = \begin{cases} 0 & t < 0, \\ \tfrac{1}{2} & 0 \le t < 1, \\ ke^{-2t} & t \ge 1, \end{cases}$$

where k is a constant. To pay for the call, 20 pence has to be inserted at the beginning, and a further 20 pence after each subsequent half-minute. Determine by how much the average cost of a call exceeds the cost of a call of average length charged at 40 pence per minute.

From the normalization of the PDF, we must have

$$1 = \int_0^\infty f(t)\, dt = \frac{1}{2} + \int_1^\infty ke^{-2t}\, dt = \frac{1}{2} + \frac{ke^{-2}}{2} \quad \Rightarrow \quad k = e^2.$$

The average length of a call is given by

$$\bar{t} = \int_0^1 t \frac{1}{2}\, dt + \int_1^\infty t\, e^2 e^{-2t}\, dt$$

$$= \frac{1}{2}\frac{1}{2} + \left[\frac{te^2 e^{-2t}}{-2}\right]_1^\infty + \int_1^\infty \frac{e^2 e^{-2t}}{2}\, dt = \frac{1}{4} + \frac{1}{2} + \frac{e^2}{2}\left[\frac{e^{-2t}}{-2}\right]_1^\infty = \frac{3}{4} + \frac{1}{4} = 1.$$

Let $p_n = \Pr\{\tfrac{1}{2}(n - 1) < t < \tfrac{1}{2}n\}$. The corresponding cost is $c_n = 20n$. Clearly, $p_1 = p_2 = \tfrac{1}{4}$ and, for $n > 2$,

$$p_n = e^2 \int_{(n-1)/2}^{n/2} e^{-2t}\, dt = e^2 \left[\frac{e^{-2t}}{-2}\right]_{(n-1)/2}^{n/2} = \frac{1}{2}e^2(e - 1)e^{-n}.$$

The average cost of a call is therefore

$$\bar{c} = 20 \left[\frac{1}{4} + 2\frac{1}{4} + \sum_{n=3}^{\infty} n\frac{1}{2}e^2(e-1)e^{-n} \right] = 15 + 10e^2(e-1) \sum_{n=3}^{\infty} ne^{-n}.$$

Now, the final summation might be recognized as part of an arithmetico-geometric series whose sum can be found from the standard formula

$$S = \frac{a}{1-r} + \frac{rd}{(1-r)^2},$$

with $a = 0, d = 1$ and $r = e^{-1}$, or could be evaluated directly by noting that as a geometric series,

$$\sum_{n=0}^{\infty} e^{-nx} = \frac{1}{1-e^{-x}}.$$

Differentiating this with respect to x and then setting $x = 1$ gives

$$-\sum_{n=0}^{\infty} ne^{-nx} = -\frac{e^{-x}}{(1-e^{-x})^2} \quad \Rightarrow \quad \sum_{n=0}^{\infty} ne^{-n} = \frac{e^{-1}}{(1-e^{-1})^2}.$$

From either method it follows that

$$\sum_{n=3}^{\infty} ne^{-n} = \frac{e}{(e-1)^2} - e^{-1} - 2e^{-2}$$

$$= \frac{e - e + 2 - e^{-1} - 2 + 4e^{-1} - 2e^{-2}}{(e-1)^2} = \frac{3e^{-1} - 2e^{-2}}{(e-1)^2}.$$

The total charge therefore exceeds that of a call of average length (1 minute) charged at 40 pence per minute by the amount (in pence)

$$15 + 10e^2(e-1)\frac{3e^{-1} - 2e^{-2}}{(e-1)^2} - 40 = \frac{10(3e-2) - 25e + 25}{e-1} = \frac{5e+5}{e-1} = 10.82.$$

16.15 A tennis tournament is arranged on a straight knockout basis for 2^n players, and for each round, except the final, opponents for those still in the competition are drawn at random. The quality of the field is so even that in any match it is equally likely that either player will win. Two of the players have surnames that begin with "Q". Find the probabilities that they play each other

(a) in the final,
(b) at some stage in the tournament.

Let p_r be the probability that *before* the rth round the two players are both still in the tournament (and, by implication, have not met each other). Clearly, $p_1 = 1$.

Before the rth round there are 2^{n+1-r} players left in. For both "Q" players to still be in before the $(r+1)$th round, Q_1 must avoid Q_2 in the draw and both must win their matches. Thus

$$p_{r+1} = \frac{2^{n+1-r} - 2}{2^{n+1-r} - 1}\left(\frac{1}{2}\right)^2 p_r.$$

(a) The probability that they meet in the final is p_n, given by

$$
\begin{aligned}
p_n &= 1 \frac{2^n - 2}{2^n - 1} \frac{1}{4} \frac{2^{n-1} - 2}{2^{n-1} - 1} \frac{1}{4} \cdots \frac{2^2 - 2}{2^2 - 1} \frac{1}{4} \\
&= \left(\frac{1}{4}\right)^{n-1} 2^{n-1} \left[\frac{(2^{n-1} - 1)(2^{n-2} - 1)\cdots(2^1 - 1)}{(2^n - 1)(2^{n-1} - 1)\cdots(2^2 - 1)}\right] \\
&= \left(\frac{1}{4}\right)^{n-1} 2^{n-1} \frac{1}{2^n - 1} \\
&= \frac{1}{2^{n-1}(2^n - 1)}.
\end{aligned}
$$

(b) The more general solution to the recurrence relation derived above is

$$
\begin{aligned}
p_r &= 1 \frac{2^n - 2}{2^n - 1} \frac{1}{4} \frac{2^{n-1} - 2}{2^{n-1} - 1} \frac{1}{4} \cdots \frac{2^{n+2-r} - 2}{2^{n+2-r} - 1} \frac{1}{4} \\
&= \left(\frac{1}{4}\right)^{r-1} 2^{r-1} \left[\frac{(2^{n-1} - 1)(2^{n-2} - 1)\cdots(2^{n+1-r} - 1)}{(2^n - 1)(2^{n-1} - 1)\cdots(2^{n+2-r} - 1)}\right] \\
&= \left(\frac{1}{2}\right)^{r-1} \frac{2^{n+1-r} - 1}{2^n - 1}.
\end{aligned}
$$

Before the rth round, if they are both still in the tournament, the probability that they will be drawn against each other is $(2^{n-r+1} - 1)^{-1}$. Consequently, the chance that they will meet at *some* stage is

$$
\begin{aligned}
\sum_{r=1}^{n} p_r \frac{1}{2^{n-r+1} - 1} &= \sum_{r=1}^{n} \left(\frac{1}{2}\right)^{r-1} \frac{2^{n+1-r} - 1}{2^n - 1} \frac{1}{2^{n-r+1} - 1} \\
&= \frac{1}{2^n - 1} \sum_{r=1}^{n} \left(\frac{1}{2}\right)^{r-1} \\
&= \frac{1}{2^n - 1} \frac{1 - (\frac{1}{2})^n}{1 - \frac{1}{2}} = \frac{1}{2^{n-1}}.
\end{aligned}
$$

This same conclusion can also be reached in the following way.

The probability that Q_1 *is not* put out of (i.e. wins) the tournament is $(\frac{1}{2})^n$. It follows that the probability that Q_1 *is* put out is $1 - (\frac{1}{2})^n$ and that the player responsible is Q_2 with probability $[1 - (\frac{1}{2})^n]/(2^n - 1) = 2^{-n}$. Similarly, the probability that Q_2 is put out and that the player responsible is Q_1 is also 2^{-n}. These are exclusive events but cover all cases in which Q_1 and Q_2 meet during the tournament, the probability of which is therefore $2 \times 2^{-n} = 2^{1-n}$.

16.17 This problem is about interrelated binomial trials.

(a) In two sets of binomial trials T and t, the probabilities that a trial has a successful outcome are P and p, respectively, with corresponding probabilities of failure of $Q = 1 - P$ and $q = 1 - p$. One "game" consists of a trial T, followed, if T is successful, by a trial t and then a further trial T. The two trials continue to alternate until one of the T-trials fails, at which point the game ends. The score S for the game is the total number of successes in the t-trials. Find the PGF for S and use it to show that

$$E[S] = \frac{Pp}{Q}, \qquad V[S] = \frac{Pp(1 - Pq)}{Q^2}.$$

(b) Two normal unbiased six-faced dice A and B are rolled alternately starting with A; if A shows a 6 the experiment ends. If B shows an odd number no points are scored, one point is scored for a 2 or a 4, and two points are awarded for a 6. Find the average and standard deviation of the score for the experiment and show that the latter is the greater.

(a) This is a situation in which the score for the game is a variable-length sum, the length N being determined by the outcome of the T-trials. The probability that $N = n$ is given by $h_n = P^n Q$, since n T-trials must succeed and then be followed by a failing T-trial. Thus the PGF for the length of each "game" is given by

$$\chi_N(t) \equiv \sum_{n=0}^{\infty} h_n t^n = \sum_{n=0}^{\infty} P^n Q t^n = \frac{Q}{1 - Pt}.$$

For each permitted Bernoulli t-trial, $X_i = 1$ with probability p and $X_i = 0$ with probability q; its PGF is thus $\Phi_X(t) = q + pt$. The score for the game is $S = \sum_{i=1}^{N} X_i$ and its PGF is given by the compound function

$$\Xi_S(t) = \chi_N(\Phi_X(t))$$
$$= \frac{Q}{1 - P(q + pt)},$$

in which the PGF for a single t-trial forms the argument of the PGF for the length of each "game".

It follows that the mean of S is found from

$$\Xi'_S(t) = \frac{QPp}{(1 - Pq - Ppt)^2} \quad \Rightarrow \quad E[S] = \Xi'_S(1) = \frac{QPp}{(1 - P)^2} = \frac{Pp}{Q}.$$

To calculate the variance of S we need to find $\Xi''_S(1)$. This second derivative is

$$\Xi''_S(t) = \frac{2QP^2 p^2}{(1 - Pq - Ppt)^3} \quad \Rightarrow \quad \Xi''_S(1) = \frac{2P^2 p^2}{Q^2}.$$

The variance is therefore

$$V[S] = \Xi''_S(1) + \Xi'_S(1) - [\Xi'_S(1)]^2$$
$$= \frac{2P^2 p^2}{Q^2} + \frac{Pp}{Q} - \frac{P^2 p^2}{Q^2}$$
$$= \frac{Pp(Pp + Q)}{Q^2} = \frac{Pp(P - Pq + Q)}{Q^2} = \frac{Pp(1 - Pq)}{Q^2}.$$

(b) For die A: $P = \frac{5}{6}$ and $Q = \frac{1}{6}$ giving $\chi_N(t) = 1/(6 - 5t)$.

For die B: $\Pr(X = 0) = \frac{3}{6}$, $\Pr(X = 1) = \frac{2}{6}$ and $\Pr(X = 2) = \frac{1}{6}$ giving $\Phi_X(t) = (3 + 2t + t^2)/6$.

The PGF for the game score S is thus

$$\Xi_S(t) = \frac{1}{6 - \frac{5}{6}(3 + 2t + t^2)} = \frac{6}{21 - 10t - 5t^2}.$$

We need to evaluate the first two derivatives of $\Xi_S(t)$ at $t = 1$, as follows:

$$\Xi_S'(t) = \frac{-6(-10 - 10t)}{(21 - 10t - 5t^2)^2} = \frac{60 + 60t}{(21 - 10t - 5t^2)^2}$$

$$\Rightarrow \quad E[S] = \Xi_S'(1) = \frac{120}{6^2} = \frac{10}{3} = 3.33,$$

$$\Xi_S''(t) = \frac{60}{(21 - 10t - 5t^2)^2} - \frac{2(60 + 60t)(-10 - 10t)}{(21 - 10t - 5t^2)^3}$$

$$\Rightarrow \quad \Xi_S''(1) = \frac{60}{36} - \frac{2(120)(-20)}{(6)^3} = \frac{215}{9}.$$

Substituting the calculated values gives $V[S]$ as

$$V[S] = \frac{215}{9} + \frac{10}{3} - \left(\frac{10}{3}\right)^2 = \frac{145}{9},$$

from which it follows that

$$\sigma_S = \sqrt{V[S]} = 4.01, \text{ i.e. greater than the mean.}$$

16.19 A point P is chosen at random on the circle $x^2 + y^2 = 1$. The random variable X denotes the distance of P from $(1, 0)$. Find the mean and variance of X and the probability that X is greater than its mean.

With O as the center of the unit circle and Q as the point $(1, 0)$, let OP make an angle θ with the x-axis OQ. The random variable X then has the value $2\sin(\theta/2)$ with θ uniformly distributed on $(0, 2\pi)$, i.e.

$$f(x)\,dx = \frac{1}{2\pi}\,d\theta.$$

The mean of X is given straightforwardly by

$$\langle X \rangle = \int_0^2 X f(x)\,dx = \int_0^{2\pi} 2\sin\left(\frac{\theta}{2}\right)\frac{1}{2\pi}\,d\theta = \frac{1}{\pi}\left[-2\cos\frac{\theta}{2}\right]_0^{2\pi} = \frac{4}{\pi}.$$

For the variance we have

$$\sigma_X^2 = \langle X^2 \rangle - \langle X \rangle^2 = \int_0^{2\pi} 4\sin^2\left(\frac{\theta}{2}\right)\frac{1}{2\pi}\,d\theta - \frac{16}{\pi^2} = \frac{4}{2\pi}\frac{1}{2}2\pi - \frac{16}{\pi^2} = 2 - \frac{16}{\pi^2}.$$

When $X = \langle X \rangle = 4/\pi$, the angle $\theta = 2 \sin^{-1}(2/\pi)$ and so

$$\Pr(X > \langle X \rangle) = \frac{2\pi - 4 \sin^{-1} \dfrac{2}{\pi}}{2\pi} = 0.561.$$

16.21 The number of errors needing correction on each page of a set of proofs follows a Poisson distribution of mean μ. The cost of the first correction on any page is α and that of each subsequent correction on the same page is β. Prove that the average cost of correcting a page is

$$\alpha + \beta(\mu - 1) - (\alpha - \beta)e^{-\mu}.$$

Since the number of errors on a page is Poisson distributed, the probability of n errors on any particular page is

$$\Pr(n \text{ errors}) = p_n = e^{-\mu} \frac{\mu^n}{n!}.$$

The average cost per page, found by averaging the corresponding cost over all values of n, is

$$c = 0\, p_0 + \alpha p_1 + \sum_{n=2}^{\infty} [\alpha + (n-1)\beta] p_n$$

$$= \alpha \mu e^{-\mu} + (\alpha - \beta) \sum_{n=2}^{\infty} p_n + \beta \sum_{n=2}^{\infty} n p_n.$$

Now, $\sum_{n=0}^{\infty} p_n = 1$ and, for a Poisson distribution, $\sum_{n=0}^{\infty} n p_n = \mu$. These can be used to evaluate the above, once the $n = 0$ and $n = 1$ terms have been removed. Thus

$$c = \alpha \mu e^{-\mu} + (\alpha - \beta)(1 - e^{-\mu} - \mu e^{-\mu}) + \beta(\mu - 0 - \mu e^{-\mu})$$

$$= \alpha + \beta(\mu - 1) + e^{-\mu}(\alpha \mu - \alpha + \beta - \mu \alpha + \mu \beta - \mu \beta)$$

$$= \alpha + \beta(\mu - 1) + e^{-\mu}(\beta - \alpha),$$

as given in the question.

16.23 The probability distribution for the number of eggs in a clutch is $\text{Po}(\lambda)$, and the probability that each egg will hatch is p (independently of the size of the clutch). Show by direct calculation that the probability distribution for the number of chicks that hatch is $\text{Po}(\lambda p)$.

Clearly, to determine the probability that a clutch produces k chicks, we must consider clutches of size n, for all $n \geq k$, and for each such clutch find the probability that exactly k of the n chicks do hatch. We then average over all n, weighting the results according to the distribution of n.

The probability that k chicks hatch from a clutch of size n is $^nC_k p^k q^{n-k}$, where $q = 1 - p$. The probability that the clutch is of size n is $e^{-\lambda}\lambda^n/n!$. Consequently, the overall probability of k chicks hatching from a clutch is

$$
\begin{aligned}
\Pr(k \text{ chicks}) &= \sum_{n=k}^{\infty} e^{-\lambda} \frac{\lambda^n}{n!} \, {}^nC_k \, p^k \, q^{n-k} \\
&= e^{-\lambda} p^k \lambda^k \sum_{n=k}^{\infty} \frac{(\lambda q)^{n-k}}{n!} \frac{n!}{k! \, (n-k)!}, \quad \text{set } n - k = m, \\
&= e^{-\lambda} \frac{(\lambda p)^k}{k!} \sum_{m=0}^{\infty} \frac{(\lambda q)^m}{m!} \\
&= e^{-\lambda} \frac{(\lambda p)^k}{k!} e^{\lambda q} \\
&= \frac{e^{-\lambda p} (\lambda p)^k}{k!},
\end{aligned}
$$

since $q = 1 - p$. Thus $\Pr(k \text{ chicks})$ is distributed as a Poisson distribution with parameter $\mu = \lambda p$.

16.25 Under EU legislation on harmonization, all kippers are to weigh 0.2000 kg and vendors who sell underweight kippers must be fined by their government. The weight of a kipper is normally distributed with a mean of 0.2000 kg and a standard deviation of 0.0100 kg. They are packed in cartons of 100 and large quantities of them are sold.

Every day a carton is to be selected at random from each vendor and tested according to one of the following schemes, which have been approved for the purpose.

(a) The entire carton is weighed and the vendor is fined 2500 euros if the average weight of a kipper is less than 0.1975 kg.
(b) Twenty five kippers are selected at random from the carton; the vendor is fined 100 euros if the average weight of a kipper is less than 0.1980 kg.
(c) Kippers are removed one at a time, at random, until one has been found that weighs *more* than 0.2000 kg; the vendor is fined $4n(n-1)$ euros, where n is the number of kippers removed.

Which scheme should the Chancellor of the Exchequer be urging his government to adopt?

For these calculations we measure weights in grammes.

(a) For this scheme we have a normal distribution with mean $\mu = 200$ and s.d. $\sigma = 10$. The s.d. for a carton is $\sqrt{100}\,\sigma = 100$ and the mean weight is 20 000. There is a penalty if the weight of a carton is less than 19 750. This critical value represents a standard variable of

$$
Z = \frac{19\,750 - 20\,000}{100} = -2.5.
$$

The probability that $Z < -2.5 = 1 - \Phi(2.5) = 1 - 0.9938 = 0.0062$. Thus the average fine per carton tested on this scheme is $0.0062 \times 2500 = 15.5$ euros.

(b) For this scheme the general parameters are the same but the mean weight of the sample measured is 5000 and its s.d. is $\sqrt{25}\,(10) = 50$. The Z-value at which a fine is imposed is

$$Z = \frac{(198 \times 25) - 5000}{50} = -1.$$

The probability that $Z < -1.0 = 1 - \Phi(1.0) = 1 - 0.8413 = 0.1587$. Thus the average fine per carton tested on this scheme is $0.1587 \times 100 = 15.9$ euros.

(c) This scheme is a series of Bernoulli trials in which the probability of success is $\frac{1}{2}$ (since half of all kippers weigh more than 200 and the distribution is normal). The probability that it will take n kippers to find one that passes the test is $q^{n-1}p = (\frac{1}{2})^n$. The expected fine is therefore

$$f = \sum_{n=2}^{\infty} 4n(n-1)\left(\frac{1}{2}\right)^n = 4\frac{2(\frac{1}{4})}{(\frac{1}{2})^3} = 16 \text{ euros}.$$

The expression for the sum was found by twice differentiating the sum of the geometric series $\sum r^n$ with respect to r, as follows:

$$\sum_{n=0}^{\infty} r^n = \frac{1}{1-r} \quad \Rightarrow \quad \sum_{n=1}^{\infty} nr^{n-1} = \frac{1}{(1-r)^2}$$

$$\Rightarrow \quad \sum_{n=2}^{\infty} n(n-1)r^{n-2} = \frac{2}{(1-r)^3}$$

$$\Rightarrow \quad \sum_{n=2}^{\infty} n(n-1)r^n = \frac{2r^2}{(1-r)^3}.$$

There is, in fact, little to choose between the schemes on monetary grounds; no doubt political considerations, such as the current unemployment rate, will decide!

16.27 A practical-class demonstrator sends his 12 students to the storeroom to collect apparatus for an experiment, but forgets to tell each which type of component to bring. There are three types, A, B and C, held in the stores (in large numbers) in the proportions 20%, 30% and 50%, respectively, and each student picks a component at random. In order to set up one experiment, one unit each of A and B and two units of C are needed. Let $\Pr(N)$ be the probability that at least N experiments can be set up.

(a) Evaluate $\Pr(3)$.
(b) Find an expression for $\Pr(N)$ in terms of k_1 and k_2, the numbers of components of types A and B, respectively, selected by the students. Show that $\Pr(2)$ can be written in the form

$$\Pr(2) = (0.5)^{12} \sum_{i=2}^{6} {}^{12}C_i \, (0.4)^i \sum_{j=2}^{8-i} {}^{12-i}C_j \, (0.6)^j.$$

(c) By considering the conditions under which no experiments can be set up, show that $\Pr(1) = 0.9145$.

(a) To make three experiments possible the 12 components picked must be three each of A and B and six of C. The probability of this is given by the multinomial distribution as

$$\Pr(3) = \frac{(12)!}{3!\,3!\,6!}\,(0.2)^3(0.3)^3(0.5)^6 = 0.06237.$$

(b) Let the numbers of A, B and C selected be k_1, k_2 and k_3, respectively, and consider when *at least* N experiments can be set up. We have the obvious inequalities $k_1 \geq N$, $k_2 \geq N$ and $k_3 \geq 2N$. In addition $k_3 = 12 - k_1 - k_2$, implying that $k_2 \leq 12 - 2N - k_1$. Further, k_1 cannot be greater than $12 - 3N$ if at least N experiments are to be set up, as each requires three other components that are not of type A. These inequalities set the limits on the acceptable values of k_1 and k_2 (k_3 is not a third independent variable). Thus $\Pr(N)$ is given by

$$\sum_{k_1 \geq N}^{12-3N} \sum_{k_2 \geq N}^{12-2N-k_1} \frac{(12)!}{k_1!\,k_2!\,(12 - k_1 - k_2)!}\,(0.2)^{k_1}\,(0.3)^{k_2}\,(0.5)^{12-k_1-k_2}.$$

The answer to part (a) is a particular case of this with $N = 3$, when each summation reduces to a single term.

For $N = 2$ the expression becomes

$$\Pr(2) = \sum_{k_1 \geq 2}^{6} \sum_{k_2 \geq 2}^{8-k_1} \frac{(12)!}{k_1!\,k_2!\,(12 - k_1 - k_2)!}\,(0.2)^{k_1}\,(0.3)^{k_2}\,(0.5)^{12-k_1-k_2}$$

$$= (0.5)^{12} \sum_{i=2}^{6} \sum_{j=2}^{8-i} \frac{(12)!\,(0.2/0.5)^i}{i!\,(12 - i)!}\,\frac{(12 - i)!\,(0.3/0.5)^j}{j!\,(12 - i - j)!}$$

$$= (0.5)^{12} \sum_{i=2}^{6} {}^{12}C_i\,(0.4)^i \sum_{j=2}^{8-i} {}^{12-i}C_j\,(0.6)^j.$$

(c) No experiment can be set up if any one of the following four events occurs: $A_1 = (k_1 = 0)$, $A_2 = (k_2 = 0)$, $A_3 = (k_3 = 0)$ and $A_4 = (k_3 = 1)$. The probability for the union of these four events is given by

$$\Pr(A_1 \cup A_2 \cup A_3 \cup A_4) = \sum_{i=1}^{4} \Pr(A_i) - \sum_{i,j} \Pr(A_i \cap A_j) + \cdots .$$

The probabilities $\Pr(A_i)$ are straightforward to calculate as follows:

$$\Pr(A_1) = (1 - 0.2)^{12}, \qquad \Pr(A_2) = (1 - 0.3)^{12},$$
$$\Pr(A_3) = (1 - 0.5)^{12}, \qquad \Pr(A_4) = {}^{12}C_1(1 - 0.5)^{12}(0.5).$$

The calculation of the probability for the intersection of two events is typified by

$$\Pr(A_1 \cap A_2) = [1 - (0.2 + 0.3)]^{12}$$
$$\text{and } \Pr(A_1 \cap A_4) = {}^{12}C_1[1 - (0.2 + 0.5)]^{11}(0.5)^1.$$

A few trial evaluations show that these are of order 10^{-4} and can be ignored by comparison with the larger terms in the first sum, which are (after rounding)

$$\sum_{i=1}^{4} \Pr(A_i) = (0.8)^{12} + (0.7)^{12} + (0.5)^{12} + 12(0.5)^{11}(0.5)$$

$$= 0.0687 + 0.0138 + 0.0002 + 0.0029 = 0.0856.$$

Since the probability of no experiments being possible is 0.0856, it follows that $\Pr(1) = 0.9144$.

16.29 The continuous random variables X and Y have a joint PDF proportional to $xy(x - y)^2$ with $0 \le x \le 1$ and $0 \le y \le 1$. Find the marginal distributions for X and Y and show that they are negatively correlated with correlation coefficient $-\frac{2}{3}$.

This PDF is clearly symmetric between x and y. We start by finding its normalization constant c:

$$\int_0^1 \int_0^1 c(x^3 y - 2x^2 y^2 + xy^3) \, dx \, dy = c \left(\frac{1}{4} \frac{1}{2} - 2 \frac{1}{3} \frac{1}{3} + \frac{1}{2} \frac{1}{4} \right) = \frac{c}{36}.$$

Thus, we must have that $c = 36$.

The marginal distribution for x is given by

$$f(x) = 36 \int_0^1 (x^3 y - 2x^2 y^2 + xy^3) \, dy$$
$$= 36 \left(\tfrac{1}{2} x^3 - \tfrac{2}{3} x^2 + \tfrac{1}{4} x \right)$$
$$= 18x^3 - 24x^2 + 9x,$$

and the mean of x by

$$\mu_X = \bar{x} = \int_0^1 (18x^4 - 24x^3 + 9x^2) \, dx = \frac{18}{5} - \frac{24}{4} + \frac{9}{3} = \frac{3}{5}.$$

By symmetry, the marginal distribution and the mean for y are $18y^3 - 24y^2 + 9y$ and $\frac{3}{5}$, respectively.

To calculate the correlation coefficient we also need the variances of x and y and their covariance. The variances, obviously equal, are given by

$$\sigma_X^2 = \int_0^1 x^2(18x^3 - 24x^2 + 9x) \, dx - \left(\tfrac{3}{5} \right)^2$$
$$= \frac{18}{6} - \frac{24}{5} + \frac{9}{4} - \frac{9}{25}$$
$$= \frac{900 - 1440 + 675 - 108}{300} = \frac{9}{100}.$$

The standard deviations σ_X and σ_Y are therefore both equal to $3/10$.

The covariance is calculated next; it is given by

$$\text{Cov}[X, Y] = \langle XY \rangle - \mu_X \mu_Y$$

$$= 36 \int_0^1 \int_0^1 (x^4 y^2 - 2x^3 y^3 + x^2 y^4)\, dx\, dy - \frac{3}{5} \frac{3}{5}$$

$$= \frac{36}{5 \times 3} - \frac{72}{4 \times 4} + \frac{36}{3 \times 5} - \frac{9}{25}$$

$$= \frac{12}{5} - \frac{9}{2} + \frac{12}{5} - \frac{9}{25}$$

$$= \frac{120 - 225 + 120 - 18}{50} = -\frac{3}{50}.$$

Finally,

$$\text{Corr}[X, Y] = \frac{\text{Cov}[X, Y]}{\sigma_X \sigma_Y} = \frac{-\frac{3}{50}}{\frac{3}{10} \frac{3}{10}} = -\frac{2}{3}.$$

16.31 Two continuous random variables X and Y have a joint probability distribution

$$f(x, y) = A(x^2 + y^2),$$

where A is a constant and $0 \le x \le a, 0 \le y \le a$. Show that X and Y are negatively correlated with correlation coefficient $-15/73$. By sketching a rough contour map of $f(x, y)$ and marking off the regions of positive and negative correlation, convince yourself that this (perhaps counter-intuitive) result is plausible.

The calculations of the various parameters of the distribution are straightforward (see Problem 16.29). The parameter A is determined by the normalization condition:

$$1 = \int_0^a \int_0^a A(x^2 + y^2)\, dx\, dy = A\left(\frac{a^4}{3} + \frac{a^4}{3}\right) \quad \Rightarrow \quad A = \frac{3}{2a^4}.$$

The two expectation values required are given by

$$E[X] = \int_0^a \int_0^a Ax(x^2 + y^2)\, dx\, dy$$

$$= \frac{3}{2a^4}\left(\frac{a^5}{4 \times 1} + \frac{a^5}{2 \times 3}\right) = \frac{5a}{8}, \qquad (E[Y] = E[X]),$$

$$E[X^2] = \int_0^a \int_0^a Ax^2(x^2 + y^2)\, dx\, dy$$

$$= \frac{3}{2a^4}\left(\frac{a^6}{5 \times 1} + \frac{a^6}{3 \times 3}\right) = \frac{7a^2}{15}.$$

Hence the variance, calculated from the general result $V[X] = E[X^2] - (E[X])^2$, is

$$V[X] = \frac{7a^2}{15} - \left(\frac{5a}{8}\right)^2 = \frac{73}{960}a^2,$$

and the standard deviations are given by

$$\sigma_X = \sigma_Y = \sqrt{\frac{73}{960}}\, a.$$

To obtain the correlation coefficient we need also to calculate the following:

$$E[XY] = \int_0^a \int_0^a Axy(x^2 + y^2)\,dx\,dy$$

$$= \frac{3}{2a^4}\left(\frac{a^6}{4 \times 2} + \frac{a^6}{2 \times 4}\right) = \frac{3a^2}{8}.$$

Then the covariance, given by Cov $[X, Y] = E[XY] - E[X]E[Y]$, is evaluated as

$$\mathrm{Cov}\,[X, Y] = \frac{3}{8}a^2 - \frac{5a}{8}\frac{5a}{8} = -\frac{a^2}{64}.$$

Combining this last result with the standard deviations calculated above, we then obtain

$$\mathrm{Corr}\,[X, Y] = \frac{-(a^2/64)}{\sqrt{\frac{73}{960}}\, a \sqrt{\frac{73}{960}}\, a} = -\frac{15}{73}.$$

As the means of both X and Y are $\frac{5}{8}a = 0.62a$, the areas of the square of side a for which $X - \mu_X$ and $Y - \mu_Y$ have the same sign (i.e. regions of positive correlation) are about $(0.62)^2 \approx 39\%$ and $(0.38)^2 \approx 14\%$ of the total area of the square. The regions of negative correlation occupy some 47% of the square.

However, $f(x, y) = A(x^2 + y^2)$ favors the regions where one or both of x and y are large and close to unity. Broadly speaking, this gives little weight to the region in which both X and Y are less than their means, and so, although it is the largest region in area, it contributes relatively little to the overall correlation. The two (equal area) regions of negative correlation together outweigh the smaller high probability region of positive correlation in the top right-hand corner of the square; the overall result is a net negative correlation coefficient.

17 Statistics

17.1 A group of students uses a pendulum experiment to measure g, the acceleration of free fall, and obtains the following values (in $m\,s^{-2}$): 9.80, 9.84, 9.72, 9.74, 9.87, 9.77, 9.28, 9.86, 9.81, 9.79, 9.82. What would you give as the best value and standard error for g as measured by the group?

We first note that the reading of $9.28\,m\,s^{-2}$ is so far from the others that it is almost certainly in error and should not be used in the calculation. The mean of the ten remaining values is 9.802 and the standard deviation of the sample about its mean is 0.04643. After including Bessel's correction factor, the estimate of the population s.d. is $\sigma = 0.0489$, leading to a s.d. in the measured value of the mean of $0.0489/\sqrt{10} = 0.0155$. We therefore give the best value and standard error for g as $9.80 \pm 0.02\,m\,s^{-2}$.

17.3 The following are the values obtained by a class of 14 students when measuring a physical quantity x: 53.8, 53.1, 56.9, 54.7, 58.2, 54.1, 56.4, 54.8, 57.3, 51.0, 55.1, 55.0, 54.2, 56.6.

(a) Display these results as a histogram and state what you would give as the best value for x.
(b) Without calculation, estimate how much reliance could be placed upon your answer to (a).
(c) Data books give the value of x as 53.6 with negligible error. Are the data obtained by the students in conflict with this?

(a) The histogram in Figure 17.1 shows no reading that is an obvious mistake and there is no reason to suppose other than a Gaussian distribution. The best value for x is the arithmetic mean of the 14 values given, i.e. 55.1.

(b) We note that 11 values, i.e. approximately two-thirds of the 14 readings, lie within ± 2 bins of the mean. This estimates the s.d. for the population as 2.0 and gives a standard error in the mean of $\approx 2.0/\sqrt{14} \approx 0.6$.

(c) Within the accuracy we are likely to achieve by estimating σ for the sample by eye, the value of Student's t is $(55.1 - 53.6)/0.6$, i.e. about 2.5. With 14 readings there are 13 degrees of freedom. From standard tables for the Student's t-test, $C_{13}(2.5) \approx 0.985$. It is therefore likely at the $2 \times 0.015 = 3\%$ significance level that the data are in conflict with the accepted value.

[Numerical analysis of the data, rather than a visual estimate, gives the lower value 0.51 for the standard error in the mean and implies that there is a conflict between the data and the accepted value at the 1.0% significance level.]

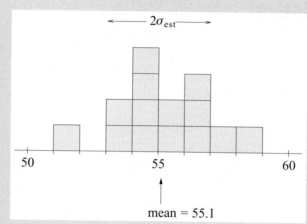

Figure 17.1 Histogram of the data in Problem 17.3.

17.5 A population contains individuals of k types in equal proportions. A quantity X has mean μ_i amongst individuals of type i and variance σ^2, which has the same value for all types. In order to estimate the mean of X over the whole population, two schemes are considered; each involves a total sample size of nk. In the first the sample is drawn randomly from the whole population, whilst in the second (*stratified sampling*) n individuals are randomly selected from each of the k types.

Show that in both cases the estimate has expectation

$$\mu = \frac{1}{k}\sum_{i=1}^{k}\mu_i,$$

but that the variance of the first scheme exceeds that of the second by an amount

$$\frac{1}{k^2 n}\sum_{i=1}^{k}(\mu_i - \mu)^2.$$

(i) For the first scheme the estimator $\hat{\mu}$ has expectation

$$\langle \hat{\mu} \rangle = \frac{1}{nk}\sum_{j=1}^{nk}\langle x_j \rangle,$$

where

$$\langle x_j \rangle = \frac{1}{k}\sum_{i=1}^{k}\mu_i \text{ for all } j,$$

since the k types are in equal proportions in the population. Thus,

$$\langle\hat{\mu}\rangle = \frac{1}{nk}\sum_{j=1}^{nk}\frac{1}{k}\sum_{i=1}^{k}\mu_i = \frac{1}{k}\sum_{i=1}^{k}\mu_i = \mu.$$

The variance of $\hat{\mu}$ is given by

$$V[\hat{\mu}] = \frac{1}{n^2k^2}\,nk\,V[x]$$

$$= \frac{1}{nk}\,(\langle x^2\rangle - \mu^2)$$

$$= \frac{1}{nk}\left(\frac{1}{k}\sum_{i=1}^{k}\langle x_i^2\rangle - \mu^2\right),$$

again since the k types are in equal proportions in the population.

Now we use the relationship $\sigma^2 = \langle x_i^2\rangle - \mu_i^2$ to replace $\langle x_i^2\rangle$ for each type, noting that σ^2 has the same value in each case. The expression for the variance becomes

$$V[\hat{\mu}] = \frac{1}{nk}\left[\frac{1}{k}\sum_{i=1}^{k}(\mu_i^2 + \sigma^2) - \mu^2\right]$$

$$= \frac{\sigma^2 - \mu^2}{nk} + \frac{1}{nk^2}\sum_{i=1}^{k}(\mu_i - \mu + \mu)^2$$

$$= \frac{\sigma^2 - \mu^2}{nk} + \frac{1}{nk^2}\sum_{i=1}^{k}\left[(\mu_i - \mu)^2 + 2\mu(\mu_i - \mu) + \mu^2\right]$$

$$= \frac{\sigma^2 - \mu^2}{nk} + \frac{1}{nk^2}\sum_{i=1}^{k}(\mu_i - \mu)^2 + 0 + \frac{k\mu^2}{nk^2}$$

$$= \frac{\sigma^2}{nk} + \frac{1}{nk^2}\sum_{i=1}^{k}(\mu_i - \mu)^2.$$

(ii) For the second scheme the calculations are more straightforward. The expectation value of the estimator $\hat{\mu} = (nk)^{-1}\sum_{i=1}^{k}\langle x_i\rangle$ is

$$\langle\hat{\mu}\rangle = \frac{1}{nk}\sum_{i=1}^{k}n\mu_i = \frac{1}{k}\sum_{i=1}^{k}\mu_i = \mu,$$

whilst the variance is given by

$$V[\hat{\mu}] = \frac{1}{n^2k^2}\sum_{i=1}^{k}V[\langle x_i\rangle] = \frac{1}{n^2k^2}\sum_{i=1}^{k}n\sigma_i^2 = \frac{1}{k^2}\frac{k\sigma^2}{n} = \frac{\sigma^2}{kn},$$

since $\sigma_i^2 = \sigma^2$ for all i.

Comparing the results from (i) and (ii), we see that the variance of the estimator in the first scheme is larger by

$$\frac{1}{nk^2}\sum_{i=1}^{k}(\mu_i - \mu)^2.$$

17.7 According to a particular theory, two dimensionless quantities X and Y have equal values. Nine measurements of X gave values of 22, 11, 19, 19, 14, 27, 8, 24 and 18, whilst seven measured values of Y were 11, 14, 17, 14, 19, 16 and 14. Assuming that the measurements of both quantities are Gaussian distributed with a common variance, are they consistent with the theory? An alternative theory predicts that $Y^2 = \pi^2 X$; are the data consistent with this proposal?

On the hypothesis that $X = Y$ and both quantities have Gaussian distributions with a common variance, we need to calculate the value of t given by

$$t = \frac{\bar{w} - \omega}{\hat{\sigma}}\left(\frac{N_1 N_2}{N_1 + N_2}\right)^{1/2},$$

where $\bar{w} = \bar{x}_1 - \bar{x}_2$, $\omega = \mu_1 - \mu_2 = 0$ and

$$\hat{\sigma} = \left[\frac{N_1 s_1^2 + N_2 s_2^2}{N_1 + N_2 - 2}\right]^{1/2}.$$

The nine measurements of X have a mean of 18.0 and a value for s^2 of 33.33. The corresponding values for the seven measurements of Y are 15.0 and 5.71. Substituting these values gives

$$\hat{\sigma} = \left[\frac{9 \times 33.33 + 7 \times 5.71}{9 + 7 - 2}\right]^{1/2} = 4.93,$$

$$t = \frac{18.0 - 15.0 - 0}{4.93}\left(\frac{9 \times 7}{9 + 7}\right)^{1/2} = 1.21.$$

This variable follows a Student's t-distribution for $9 + 7 - 2 = 14$ degrees of freedom. Interpolation in standard tables gives $C_{14}(1.21) \approx 0.874$, showing that a larger value of t could be expected in about $2 \times (1 - 0.874) = 25\%$ of cases. Thus no inconsistency between the data and the first theory has been established.

For the second theory we are testing Y^2 against $\pi^2 X$; the former will not be Gaussian distributed and the two distributions will not have a common variance. Thus the best we can do is to compare the difference between the two expressions, evaluated with the mean values of X and Y, against the estimated error in that difference.

The difference in the expressions is $(15.0)^2 - 18.0\pi^2 = 47.3$. The error in the difference between the functions of Y and X is given approximately by

$$V(Y^2 - \pi^2 X) = (2Y)^2\, V[Y] + (\pi^2)^2\, V[X]$$

$$= (30.0)^2 \frac{5.71}{7 - 1} + (\pi^2)^2 \frac{33.33}{9 - 1}$$

$$= 1262 \quad \Rightarrow \quad \sigma \approx 35.5.$$

The difference is thus about $47.3/35.5 = 1.33$ standard deviations away from the theoretical value of 0. The distribution will not be truly Gaussian but, if it were, this figure would have a probability of being exceeded in magnitude some $2 \times (1 - 0.908) = 18\%$ of the time. Again no inconsistency between the data and theory has been established.

17.9 During an investigation into possible links between mathematics and classical music, pupils at a school were asked whether they had preferences (a) between mathematics and English, and (b) between classical and pop music. The results are given below.

	Classical	None	Pop
Mathematics	23	13	14
None	17	17	36
English	30	10	40

Determine whether there is any evidence for

(a) a link between academic and musical tastes, and
(b) a claim that pupils either had preferences in both areas or had no preference.

You will need to consider the appropriate value for the number of degrees of freedom to use when applying the χ^2 test.

We first note that there were 200 pupils taking part in the survey. Denoting no academic preference between mathematics and English by NA and no musical preference by NM, we draw up an enhanced table of the actual numbers m_{XY} of preferences for the various combinations that also shows the overall probabilities p_X and p_Y of the three choices in each selection.

	C	NM	P	Total	p_X
M	23	13	14	50	0.25
NA	17	17	36	70	0.35
E	30	10	40	80	0.40
Total	70	40	90	200	
p_Y	0.35	0.20	0.45		

(a) If we now assume the (null) hypothesis that there are no correlations in the data and that any apparent correlations are the result of statistical fluctuations, then the expected number of pupils opting for the combination X and Y is $n_{XY} = 200 \times p_X \times p_Y$. A table of n_{XY} is as follows:

	C	NM	P	Total
M	17.5	10	22.5	50
NA	24.5	14	31.5	70
E	28	16	36	80
Total	70	40	90	200

Taking the standard deviation as the square root of the expected number of votes for each particular combination, the value of χ^2 is given by

$$\chi^2 = \sum_{\text{all XY combinations}} \left(\frac{n_i - m_i}{\sqrt{n_i}} \right)^2 = 12.3.$$

For an $n \times n$ correlation table (here $n = 3$), the $(n-1) \times (n-1)$ block of entries in the upper left can be filled in arbitrarily. But, as the totals for each row and column are predetermined, the remaining $2n - 1$ entries are not arbitrary. Thus the number of degrees of freedom (d.o.f.) for such a table is $(n-1)^2$, here 4 d.o.f. From tables, a χ^2 of 12.3 for 4 d.o.f. makes the assumed hypothesis less than 2% likely, and so it is almost certain that a correlation between academic and musical tastes does exist.

(b) To investigate a claim that pupils either had preferences in both areas or had no preference, we must combine expressed preferences for classical or pop into one set labeled PM meaning "expressed a musical preference"; similarly for academic subjects. The correlation table is now a 2×2 one and will have only one degree of freedom. The actual data table is

	PM	NM	Total	p_X
PA	107	23	130	0.65
NA	53	17	70	0.35
Total	160	40	200	
p_Y	0.80	0.20		

and the expected ($n_{XY} = 200 p_X p_Y$) one is

	PM	NM	Total
PA	104	26	130
NA	56	14	70
Total	160	40	200

The value of χ^2 is

$$\chi^2 = \frac{(-3)^2}{104} + \frac{(3)^2}{26} + \frac{(3)^2}{56} + \frac{(-3)^2}{14} = 1.24.$$

This is close to the expected value (1) of χ^2 for 1 d.o.f. and is neither too big nor too small. Thus there is no evidence for the claim (or for any tampering with the data!).

17.11 A particle detector consisting of a shielded scintillator is being tested by placing it near a particle source whose intensity can be controlled by the use of absorbers. It might register counts even in the absence of particles from the source because of the cosmic ray background.

The number of counts n registered in a fixed time interval as a function of the source strength s is given as:

source strength s:	0	1	2	3	4	5	6
counts n:	6	11	20	42	44	62	61

At any given source strength, the number of counts is expected to be Poisson distributed with mean

$$n = a + bs,$$

where a and b are constants. Analyze the data for a fit to this relationship and obtain the best values for a and b together with their standard errors.

(a) How well is the cosmic ray background determined?

(b) What is the value of the correlation coefficient between a and b? Is this consistent with what would happen if the cosmic ray background were imagined to be negligible?

(c) Do the data fit the expected relationship well? Is there any evidence that the reported data "are too good a fit"?

Because in this problem the independent variable s takes only consecutive integer values, we will use it as a label i and denote the number of counts corresponding to $s = i$ by n_i. As the data are expected to be Poisson distributed, the best estimate of the variance of each reading is equal to the best estimate of the reading itself, namely the actual measured value. Thus each reading n_i has an error of $\sqrt{n_i}$, and the covariance matrix N takes the form $N = \mathrm{diag}(n_0, n_1, \ldots, n_6)$, i.e. it is diagonal, but not a multiple of the unit matrix.

The expression for χ^2 is

$$\chi^2(a, b) = \sum_{i=0}^{6} \left(\frac{n_i - a - bi}{\sqrt{n_i}} \right)^2 \qquad (*).$$

Minimization with respect to a and b gives the simultaneous equations

$$0 = \frac{\partial \chi^2}{\partial a} = -2 \sum_{i=0}^{6} \frac{n_i - a - bi}{n_i},$$

$$0 = \frac{\partial \chi^2}{\partial b} = -2 \sum_{i=0}^{6} \frac{i(n_i - a - bi)}{n_i}.$$

As is shown more generally in textbooks on numerical computing (e.g. William H. Press *et al.*, *Numerical Recipes in C*, 2nd edn (Cambridge: Cambridge University Press, 1996), Sect. 15.2), these equations are most conveniently solved by defining the quantities

$$S \equiv \sum_{i=0}^{6} \frac{1}{n_i}, \quad S_x \equiv \sum_{i=0}^{6} \frac{i}{n_i}, \quad S_y \equiv \sum_{i=0}^{6} \frac{n_i}{n_i},$$

$$S_{xx} \equiv \sum_{i=0}^{6} \frac{i^2}{n_i}, \quad S_{xy} \equiv \sum_{i=0}^{6} \frac{in_i}{n_i}, \quad \Delta \equiv SS_{xx} - (S_x)^2.$$

With these definitions (which correspond to the quantities calculated and accessibly stored in most calculators programmed to perform least-squares fitting), the solutions for the best estimators of a and b are

$$\hat{a} = \frac{S_{xx}S_y - S_x S_{xy}}{\Delta},$$

$$\hat{b} = \frac{S_{xy}S - S_x S_y}{\Delta},$$

with variances and covariance given by

$$\sigma_a^2 = \frac{S_{xx}}{\Delta}, \quad \sigma_b^2 = \frac{S}{\Delta}, \quad \mathrm{Cov}(a, b) = -\frac{S_x}{\Delta}.$$

The computed values of these quantities are: $S = 0.38664$; $S_x = 0.53225$; $S_y = 7$; $S_{xx} = 1.86221$; $S_{xy} = 21$; $\Delta = 0.43671$.

From these values, the best estimates of $\hat{a}$, $\hat{b}$ and the variances σ_a^2 and σ_b^2 are

$$\hat{a} = 4.2552, \quad \hat{b} = 10.061, \quad \sigma_a^2 = 4.264, \quad \sigma_b^2 = 0.8853.$$

The covariance is $\mathrm{Cov}(a, b) = -1.2187$, giving estimates for a and b of

$$a = 4.3 \pm 2.1 \quad \text{and} \quad b = 10.06 \pm 0.94,$$

with a correlation coefficient $r_{ab} = -0.63$.

(a) The cosmic ray background must be present, since $n(0) \neq 0$, but its value of about 4 is uncertain to within a factor of 2.

(b) The correlation between a and b is negative and quite strong. This is as expected since, if the cosmic ray background represented by a were reduced towards zero, then b would have to be increased to compensate when fitting to the measured data for non-zero source strengths.

(c) A measure of the goodness-of-fit is the value of χ^2 achieved using the best-fit values for a and b. Direct resubstitution of the values found into (∗) gives $\chi^2 = 4.9$. If the weight of a particular reading is taken as the square root of the predicted (rather than the measured) value, then χ^2 rises slightly to 5.1. In either case the result is almost exactly that "expected" for 5 d.o.f. – neither too good nor too bad. There are five degrees of freedom because there are seven data points and two parameters have been chosen to give a best fit.

17.13 The following are the values and standard errors of a physical quantity $f(\theta)$ measured at various values of θ (in which there is negligible error):

θ	0	$\pi/6$	$\pi/4$	$\pi/3$
$f(\theta)$	3.72 ± 0.2	1.98 ± 0.1	-0.06 ± 0.1	-2.05 ± 0.1

θ	$\pi/2$	$2\pi/3$	$3\pi/4$	π
$f(\theta)$	-2.83 ± 0.2	1.15 ± 0.1	3.99 ± 0.2	9.71 ± 0.4

Theory suggests that f should be of the form $a_1 + a_2 \cos\theta + a_3 \cos 2\theta$. Show that the normal equations for the coefficients a_i are

$$481.3a_1 + 158.4a_2 - 43.8a_3 = 284.7,$$
$$158.4a_1 + 218.8a_2 + 62.1a_3 = -31.1,$$
$$-43.8a_1 + 62.1a_2 + 131.3a_3 = 368.4.$$

(a) If you have matrix inversion routines available on a computer, determine the best values and variances for the coefficients a_i and the correlation between the coefficients a_1 and a_2.
(b) If you have only a calculator available, solve for the values using a Gauss–Seidel iteration and start from the approximate solution $a_1 = 2$, $a_2 = -2$, $a_3 = 4$.

Assume that the measured data have uncorrelated errors. The quoted errors are not all equal and so the covariance matrix N, whilst being diagonal, will not be a multiple of the unit matrix; it will be

$$N = \text{diag}(0.04,\ 0.01,\ 0.01,\ 0.01,\ 0.04,\ 0.01,\ 0.04,\ 0.16).$$

Using as base functions the three functions $h_1(\theta) = 1$, $h_2(\theta) = \cos\theta$ and $h_3(\theta) = \cos 2\theta$, we calculate the elements of the 8×3 response matrix $R_{ij} = h_j(\theta_i)$. To save space we display its 3×8 transpose:

$$R^{\text{T}} = \begin{pmatrix} 1 & 1 & 1 & 1 & 1 & 1 & 1 & 1 \\ 1 & 0.866 & 0.707 & 0.500 & 0 & -0.500 & -0.707 & -1 \\ 1 & 0.500 & 0 & -0.500 & -1 & -0.500 & 0 & 1 \end{pmatrix}.$$

Then

$$R^{\text{T}}N^{-1} = \begin{pmatrix} 25 & 100 & 100 & 100 & 25 & 100 & 25 & 6.25 \\ 25 & 86.6 & 70.7 & 50 & 0 & -50 & -17.7 & -6.25 \\ 25 & 50.0 & 0 & -50.0 & -25 & -50 & 0 & 6.25 \end{pmatrix}$$

and

$$R^TN^{-1}R = R^TN^{-1} \begin{pmatrix} 1 & 1 & 1 \\ 1 & 0.866 & 0.500 \\ 1 & 0.707 & 0 \\ 1 & 0.500 & -0.500 \\ 1 & 0 & -1 \\ 1 & -0.500 & -0.500 \\ 1 & -0.707 & 0 \\ 1 & -1 & 1 \end{pmatrix}$$

$$= \begin{pmatrix} 481.25 & 158.35 & -43.75 \\ 158.35 & 218.76 & 62.05 \\ -43.75 & 62.05 & 131.25 \end{pmatrix}.$$

From the measured values,

$$f = (3.72,\ 1.98,\ -0.06,\ -2.05,\ -2.83,\ 1.15,\ 3.99,\ 9.71)^T,$$

we need to calculate $R^TN^{-1}f$, which is given by

$$\begin{pmatrix} 25 & 100 & 100 & 100 & 25 & 100 & 25 & 6.25 \\ 25 & 86.6 & 70.7 & 50 & 0 & -50 & -17.7 & -6.25 \\ 25 & 50.0 & 0 & -50 & -25 & -50 & 0 & 6.25 \end{pmatrix} \begin{pmatrix} 3.72 \\ 1.98 \\ -0.06 \\ -2.05 \\ -2.83 \\ 1.15 \\ 3.99 \\ 9.71 \end{pmatrix},$$

i.e. $(284.7,\ -31.08,\ 368.44)^T$.

The vector of LS estimators of a_i satisfies $R^TN^{-1}R\hat{a} = R^TN^{-1}f$. Substituting the forms calculated above into the two sides of the equality gives the set of equations stated in the question.

(a) Machine (or manual!) inversion gives

$$(R^TN^{-1}R)^{-1} = 10^{-3} \begin{pmatrix} 3.362 & -3.177 & 2.623 \\ -3.177 & 8.282 & -4.975 \\ 2.623 & -4.975 & 10.845 \end{pmatrix}.$$

From this (covariance matrix) we can calculate the standard errors on the a_i from the square roots of the terms on the leading diagonal as ± 0.058, ± 0.091 and ± 0.104. We can further calculate the correlation coefficient r_{12} between a_1 and a_2 as

$$r_{12} = \frac{-3.177 \times 10^{-3}}{0.058 \times 0.091} = -0.60.$$

The best values for the a_i are given by the result of multiplying the column matrix $(284.7,\ -31.08,\ 368.44)^T$ by the above inverted matrix. This yields $(2.022,\ -2.944,\ 4.897)^T$ to give the best estimates of the a_i as

$$a_1 = 2.02 \pm 0.06, \qquad a_2 = -2.99 \pm 0.09, \qquad a_3 = 4.90 \pm 0.10.$$

(b) Denote the given set of equations by $Aa = b$ and start by dividing each equation by the quantity needed to make the diagonal elements of A each equal to unity; this produces $Ca = d$. Then, writing $C = I - F$ yields the basis of the iteration scheme,

$$a_{n+1} = Fa_n + d.$$

We use only the simplest form of Gauss–Seidel iteration (with no separation into upper and lower diagonal matrices).

The explicit form of $Ca = d$ is

$$\begin{pmatrix} 1 & 0.3290 & -0.0909 \\ 0.7239 & 1 & 0.2836 \\ -0.3333 & 0.4728 & 1 \end{pmatrix} \begin{pmatrix} a_1 \\ a_2 \\ a_3 \end{pmatrix} = \begin{pmatrix} 0.5916 \\ -0.1421 \\ 2.8072 \end{pmatrix}$$

and

$$F = \begin{pmatrix} 0 & -0.3290 & 0.0909 \\ -0.7239 & 0 & -0.2836 \\ 0.3333 & -0.4728 & 0 \end{pmatrix}.$$

Starting with the approximate solution $a_1 = 2$, $a_2 = -2$, $a_3 = 4$ gives as the result of the first ten iterations

a_1	a_2	a_3
2.000	-2.000	4.000
1.613	-2.724	4.419
1.890	-2.563	4.633
1.856	-2.824	4.649
1.943	-2.804	4.761
1.947	-2.899	4.781
1.980	-2.907	4.827
1.987	-2.944	4.842
2.000	-2.953	4.861
2.005	-2.969	4.870
2.011	-2.975	4.879

This final set of values is in close agreement with that obtained by direct inversion; in fact, after 18 iterations the values agree exactly to three significant figures. Of course, using this method makes it difficult to estimate the errors in the derived values.

BUILDING COMMUNITY

A SPECIAL REPORT

Building Community

A NEW FUTURE FOR
ARCHITECTURE EDUCATION AND PRACTICE

ERNEST L. BOYER AND LEE D. MITGANG

THE CARNEGIE FOUNDATION
FOR THE ADVANCEMENT OF TEACHING

5 IVY LANE, PRINCETON, NEW JERSEY 08540

Copyright © 1996

The Carnegie Foundation
for the Advancement of Teaching

This report is published as part of the effort by The Carnegie Foundation for the Advancement of Teaching to explore significant issues in education. The views expressed should not necessarily be ascribed to individual members of the Board of Trustees of The Carnegie Foundation.

LIBRARY OF CONGRESS CATALOGING-IN-PUBLICATION DATA

Boyer, Ernest L.
   Building community : a new future for architecture education and
   practice : a special report / Ernest L. Boyer and Lee D. Mitgang.
       p.   cm.
   Includes bibliographical references and index.
   ISBN 0-931050-59-6
     1. Architecture—Study and teaching—United States.
   2. Architectural surveys—United States.   I. Mitgang, Lee D., 1949-
   II. Title.
NA2105.B69   1996
720'.7'1173—dc20   96-13832
                                                CIP

Second printing, 1996

Copies are available from

CALIFORNIA PRINCETON FULFILLMENT SERVICES
1445 Lower Ferry Road
Ewing, New Jersey 08618

TOLL-FREE—U.S. & CANADA (800) 777-4726 FAX (800) 999-1958

PHONE (609) 833-1759 FAX (609) 883-7413

CONTENTS

ACKNOWLEDGMENTS

WE MUST THANK, first of all, the leaders and membership of the five national architectural organizations that coalesced around the need for an independent study of professional education and practice: the American Institute of Architects, the Association of Collegiate Schools of Architecture, the American Institute of Architecture Students, the National Council of Architectural Registration Boards, and the National Architectural Accrediting Board, Inc. Throughout this project, they provided encouragement and generous funding for the research, met every request for information or assistance, and preserved the independence and integrity of the process.

We especially thank, personally and professionally, Norman L. Koonce, FAIA, president of the American Architectural Foundation, and Alan Sandler, the Foundation's senior director of education programs. From beginning to end, their friendship and endless diligence in making the intentions of our project better known and in easing our access to the world of architecture lifted countless burdens from our shoulders. The report simply could not have been completed without their support.

Additional support for the research was generously provided by the Otis Elevator Company, and we thank them as well.

The idea for this project dates to 1987, when leaders of the five organizations first broached it to Ernest L. Boyer, president of The Carnegie Foundation. Many past and present leaders of the five groups deserve our gratitude for supporting the study, but if there is a true godfather, it is W. Cecil Steward, FAIA, past president of the American Institute of Architects and dean of the University of Nebraska College of Architecture. Cecil worked tirelessly with the five organizations to make this project a reality, and in the end provided valuable editorial critique that greatly enriched the final report.

We must also thank the board of trustees of The Carnegie Foundation for recognizing the broader relevance of architecture to educational excellence and for generously agreeing, shortly before Dr. Boyer's death, to assume editorial and financial responsibility for the completion and publication of the report. And we gratefully acknowledge Dr. Charles Glassick, interim president of the Foundation, and Robert Hochstein, assistant to the president and long-time director of communications, for providing the keen-eyed editorial comments and the infectious confidence that saw this project through its final months.

IN THE CONDUCT OF THIS STUDY, many of the nation's prominent architectural scholars and practitioners advised us on the crucial issues of our project, and we now thank all who graciously gave their time and expertise.

The five national architecture organizations established an Architectural Study Advisory Group which met with us twice a year in Washington. Along with Norman Koonce and Alan Sandler, those who served included: Virgil Carter, FAIA, vice president of the American Institute of Architects Education Department; L. William Chapin II, FAIA, Chapin and Tomaselli Architects; John C. Gaunt, FAIA, dean, University of Kansas School of Architecture & Urban Design; Chuck Hamlin, vice president, AIA public affairs; Terrence M. McDermott, executive vice president and chief executive officer of the AIA; Christine Malecki-West, past vice president of the American Institute of Architecture Students; Dee Christy Briggs, past president of AIAS; Lawrence Guillemette, executive director of AIAS; Irene Tyson, past executive director of AIAS; Kent L. Hubbell, chairman, Cornell University Department of Architecture; Linda Groat, University of Michigan College of Architecture & Urban Planning; Richard E. McCommons, FAIA, past executive director of ACSA; Walter Bank, past executive director of ACSA; Martin Moeller, executive director of ACSA; Gregory S. Palermo, FAIA, Iowa State University Department of Architecture; George B. Terrien, president, Boston Architectural Center; John M. Maudlin-Jeronimo, FAIA, executive director, National Architectural Accrediting Board; Richard W. Quinn, FAIA, Stecker La Bau Arneill McManus Architects, Inc.;

Darrell L. Smith, TBG Architects and Planners; and Samuel T. Balen, FAIA, executive vice president, the National Council of Architectural Registration Boards.

On three occasions, we invited prominent figures in the architecture field to the Foundation's Princeton headquarters to discuss our study. Those who attended one or more of those gatherings included: Kathryn Anthony, of the University of Illinois at Urbana-Champaign School of Architecture; Douglas Bailey, past president of American Institute of Architecture Students; James F. Barker, dean, Clemson University College of Architecture; Robert M. Beckley, FAIA, dean, University of Michigan College of Architecture & Urban Planning; Walter Carry, FAIA, of Cooper, Carry & Associates, Inc.; L. William Chapin, Alethea Cheng, and Meghan Corwin, students at Pratt Institute; Dr. Michael J. Crosbie, past editor of *Progressive Architecture;* Sam Davis, FAIA, chairman, University of California–Berkeley Department of Architecture; Joseph Esherick, FAIA, of Esherick, Homsey Dodge & Davis; Helen Ferguson, student at Princeton University; Jerry Finrow, dean, University of Washington Department of Architecture; Harrison Fraker, FAIA, dean, University of California–Berkeley College of Environmental Design; Robert Gutman, Princeton University School of Architecture; Leigh Hubbard, past vice president, American Institute of Architecture Students; and again Kent L. Hubbell.

Also, Steven W. Hurtt, dean, University of Maryland School of Architecture; Douglas Kelbaugh, University of Washington Department of Architecture; Laurie Maurer, of Maurer & Maurer Architects; Susan Maxman, of Susan Maxman Architects; Beth McLendon, student at Princeton University; Tom McPeek, student at Ball State University; William G. McMinn, FAIA, dean, Cornell University College of Architecture, Art & Planning; Courtney Miller, past president, American Institute of Architecture Students; Ted Pappas, Pappas Associates Architects; Alan Paradis, APS, Inc.; Charles Pearman, Cornell University; Elizabeth Plater-Zybeck, dean, University of Miami School of Architecture; Alan J. Plattus, associate dean, Yale University School of Architecture; William Porter, past head, Massachusetts Institute of Technology Department of Architecture; Donna V. Robertson, dean, Tulane University; Harry G. Robinson III, FAIA, dean, Howard University School of Architecture and Planning; Peter Samton, Gruzen Samton Architects, Planners; Carl M. Sapers, of the law firm Hill and Barlow; Donald Schön, Massachusetts Institute of Technology; Edward Sovik, FAIA, of Sovik, Mathre, Sathrum; W. Cecil Steward; George Terrien; Suzanna Torre, director, Art Academy, Cranbrook Academy of Art; and Cynthia Weese, FAIA, dean, Washington University School of Architecture.

We are also grateful to William Porter, past head of MIT's architecture program, for his suggestions and critique of our initial project plan; Mina Marefat, of the Smithsonian Institution, for sharing with us information about architecture education in non-Western cultures; and Denis Jesson, of the University of Manitoba, who generously shared with us his continuing work on architectural research.

We owe a special debt to the hundreds of students, faculty members, administrators, and practicing architects we met during our field work, and the hundreds of others who participated in our 1994 Survey on the Education of Architects. Unnamed here for reasons of confidentiality, they nonetheless have our gratitude for their hospitality, courtesy, and cooperation. Their words and experiences are the backbone of this report, and we can't thank them enough for their support and candor.

MANY MEMBERS OF THE CARNEGIE STAFF contributed to this study from beginning to end. We especially thank Gene Maeroff, senior fellow at the Foundation. Gene helped conduct the fieldwork, wrote perceptive site visit reports, refined the surveys, critiqued drafts, and, throughout, provided invaluable counsel. We are also indebted to Michael Timpane, vice president of the Foundation, for his insightful review of final chapter drafts. Mary Huber, also a senior fellow at the Foundation, gave us detailed and altogether superb editorial suggestions that resulted in a greatly improved report. Along with Mary Huber, former Carnegie scholars Warren Bryan Martin and Eugene Rice took part in crucial early discussions with the architecture groups, which paved the way for the project.

For the Foundation's Survey on the Education of Architects, we acknowledge the expert work of Mary Jean Whitelaw, who helped develop the survey instrument, edited the final tables, and drafted the technical notes at the end of this report. Craig Wacker spent many hours organizing into usable form the hundreds of written responses to the surveys. Lois Harwood assisted in reviewing the surveys, and Pat Novak, Laura Bell, Kelli-Ann Lanino, and Louise Underwood handled the mailing and distribution.

Jan Hempel worked tirelessly, under tight deadline, to carry the project through final editing and production, refining the structure and preparing the manuscript for bound books. Carol Tate and Johanna Wilson assisted in the final editing phase and performed the essential job of proofing citations. Hinda Greenberg, director of the Foundation's information center, assisted by Torin Dilley, came through unfailingly in meeting many research requests.

Carol Duryea, Robert Lucas, and Arlene Hobson Gundrum smoothed out our travel schedules and expenses. Jacquelyn Minning handled phone calls, and Jeanine Natriello provided valuable help in publicity work. David Walter, the Foundation's treasurer, oversaw the budget and went beyond the call in countless ways. Louise Underwood, long-time secretary to the Foundation president, provided constant day-to-day support and much wise counsel, through every phase. Dawn Berberian cheerfully typed thousands of our words, deciphered impossible handwriting, and met impossible deadlines.

Our final appreciation and blessings go to the Boyer family, and most especially to Kay, wife of the late Ernest Boyer. She was the rock who kept Ernie anchored and miraculously energetic through his final days. And even after he was gone, Kay found the generosity of spirit to cheer this project on to completion.

Ernest L. Boyer, 1928–1995

I T SEEMS NATURAL that the architecture community and Ernest L. Boyer would eventually discover each other. In architects and architecture educators, Dr. Boyer found kindred spirits: humanists, idealists, artists, communicators in many media. He often said that "the future belongs to the integrators," and architecture, which involves so centrally the integration of many disciplines through the design act, deeply intrigued him. Most essentially, he saw in architecture education the epitome of his vision of "engaged scholarship," which, if combined with civic commitment, holds virtually endless possibilities for renewing communities and even nations.

What first brought leaders of the architecture world to Ernie's door nearly ten years ago were anxieties surrounding the future of their profession and architecture education. Still, it was the *possibilities*, far more than the *problems*, that excited him. In 1993, after lengthy discussions, he was finally persuaded to undertake this thirty-month, comprehensive study of architecture education and practice.

In Dr. Boyer, the architecture world found one of this century's great educational leaders, a scholar with the lively intellect, vast experience, moral authority, and generosity of spirit who might help a profession riven by self-doubt find common ground and purpose. Indeed, his singular ability to find simple truths in hopeless thickets of complexity shines through, again and again, in this report.

In a career spanning four decades, Dr. Boyer was a campus leader, distinguished teacher, and a U.S. Commissioner of Education, and finally, one of this nation's most trusted voices of educational reform. As president of The Carnegie Foundation from 1979 to 1995, he was the author of a series of publications that literally transformed the national debate about primary and secondary schools, undergraduate college education, and the nature of scholarship itself.

His unparalleled reputation as an objective but loving critic of American education undoubtedly contributed to the bold decision by the five national architecture organizations representing educators, practitioners, students, accreditors, and regulators to entrust the conduct of this report to nonarchitects. Dr. Boyer had, above all, an unwavering faith in the willingness and capacity of educators to renew themselves, given the right ideas and encouragement. A devout Quaker, he was universally revered as a peacemaker in an age of conflict and cynicism. More than once we heard it said that perhaps no one in America but Ernie Boyer could have gotten the often fractious worlds of architecture education and practice to join forces to make this study possible.

To our sorrow, Dr. Boyer passed away in December 1995. Though prevented from writing the final report, Dr. Boyer was enthusiastically engaged even in his last days in shaping the core arguments we make here. Our hope is that, while the lyrics are not precisely his, readers will be able to discover in these pages the melodies of Ernie's ideas and spirit.

LEE D. MITGANG
Senior Fellow
The Carnegie Foundation
for the Advancement of Teaching

Making the Connections

I N THE FALL OF 1993, when work began on this thirty-month study of architecture education and practice, my daughter's fifth-grade class was busy building toothpick bridges. The goal was to make the strongest bridge possible, capable of supporting the most metal weights. The added hitch was that each child's bridge had to be constructed within an imaginary budget of $1.4 million. Each toothpick, each glob of glue cost money, and to succeed in the assignment, the children had to learn and apply some basics about physics and tectonics in order to stay within budget and make each toothpick support as much weight as possible. As each child's bridge was ready for stress-testing, the class gathered excitedly and considered, collectively, how the next child's bridge might be made even stronger.

In weighing the problems, possibilities, and larger meanings of the education of architects, my co-author and departed friend, Ernest Boyer, and I reflected on that simple-sounding elementary school exercise. It was intriguing that, wittingly or not, the tasks, thought processes, and goals we found in visits to design studios at architecture schools and those at work in this grade school project were so strikingly similar.

The message couldn't be more compelling: architecture education centers, appropriately enough, on preparing future architects capable of designing sturdy, beautiful, and useful structures that serve users, strengthen communities, and enhance the environment. We became convinced, however, that the core elements of architecture education— learning to design within constraints, collaborative learning, and the refining of knowledge through the reflective act of design—have relevance and power far beyond the training of future architects. The basic canons of design education to be found at the nation's 103 accredited architecture programs could be as enriching for students of all ages and interests as they are for aspiring architects, if only they were better known and more widely appreciated.

We concluded, in short, that architecture education is really about fostering the learning habits needed for the discovery, integration, application, and sharing of knowledge over a lifetime.

Along with the vast potential, however, what this also points up is the architecture community's long history of failure to connect itself firmly to the larger concerns confronting families, businesses, schools, communities, and society. Those fifth graders had no idea, and were never told, that they were actually engaged in the rudiments of design work. While few of those youngsters will ever go on to be practicing architects, all will be lifetime users of buildings, and beholders and inhabitants of the built environment. But, like the rest of the public, most spend their school days and indeed their lives seldom if ever considering the permanent and profound impact architecture has on their own personal health, productivity, and happiness, and on community life. In short, too many Americans will spend their lives as architectural illiterates. Unless those connections can be more clearly established in schools and in public discourse, architecture will remain omnipresent yet underappreciated and shrouded in mystery.

Even on the college campuses where most architecture programs are located, the potential of design education to enrich learning and life has been inadequately explored. We discovered that architecture students and faculty are too often disconnected from other disciplines, and distant from the social and cultural mainstream of campus life. At some schools, the curriculum seems remote from the concerns of clients, communities, or the larger challenges of the human condition. And at hardly any campus is knowledge of architecture considered a basic element of the liberal education of all students, whatever their future plans.

This report, then, is directed at two ends. It is concerned, first, with examining the problems and the possibilities of architecture education as it has evolved in this country. In the course of our research, we discovered a rich storehouse of promising practices on campuses and in architecture firms across the country, including outstanding examples of teaching excellence and integrative learning. We were especially inspired by the design studio, the distinctive holy-of-holies of architecture education where generations of bleary-eyed students have hunched over drafting tables til all hours working on balsa or cardboard creations amid old sofas and soda cans. We are convinced that these studios, scruffy though they may look, are nonetheless models for creative learning that others on campus might well think about.

At the same time, we also found a sense of disconnection—between architects and other disciplines on campus, and between the two separate worlds of architecture education and practice. Architecture students and faculty at many schools seem isolated, socially and intellectually, from the mainstream of campus life. And if the most pessimistic observers are correct, the gulf dividing architecture schools and the practice world has grown perilously wide. It is our hope that this report will help promote more constructive engagement between educators and practicing architects. We offer a new, comprehensive framework for sustained dialogue aimed at establishing not only a more unified and prosperous profession, but a healthier physical environment and more wholesome communities.

But this report isn't just for present and future architects. It is for everyone who worries, with us, about whether beauty still has a valued place in our society, whether our cities will be centers of civilization or decay, and whether our children will inherit a wholesome physical environment that uplifts the spirit and promotes health and prosperity. The second goal of this study, then, is to take a modest step toward demystifying the purposes and potentials of architecture education and practice. The knowledge of the architecture discipline and the work of practicing architects are simply too important, their power to bestow beauty or inflict damage too permanent and profound, to remain in the shadows.

In the conduct of our research, we first reviewed significant past studies of architecture education written in this century, as well as scores of books, journal articles, and scholarly papers. We read the accreditation reports of approximately fifty architecture schools, and participated as observers on an accreditation team visit to a midwestern school of architecture.

Most of all, as outsiders to the architecture field, we listened—to the satisfactions, frustrations, goals, and ideals expressed to us, in person and in writing, by hundreds of present and future members of the profession. Throughout our study, we consulted broadly with eminent practitioners and scholars in architecture and in allied fields, and attended meetings and symposia sponsored by national and regional architecture groups.

During the spring and fall of 1994, we spent several days each at fifteen accredited schools of architecture. Campuses were carefully selected to represent the rich geographic, institutional, and programmatic diversity among the nation's 103 accredited schools. Our visits included, for example, a program in an inner-city art institute, a large research-oriented program with more than one thousand students in the South, a highly selective program in the East with only 130 students, an upper midwestern institution with relatively open enrollment, a historically Black institution, a free-standing private institute in the West, a New England program that blends work and study, and two programs within schools of technology, one on the East Coast, one on the West Coast.

During our visits, we interviewed, separately and in small groups, approximately three hundred students, one hundred and fifty faculty members, and twenty-five deans and department heads. We came away with a powerful appreciation of the breadth and depth of architecture scholarship. We sat in on lecture classes covering, among other things, architectural theory, computer-aided design, zoning and building codes, acoustics, lighting, plumbing, sewage and drainage systems, pre-cast concrete, the history of Chinese architecture, and an ominously titled course in "Female Fetish Urban Form." We visited architectural laboratories where faculty and students were conducting research in daylighting, building technology, psychophysiology, urban development, computer animation, and virtual environments.

And we spent many hours in design studios and watched students design desert homes, silos, airport terminals, and a host of other creations. We visited several college-operated studios in inner-city storefronts whose focus was community development. A "Cultural Intervention Studio" was taught by an expert in Moslem architecture, and the goal was to help students relate their design work to a variety of Western and non-Western cultures. We observed several "design-build studios," in which students shepherd their projects through every phase from conception and design to contract negotiation to construction.

We also visited two dozen architectural firms of varying sizes, specialties, and locations, selected with the help of leading architects and educators. Those visits featured lengthy conversations with principal partners about the realities of practice, the economic, social, and political challenges ahead for the profession, and their thoughts about the education of architects. We asked practicing architects to imagine ways to strengthen the connections between the worlds of education and practice. And we met with approximately sixty interns, exploring how effectively internship works, and how well recent graduates of architecture schools around the country feel their educations have served them in the office environment. (Schools and firms that cooperated in site visits were promised anonymity. Throughout the text, however, we have identified, by name, schools and firms that we learned about through various means in our research which exemplify, in our view, promising practices.)

To broaden our perspective further, we conducted surveys of architecture students, faculty, and alumni connected with the fifteen campuses we visited, to which we received nearly five hundred replies. We distributed a separate mailed survey to the deans, department heads, and chairs of all 103 accredited architecture programs in this country. We received eighty replies which gave us not only their wisdom, but valuable information about the curricula and practices of approximately sixty-five schools. Results from our four surveys can be found throughout this report.

Our surveys also asked students, faculty, alumni, and administrators to write, at length, about the strengths and shortcomings of architecture education, the larger purposes and priorities of the profession itself, and how architecture might better help address society's challenges in the future. The result was an enormous outpouring of perspectives that greatly enriched our report.

THE SPIRIT BEHIND THIS REPORT, then, is one of optimism. We discuss, candidly, the very real problems of the education of architects. Along with our concern over the state of relations between the academy and the profession, we worry about the stresses on students, the scholarly lives of faculty, the paucity of women and minorities in both the professional and academic ranks, the inflexibility, imbalance, and narrowness of curricula in some programs, and the still-unmet challenges of internship and continuing education.

Still, in our travels to campuses and architectural offices around the country, we found a readiness for new ideas and directions. Some schools are already actively reviewing their programs. It was especially encouraging to meet architects and students eager for a greater voice in the political and economic decisions shaping the physical environment. If the architecture world succeeds in this, the nation and the world will be richer for it. Educating architects not only for competence but also for civic engagement is surely one of the highest priorities for architecture schools in the coming years.

We offer no one-size-fits-all formula for schools of architecture to embrace. We propose, instead, a broad blueprint for educational renewal, offering new language and specific priorities that might help architects decide for themselves how to define their larger purposes and connect them to the most consequential needs of society. In our wilder imaginings, Dr. Boyer and I hoped that what we learned about the possibilities of architecture education might have implications for other professions, higher education as a whole, and, perhaps in time, even fifth-grade classes.

"We in no way imagine that what we might suggest would become lockstep mandates," Dr. Boyer told a group of architectural leaders two months before he died. "But getting the right issues on the table is the first step toward renewal and so our effort, always, is not only to critique the current situation, but more importantly, to identify an agenda for the future and shape a debate around issues of greater consequence."

A PROFESSION IN PERSPECTIVE

A Profession in Perspective

THE EDUCATION OF ARCHITECTS—the men and women who design our skyscrapers and plazas, churches and museums, our schools and our homes—rests on traditions as old as history. From one generation of architects to the next, there is a rich legacy of principles and personalities that creates a common bond among veterans and novices alike. During one campus visit, we heard an eighteen-year-old student and a seasoned professor discussing intently the trinity of core values composed more than two millennia ago by the Roman architectural writer Vitruvius: "Firmness, Commodity, and Delight." Often, we heard students and faculty swap remembrances of architectural deities like "H.H." (Richardson), "Corbu," (Le Corbusier) or "Mies" (van der Rohe)—as if reminiscing fondly about familiar friends. And students throughout this country soon learn to converse comfortably in the arcane but richly fascinating language of architecture that has developed over centuries. Alongside technical twentieth-century terms such as "CAD" (computer-aided design), or "HVAC," (heat, ventilation, and air conditioning systems), aspiring architects quickly become acquainted with eighteenth-century French Beaux Arts terms such as "en charette," still the common parlance for the grueling work marathons students and practitioners engage in to solve design problems.

This sense of kinship with centuries of traditions, thoughts, and personalities is, in fact, the true tie that binds those who practice architecture with those who teach it and study it. The nobility of architecture has always rested on the idea that it is a *social art*—whose purposes include, yet transcend, the building of buildings. Architects, in short, are engaged in designing the physical features and social spaces of our daily lives, which can shape how productive, healthy, and happy we are both individually and collectively. The profound and permanent impact of the architecture profession demands an education not only highly technical and practical, but broad and intellectually liberating as well. A nineteen-year-old student told us: "Architects must use their unique skills and talents to revitalize both our communities and the world. Our purpose must be noble, to create sustainable environments which are healthy for nature and man, environments that can be improved by our intervention."

An architect in Michigan put it this way: "Architects must seek a balance between economic construction, safety and welfare of the general public, long-term value of the products they help produce—all while remaining sensitive to the aesthetic concerns of the owner, neighbors and the architect him/herself."

The essential purpose of architecture education, then, is not only the basic training of beginning practitioners, but also the initiation of students into this common legacy of knowledge, skills, and language, while instilling a sense of connectedness to the human needs that architecture, as a profession, must continually address. Architecture education, if it is to fulfill those ends, must celebrate and support, and also challenge, the profession and society as a whole.

DURING THE CONDUCT OF THIS STUDY, we discovered that despite serious concerns about the education of architects, schools of architecture draw at least as much praise as criticism from students, educators, and practicing architects. In hundreds of conversations, we met only a handful who proposed abandoning altogether the methods and traditions which have shaped architecture education for much of this century. What we heard most often were calls for *reform*, not *revolution*, and we were, in fact, impressed by the number of schools who are engaged in candid review of their programs. Reflecting on all we heard and read, we concluded that the fascination of architecture education lies far more in its *possibilities* than in its problems.

This is not to suggest that changes, even bold changes, should not be carefully considered. Still, we are convinced that architecture education, at its best, is a model that holds valuable insights and lessons for all of higher education as a new century approaches. We agree with a faculty member at a southeastern campus who called architecture education "one of the best systems of learning and personal development that has been conceived." The professor went on to say: "I would hate to see us abandon a system with so many wonderful qualities and successes for the sake of some 'bold experiment in education.'"

For the nation's 35,524 full- and part-time students who attended 103 accredited degree programs in 1994–95, the study of architecture is among the most demanding and stressful on campus, but properly pursued it continues to offer unparalleled ways to combine creativity, practicality, and idealism. Coming from public school environments where aesthetic and three-dimensional learning experiences are frequently downplayed and even disparaged, it's little surprise that eighteen- and nineteen-year-old students seeking an artistic avenue with a practical bent soon become fiercely devoted to the design studio culture.

There are several available educational paths to a professional degree. Most students pursue either a five-year undergraduate program directly from high school or a six-year master's program that combines undergraduate education and two years of graduate professional study. Some, particularly those who decide on architecture later in their college years, pursue a three- or four-year graduate degree. With few exceptions, students in practically every state are required to earn a professional degree, spend at least three years as interns, and gain a specific set of professional experiences before they can take the Architecture Registration Examination that must be passed to become licensed as an architect. (It even took Philip Johnson three tries to pass all ten parts of that dreaded test.) The path to licensure, then, is not just demanding, but long: at least eight years, and often more.

How satisfied, in fact, are students with their experiences in architecture programs? When we surveyed students and alumni from fifteen representative campuses whether they would attend their school of architecture again if they had it to do over, about eight out of ten said they would (table 1).

Table 1

IF YOU HAD IT TO DO OVER, WOULD YOU
STILL ATTEND THIS SCHOOL OF ARCHITECTURE?

	YES	NO	NOT SURE
Students	82%	11%	6%
Alumni	79	15	6

SOURCE   The Carnegie Foundation for the Advancement of Teaching, Survey on the Education of Architects, 1994.

Table 2

HOW WOULD YOU RATE MORALE
AMONG ARCHITECTURE STUDENTS?
(SURVEY OF STUDENTS)

	PERCENTAGE
Very high	10%
Fairly high	64
Fairly low	17
Very low	3
Not sure	6

SOURCE   The Carnegie Foundation for the Advancement of Teaching, Survey on the Education of Architects, 1994.

And even with the demands of school and the often dire forecasts about job prospects, we found morale among architecture students to be remarkably high. Nearly three out of four students we surveyed rated morale at their schools as either "very high" or "fairly high" (table 2).

Revealingly, more than 90 percent of administrators, students, and alumni and more than 80 percent of faculty surveyed believe that graduates of architecture school leave "well prepared" as problem solvers—a quality that leading practitioners repeatedly told us was among

Table 3

ARCHITECTURE STUDENTS LEAVE THIS SCHOOL
WELL PREPARED AS PROBLEM SOLVERS

	AGREE	DISAGREE	DON'T KNOW
Administrators	96%	3%	1%
Faculty	82	14	4
Students	91	7	2
Alumni	96	4	0

SOURCE   The Carnegie Foundation for the Advancement of Teaching, Survey on the Education of Architects, 1994.

the most prized in new employees (table 3). A twenty-two-year-old student at a midwestern campus told us her school "does an excellent job of creating problem solvers. With this talent and skill, architecture graduates can go on to be leaders and innovators in an unlimited number of areas, as well as in the profession of architecture."

Despite the evident stresses of student life, nearly 4,500 architecture students earned professional bachelor's or master's degrees in 1994–95, and most seem quite aware of the gloom-and-doom job scenarios that may await them. When we asked students what their salary expectations were five years after earning their professional degree, 24 percent said less than $30,000 and another 44 percent said in the $30,000 range (table 4). As one student put it, freshmen are routinely greeted at the portals of architecture school with the message: "I won't be rich, I won't be famous, and I have to work a lot."

We then asked students what motivated them to enter the architecture field. The most common response, given by 44 percent of those surveyed, was "putting their creative abilities to use." Intriguingly, the other most frequent responses—given by a combined 39 percent—were a desire to improve the quality of life in communities or improve the built environment as a whole. Hardly anyone cited the prestige of the

Table 4

LOOKING AHEAD, WHAT KIND OF SALARY DO
YOU EXPECT FIVE YEARS AFTER EARNING YOUR
PROFESSIONAL DEGREE IN ARCHITECTURE?
(SURVEY OF STUDENTS)

	PERCENTAGE
$20,000–29,999	24%
$30,000–39,999	44
$40,000–49,999	22
$50,000 or more	10

SOURCE   The Carnegie Foundation for the Advancement of Teaching, Survey on the Education of Architects, 1994.

profession or good salary prospects (table 5).

For most, then, the intrinsic appeal of architecture education seems to transcend its practical uncertainties, and many students cling, against the odds, to ideal visions of their potential. For many, architecture school is an opportunity to be part of a tight-knit community on campus that is defiantly proud of its distinctive methods and its reputation for long hours and hard work. "We all have a dream," said one thirty-four-year-old student. "We all want to be recognized architects. You have to pay your dues. When we enter this program they tell us that it is not high paying. But I assume money will follow if you gain recognition."

In our survey, 71 percent of students agreed that "this is an excellent time to be entering the architecture field"—a level of optimism exceeding that of both faculty and administrators (table 6).

Again, this is not to suggest that all is well with architecture education. Repeatedly in our travels, we witnessed the estrangement of the academy and the profession, the isolation and stress of student life, the disconnection of architecture from other disciplines, and the inflexibility of the curriculum on many campuses. We also saw the great unevenness of experiences in design studio, and the autocratic, one-way communication that often marks "design juries"—the tension-packed

Table 5

PLEASE RANK THE THREE MOST IMPORTANT REASONS
FOR ENTERING THE ARCHITECTURE PROFESSION
(SURVEY OF STUDENTS)

	PERCENTAGE RANKING ITEM FIRST
Putting creative abilities to practical use	44%
Improving quality of life in communities	22
Improving the built environment	17
The prestige of the profession	2
Good salary prospects	1
Other	14

SOURCE   The Carnegie Foundation for the Advancement of Teaching,
Survey on the Education of Architects, 1994.

ritual during which invited critics, including faculty, practicing architects, and more rarely clients, review and critique student work.

Further, we frequently heard that many architecture schools fit uncomfortably within the campus culture. In 1900, almost all architects served apprenticeships in offices.[1] Over the last century, professional education was brought fully into academic life, but the values and rewards of these two cultures have never been fully reconciled. Compounding those difficulties, architecture programs are generally small compared with other academic units, expensive in terms of per-pupil staffing and physical space, and for the most part, produce little research revenue for the university.

As W. Cecil Steward, dean of the University of Nebraska's College of Architecture, has noted, many university administrators, especially those on research-driven campuses, tend to see the architecture field as splintered and disputatious, and the design orientation of architecture faculty places architecture among "the 'soft,' 'fuzzy,' and undervalued disciplines in the comprehensive universities."[2]

And in his memorable Walter Gropius Lecture in 1985, Henry N.

Table 6

OVERALL, I FEEL THIS IS AN EXCELLENT
TIME TO BE ENTERING THE ARCHITECTURE FIELD

	AGREE	DISAGREE	DON'T KNOW
Administrators	67%	31%	3%
Faculty	47	47	6
Students	71	29	0

SOURCE   The Carnegie Foundation for the Advancement of Teaching, Survey on the Education of Architects, 1994.

Cobb, FAIA, then-chairman of Harvard's Department of Architecture, said that his own architecture school:

> . . . with its curious studio-based teaching methods, with its paucity of scholarly research, and its dedication to serving the highly "contaminated" professions of architecture, landscape architecture and urban planning, must appear, to borrow the language of "Peanuts," as a kind of 'Pig-Pen' character in the university family—that is to say disreputable and more or less useless, but to be tolerated with appropriate condescension and frequent expressions of dismay.[3]

Repeatedly, administrators of schools of architecture also told us that lack of financial resources was their biggest problem. The dean of one private, East Coast school of architecture told us his program has been hit with a 25 percent budget cut over the last five years, with further reductions likely. Thomas R. Wood, director of Montana State University's School of Architecture, put it bluntly: "The biggest threats I see are financial survival and, related to that issue, student access to higher education. The past five years have been disastrous for higher education. The budget reductions damage program quality, faculty morale, and limit the opportunities of people seeking an architectural education."

Table 7

THIS SCHOOL/DEPARTMENT HAS HAD TO LIVE WITH
MORE THAN ITS FAIR SHARE OF BUDGET RESTRAINTS
OVER THE PAST SEVERAL YEARS

	AGREE	DISAGREE	DON'T KNOW
Administrators	73%	27%	0%
Faculty	77	13	10

SOURCE   The Carnegie Foundation for the Advancement of Teaching,
Survey on the Education of Architects, 1994.

In our surveys, architecture administrators and faculty agreed overwhelmingly that their schools have been living with more than their fair
share of budget restraints over the past several years (table 7).

At one school we visited, not a single wall clock worked, windows
were broken, and paint was peeling off the walls. There were no electrical
outlets in the studios—a costly problem if and when the school attempts
to move decisively into the computer age. At present, 250 students at this
school share a dozen outdated computer terminals—at a time when
knowledge of computers has become practically a prerequisite for
employment.

Such problems are by no means universal. We found, in fact, quite
a number of schools of architecture which have accommodated themselves quite comfortably within the university culture, and in our surveys,
more than eight out of ten architecture school administrators and alumni
and nearly seven out of ten faculty disagreed strongly that their programs
"might be better off if they were not part of university campuses at all."
More than 75 percent of all groups surveyed also considered their schools
well regarded on campus (table 8).

Along with the challenges posed by the institutional context of many
architecture programs, the profession of architecture itself has become
many professions, some connected only tenuously, if at all, to the work
of designing buildings. Architectural practice, furthermore, is rapidly

Table 8

HOW WELL REGARDED IS THIS
SCHOOL OF ARCHITECTURE ON CAMPUS?

	WELL REGARDED	NOT WELL REGARDED	DON'T KNOW/ NOT APPLICABLE
Administrators	89%	9%	2%
Faculty	77	14	10
Students	85	11	4
Alumni	78	7	15

SOURCE   The Carnegie Foundation for the Advancement of Teaching, Survey on the Education of Architects, 1994.

becoming more global. In 1994, nearly one-third of u.s. firms with at least twenty employees, and 15 percent of firms with ten to nineteen employees were engaged in international work, according to the American Institute of Architects.[4] With virtually all architectural firms now computerized, markets halfway around the globe are quickly becoming as accessible as those around the corner. These emerging trends—globalization and computerization—have implications for architecture education that many schools are only beginning to confront.

A profession that for much of its history prized the new and unique now finds more and more of its members engaged in preservation and renovation. A profession once populated almost exclusively by wealthy, white males now must struggle to connect to a more diverse range of clientele and communities. A profession long regarded as the orchestrator of the building process now finds itself increasingly competing with general contractors, developers, interior designers, construction firms, and others. Practitioners and students who once revered that lonest of lone architectural wolves, the fictional Howard Roark from *The Fountainhead*, now talk of replacing him with a new icon: the "team player." The field has become increasingly varied, yet there remains fixed in the minds of many a single image of the architect that may well be an antique.

"We are operating a 1900-year-old education program directed toward delivering a 500-year-old model architect as we head into the 21st Century," Professor Gregory Palermo, FAIA, of Iowa State University, wrote recently.[5]

We found, in short, a profession struggling both to fit in, and if possible, to *lead*, within a social and economic context that in a number of crucial respects has been dramatically altered. We also found a profession whose faith in its own future has been shaken. What seems missing, we believe, is a sense of common purpose connecting the practice of architecture to the most consequential issues of society—and that same sense of unease permeates architecture education as well.

Urs Peter Gauchat, dean of the New Jersey Institute of Technology's School of Architecture, summarized the situation this way: "The rapidly changing context for architectural services has created a host of problems for architectural education. One problem stands out: the seed of self-doubt and the lack of a clear vision of what the architect can and should do. Many schools are permeated by a considerable lack of conviction about the future of architecture. This is often accompanied by the lack of a clear agenda about how to prepare students for a fulfilling professional life."

The atmosphere surrounding architecture education and the profession, then, is one of enormous promise and possibilities, but also anxiety perhaps unmatched in the history of the field. The mood resembles, in a fascinating way, that of public education in this country. Everyone agrees that elementary and secondary schools have serious problems. Yet at times of national self-doubt, there is a tendency to overlook the strengths of education. We expect the world of schools—demanding that they solve all of our complex social, economic and spiritual ills. When they inevitably fall short, we condemn them for failing to meet our high-minded expectations. The point is this: In the search for educational renewal at *all* levels, including the professional, care should be taken not to ascribe to schools alone what must, in the end, be a *shared* responsibility.

THE MOVE TO THE ACADEMY

In searching for common purpose and a more promising future, we begin by looking back—at the traditions, principles, and paradoxes that have, for so long, shaped the education and the professional lives of architects in the United States, and to a large degree, continue to do so.

Early in the nineteenth century, Thomas Jefferson and others with an interest in the future of architecture in a young and expanding nation sensed that more formal methods were needed to address the acute shortage of trained architects. Technical schools filled that need to some extent during the first half of the century, and a few university-based engineering schools began offering courses in architectural drawing.[6] But the dominant model of training architects was apprenticeship, at best a hit-or-miss proposition educationally.

It wasn't until the mid-nineteenth century that the momentous relocation of architecture education from office to campus began to take shape. In 1865, three years after Congress passed the Morrill Land Grant Act and eight years after the founding of the American Institute of Architects, the first formal, campus-based architecture courses in the United States were offered at MIT. The University of Illinois began its program in 1868, Cornell University in 1871, Syracuse University in 1873, and Columbia University in 1881.

Thus began a century-long process in architecture that paralleled, in many respects, the growing professionalism shaping many other fields. As Dana Cuff related in her recent study of architectural practice, training and the profession were largely unregulated during the nineteenth century. It wasn't until 1897 that Illinois became the first state to require a license to practice architecture. Most who called themselves architects were white, well-off and male, and got their training not at school but in apprenticeships of widely varying quality.[7] The rise of the city, the advent of new building materials and techniques, the birth of skyscrapers and new transportation systems—all made even more urgent the need for professionalization in architecture.

What was emerging in those early days of campus-based education, then, was a growing impulse to standardize architecture's expertise

through a more specialized, organized, and regulated educational program, and a related need to assure the public that practitioners could
assume legal responsibility for the quality and safety of their work.[8] These
needs were decisively addressed by the arrival in America of a philosophy
of education and practice developed at the Ecole des Beaux Arts, the
leading center of architecture education in France. William Robert Ware,
founder of both MIT's and Columbia University's architecture programs,
was instrumental in adapting that French philosophy to American
schools, and many of Ware's curricular precepts remain influential to this
day.

Ware's principles were:

 • that details of a practical nature that can be learned in the
office should be postponed until after formal education;
 • that courses in construction and history can be taught by
means of "cooperative student investigation . . .";
 • that architectural design should be conducted by a competitive method, with judgments by jury;
 • that the study of design should be continuous through
school, and design problems should not be overly practical, but
rather should stimulate the imagination through the study of
great masters;
 • that the study of construction should be stressed; and
 • that architectural curriculum should include as broad a
cultural background as time permitted.[9]

The Beaux Arts philosophy has remained, for a century, perhaps the
single greatest influence on how architecture education is conducted and
thought about. As Professor Kathryn Anthony, of the University of
Illinois at Champaign–Urbana, states: "The values behind the accreditation process, the way in which the curricula are structured, the jury
system, and other key aspects of architectural education continue to
primarily reflect an old Ecole des Beaux Arts model."[10]

The Beaux Arts school established, first of all, the hallowed place of
European and classical traditions in design. While that had the advantage
of distinguishing architecture from more technically driven approaches

used by engineers and others, it also placed the profession more firmly in an aesthetic realm that made it seem, to many in the public, less vital to community concerns than some other professions. The Beaux Arts philosophy also elevated design above technical aspects of the curriculum, emphasized one-on-one teaching methods, created the present-day system of design juries using outside evaluators to judge student projects, and stressed the learning of past architecture to inform present design.[11] At the same time, the ascendance of design studio under the Beaux Arts system brought with it an atmosphere of individualism, criticism, and competition among faculty and students.

By the turn of the twentieth century, there were still only nine professional schools of architecture in the United States, enrolling slightly under four hundred students.[12] For the most part, those early schools offered similar four-year programs which attempted to strike a balance between design and structures, but given the weighty contents, schools gave only "meager treatment" to either.[13] By 1912, the number of schools had grown to thirty-two, enrollments had tripled, and the Association of Collegiate Schools of Architecture (ACSA) was founded in order to foster communication among schools and establish minimal standards. By 1932, there were fifty-two institutions of collegiate rank offering professional courses in architecture.[14] Teaching, meanwhile, was carried on by "technicians" rather than scholars, and thus research, in the academic sense, was almost totally lacking.[15]

An early history of architecture education by Arthur Clason Weatherhead noted that the typical curriculum in the early twentieth century lacked cohesiveness, the business side of architecture was neglected, and there was little effort to aid the transition of students between the academy and the office. Graduates gained a strong sense of design, were resourceful in solving problems in artificial settings, and well versed in Beaux Arts logic, but their thinking tended to be two-dimensional.[16]

Following World War I, the competitive precepts of Beaux Arts were increasingly questioned. By the 1920s and 1930s, the coming of modernism, the growing rejection of Neoclassicism, and the emergence of urbanism led to a fundamental reconsideration of architecture education. Modernism was further refined by the Bauhaus philosophy imported from Germany in the 1930s by Walter Gropius at Harvard, and Ludwig Mies van der Rohe at the Illinois Institute of Technology. Under their influence, greater stress was placed in U.S. schools on draftsmanship, structural logic, appreciation of the properties of materials, and an aesthetic that derived from the exploration of geometric forms.[17] The period was marked by a shift to a more uniform, well-defined training system in which the Association of Collegiate Schools of Architecture also acted as a unifying influence.

Modernism powerfully affected the curricula at many U.S. schools. Stiffer course requirements were introduced, the study of materials and construction was emphasized, more schools related themselves to allied professional fields, and schools began to explore the relationship of architecture to human and community needs. The University of Oregon rejected the atelier, "master-pupil" model of Beaux Arts studio education and permitted students more freedom to choose their projects according to their interests.[18] The University of Cincinnati began its practice of having students alternate school studies and field work, which has continued into the 1990s.[19]

National organizations concerned with architecture education, meanwhile, were considering ways to assure more complete preparation for practice.[20] Along with the rise of professionalism and the growth of architectural knowledge came a drive for monitoring and accountability. At first, the Association of Collegiate Schools of Architecture itself performed the accreditation function by establishing minimum standards for membership. Those early standards were abandoned in 1932, however, amid criticism that they were stultifying curricula. The result was a regulatory vacuum. In 1940, a coalition of education and professional organizations founded the National Architectural Accrediting Board (NAAB). The establishment of NAAB marked a milestone in efforts to create clearer standards and regulatory safeguards in professional degree programs while recognizing the differing resources, geographic circumstances, and missions of individual schools.[21]

By the end of World War II, the nearly century-long shift away from apprenticeship to a university-based system of architecture education was virtually complete. In 1953, architecture enrollments nationwide had reached nearly ten thousand, a 150 percent increase over 1930.[22]

All along, the rationale behind the move to a campus-based architectural education system was that certain kinds of intellectual growth and learning needed for professional education was more likely to occur on campus than in an office. Still, the shift from the old apprenticeship system planted the seeds of separation between education and practice. Because education, organizationally, has been divorced from the practice world, schools have had to make extraordinary and costly efforts to help students gain real-life experiences, such as field trips, undergraduate internships, or preceptorships. Graduates, for their part, have faced the challenge of acquiring certain skills needed to make the transition from school to office not readily available in most standard curricula.

In 1956, the American Institute of Architecture Students was founded, an outgrowth of the student chapters of the AIA that had already spread to many campuses. In 1962, the membership of the National Council of Architectural Registration Boards (NCARB) voted to establish a single, national registration examination, replacing the many different exams given by states.[23]

During the 1960s, powerful new forces buffeted architecture education and campus life. Many schools sought to include social and political issues in their curricula, and a number also tried to connect students to underserved urban and rural communities by opening, for example, "Community Design Centers." The 1960s and 1970s, however, also saw increased questioning of key precepts of modernism. Changing economic and political tides led architects and architecture educators to be less optimistic about their prospects to be leaders in shaping the built environment. The resulting "postmodern" movement of the 1970s and 1980s was not simply a stylistic revolt against the alleged "glass box" sameness of modernism, but also reflected a feeling that modernist architecture had led both practitioners and educators away from some of the profession's more interesting and vital historical precepts. However, as the name suggests, postmodernism seemed more a move *away* from modernism than *toward* any specific unified approach.

By the 1990s, enthusiasm for postmodernism seemed to be evolving into an eclecticism in which many now seem inclined to disavow allegiance to any single style or philosophy. Indeed, the very notion of "style" has fallen into disfavor in many circles. Philosophically, however, a sense of modernist engagement in political, social, economic, and environmental issues appears to be reviving among at least some architects, educators, and students.

What has emerged, then, over the past century, is an ordered, university-based educational system marked by a high degree of professionalization and regulation. Together, the 103 accredited schools of architecture, the five national architecture organizations, their publications, and leading architectural journals act as the guardians and arbitors of the language and traditions of the profession, the awards and recognition programs, the educational performance objectives, and the legal terms for admission to practice.

"All of which is to say," writes Professor Palermo, "architecture and architects have perhaps never been so systematically ordered, internally defined, and—well, professional."[24]

A TRADITION OF SELF-EXAMINATION

As much as any profession, architecture has been engaged in continuous self-reflection, and over the last sixty-five years, as architecture education shifted from offices to campuses, at least half a dozen comprehensive studies have been written examining current conditions and future directions of education and practice. Those reports were, for the most part, *internal* studies, sponsored by groups within the profession itself and conducted by investigators drawn from the ranks of the profession or from schools of architecture.[25]

In reviewing those studies, what is most striking is the continued force of so many of their themes and observations, even those written decades ago. Viewed another way, however, the reports are also powerful testimony to the depth and durability of the dilemmas that continue to confront architecture education and practice.

What are the most critical issues identified in those past studies? Which still resonate today?

A Study of Architectural Schools, 1929–1932, sponsored by the Association of Collegiate Schools of Architecture (ACSA) and funded by the Carnegie Corporation, criticized the dominance of design faculty over those specializing in "construction." Design projects at many schools, it said, resulted in "paper architecture" whose real purposes and functions are often unclear.[26] The report noted the scarcity of "real research" in architectural schools,[27] and described the difficulty architecture schools often have fitting into the university culture: "The insistence of the architect on the problem method in design, with its emphasis on accomplishment rather than time spent, on student freedom rather than regimentation, completely upsets the machinery of marks, semesters, and quantitative measures. . . ."[28]

The report's description of school life more than sixty years ago also rings familiar. "Go through, of an evening, any university campus containing an architectural school. That school can be spotted without fail. It is the one brilliantly lighted attic. It is always an attic, usually in the oldest and least desirable building."[29]

Finally, architecture school was very much a man's world at the time of the Bosworth and Jones report. In 1930, just 271 female architecture students were enrolled in 34 U.S. schools of architecture.[30]

Almost twenty-five years later, in 1954, an American Institute of Architects commission produced a two-volume report, *The Architect at Mid-Century*, which presented forty-three recommendations, including an aptitude test for prospective architects, more concerted support for architectural research, establishment of "study institutes" for architecture faculty, uniform registration laws, and a single registration exam.[31]

Looking ahead to the second half of the twentieth century, the report admonished educators and practitioners to close the growing gap between them—advice we heard repeated in our own travels. Schools, the AIA report said:

> will do well to maintain the closest liaison with the profession
> in order to adjust content and method to the changing needs

of practice. And, by the same token, the profession, too, must apply its highest wisdom, most sympathetic understanding, and most penetrating vision to the problems of education. The very term "professional education" reveals by its compound form, the necessity of enlightened and harmonious cooperation.[32]

Perhaps of all major studies of architecture education, the most widely cited yet frequently misconstrued was the 1967 *Study of Education for Environmental Design*—sponsored by the American Institute of Architects and widely referred to as the "Princeton Report." Ironically, the Princeton Report owes much of its renown to a recommendation it never explicitly made. The report is often credited with proposing the "four-plus-two" master's program—four years of undergraduate study, followed by two years of graduate professional study. While the report did, in fact, propose a dramatic restructuring of architecture education that lent support to the idea, the spread of the four-plus-two programs owes more directly to two earlier reports—*Report of the Special Committee on Education, AIA*, and "*Blueprint '65: Architectural Education and Practice.*

The more visionary aspects of the Princeton Report lay elsewhere. Most essentially, it hypothesized three goals for design education: first, helping students develop the competence to work within the realities of actual practice; second, preparing graduates to be adaptable enough to grasp, and work within, "the continuing changes in the social, economic, scientific and technological setting of our society"; and finally, preparing students to develop their own analytical framework in which to envision a better society and built environment, "beyond present day constraints. . . ."[33]

To realize those objectives, the study stressed the importance of ending the isolation of the architectural discipline. It called for making connections—"building ladders and bridges," as co-author Robert Geddes put it recently—between all professions engaged in environmental design: "We needed to make working connections with others—engineers, planners, landscape architects, and an array of non-designers—who participate in the making of the built environment."[34]

To build those bridges, the report called for a flexible architecture curriculum, a wide range of teaching methods, and diverse architecture programs. Rather than proposing a "core curriculum" for all schools, the report suggested an intricate "modular, jointed framework for environmental design education" aimed at allowing students to tailor their studies to prepare them for more than nine hundred possible design-related careers.

The Princeton Report stressed the need for better understanding of architecture and the built environment among *nonarchitects*.[35] And it emphasized the importance of *continuing education*—an issue that has moved in the 1990s to the top of the academic and professional agenda:

> The idea is still accepted implicitly at many schools that what is learned in a period of four years to six years must carry a student through his entire professional career. To the extent that we are capable of developing more powerful planning and design concepts and methods through research and experimentation in the field, this idea of a termination point for education becomes less and less tenable.[36]

The Princeton Report was not immediately welcomed by the architecture education community. As Professor Geddes recounted recently, the report was issued at a time of growing disenchantment between the profession and the academy, and the failure by the AIA to invite more active participation from the collegiate schools association and other architectural and nonarchitectural organizations made the report instantly suspect to many educators.[37] Still, the wisdom and foresight of many of its observations survived the initial cool reception, and the document is today referred to by many with renewed admiration.

In 1981, a two-volume, 1,448-page analysis of the design studio was produced by a consortium of eight East Coast schools of architecture. The first volume of the *Architecture Education Study*, often called the "MIT Study," consisted of papers by six eminent scholars, each focused mainly on the methods and culture of the design studio. The second volume offered detailed case studies of design studios and student experiences.

The study introduced a topic of enduring relevance: how design education can shape attitudes about clients, users of buildings, and even fellow architects. MIT architecture professor Julian Beinart wrote that design students often held a pejorative view of architects—considering them egotists, elitists, insecure and indecisive.[38] And he discussed the disturbing attitudes architecture students harbor about clients. Some students advocated cooperation. Others suggested "educating" the client, while still others wished they could "eliminate" the client altogether—in effect, designing without compromise or constraint. Finally, some students spoke of "beguiling the client," with various kinds of "sneaky behavior" to deceive clients into thinking their wishes were being fulfilled.[39]

Lee Bolman, a professor of education at Harvard University, wrote that many practicing architects felt their schools shortchanged them in nondesign topics: 43 percent of those Bolman interviewed said they hadn't learned enough about how buildings actually get built, 39 percent wished they had been taught management skills, and 22 percent regretted not having learned how to deal better with other people. No one thought that school had provided too little training in design.[40] And Bolman argued that these curricular imbalances were not correctable simply by adding a few courses, but related instead to the more fundamental question of whether the methods and climate at most schools might be contributing to a disdain for technical and practice-oriented topics.

Asked to weigh the prospects of the profession itself, architects told Bolman in 1981 that they were fond of their work, but some believed, nonetheless, that the architecture field was "a dinosaur doomed to extinction."

> One in four felt the profession was clinging to an obsolete model of the architect's role. . . . A similar number felt that the profession did not know how to convince others of the architect's value. One-fifth of the architects suggested that the profession was dying, and an equal number felt that society

undervalued architects. . . . There were few optimistic pre-
dictions although some predicted that architects would broaden
their functions by involving themselves in planning, land use,
development, and energy-efficient design. . . .[41]

Bolman concluded with this challenge: "Many practitioners blame
the schools for failure to address such issues. They ask, in effect,
'Shouldn't the schools be doing the research and theory-building to help
in charting the profession's future? If they do not engage in that inquiry,
who will?'"[42]

In the fifteen years since the MIT study, other notable books have
been published on aspects of architectural practice or education. Two
deserve special mention. *Architectural Practice: A Critical View*, published
in 1988 by Robert Gutman, a sociologist who has taught at Princeton's
architecture program for more than twenty-five years, identified critical
issues for the profession generally:

- how to bring the supply of architects more in line with
demand for services;
- how to develop, industry-wide, a consistent "philosophy of
practice" and a clearer self-image that conforms to the realities
of the building industry;
- how architecture could pursue both educational and professional strategies to cope with increased competition from
other building professions;
- how to maintain the profitability of design businesses at a
time of rising costs and growing competition; and
- how to balance the desires and aspirations of employees with
the need for effective management and teamwork within design
firms.[43]

Those challenges, aimed principally at practitioners, raised disturbing
questions for educators as well. How effectively, for example, do teaching
methods and curricular content at most architecture schools prepare
students for a professional climate in which cooperation, management
acumen, and specialization of services are increasingly valued? Have the

changes in professional climate described by Gutman dangerously widened the gap between education and practice, and if so, what should be done?

University of Southern California Professor Dana Cuff's 1991 study, *Architecture: The Story of Practice*, employs the language of social science to describe the role of schools as "socializers" of young architects into the "culture" of the profession. Tracing the development of this culture, Cuff probed how schools of architecture help create in students, through their curricula and methods, a professional "ethos": what is valued, what is devalued.[44]

Cuff's telling portrait of life in architecture school adds modern insights to the picture presented sixty years earlier in the first ACSA-sponsored report. Students, she wrote, "stay up late, are never home, spend all their time in studio, and belong to a clique of other architecture students. . . . Here, in this earliest phase of becoming an architect, we see kernels of architects' later values, such as the principle of peer review and a developing segregation from the general public."[45]

Cuff especially notes the difficulties women face in entering academies where the curriculum and classroom culture reflect a long history of male dominance. Whatever their virtues, the design jury, the desk-side critique of student design work known as the "desk crit," and the grueling string of all-nighters known as the "charette"—reinforce a macho, boot camp atmosphere. Adding to the discomfort of female students is the fact that role models in studio and in the curriculum remain predominantly male.[46] Most essentially, Cuff argued, schools should do more to extend education "beyond context-free design," and she urged both the academy and the profession to pay more attention to "the social aspects of architecture."[47]

TO SUMMARIZE, our review of the history of American architectural education and of past efforts to assess it reveals as many possibilities as problems and as much continuity as change. A rich legacy of traditions continues to shape the methods and content of architecture education as well as professional practice, and an understanding of the historical roots of those practices is surely a necessary prelude to considering current conditions or future directions.

The distinguished scholar and practitioner Gerald McCue, of Harvard University, notes that the common pitfalls facing *any* study of architecture education include a tendency to exaggerate "newness and change," while downgrading that which is traditional, fundamental, and likely to continue. There is also, McCue says, a tendency to assume that problems, once identified, will necessarily be solved rationally. "It's like having a toothache forever, and yet we did. We're still dealing with some of these toothaches thirty years later."

With those wise precautions in mind, we begin our examination of current conditions and future prospects of architecture education by recalling, once more, the two-thousand-year-old Vitruvian trinity: Firmness, Commodity, and Delight. Throughout history, what has distinguished "architecture" from the mere building of buildings is the insight and skill to blend the useful with the timeless, the technically sound, with the beautiful. The challenge that has always faced both the academy and the profession has been discovering the right balance of those three ancient ideals, each so indispensable to successful architecture. That challenge continues today.

A New Vision

In the chapters that follow, we recommend a new framework for renewing architecture education and practice based on seven separate but interlocking priorities. The new vision builds on the traditions, history, and critiques already described, as well as best practices we discovered in campuses and offices around this country. If widely embraced, these "designs for renewal" would, we believe, help promote a more fruitful partnership between educators and practitioners that would not only enhance the competence of future architects but also lead the profession into more constructive engagement with the most pressing problems of our communities, our nation, and our planet.

We propose, as the first and most essential goal, *an enriched mission*—connecting schools and the profession more effectively to the changing social context. Specifically, we recommend that schools of architecture should embrace, as their primary objectives, the education of future practitioners trained and dedicated to promoting the value of beauty in our society; the rebirth and preservation of our cities; the need to build for human needs and happiness; and the creation of a healthier, more environmentally sustainable architecture that respects precious resources.

Second, we propose a more inclusive institutional context for the scholarly life of architects based on the principle of *diversity with dignity*. We imagine a landscape of architecture programs in which the multiple missions of schools are celebrated, and the varied talents of architecture faculty are supported and rewarded in a scholarly climate that encourages excellence in research, teaching, the application of knowledge, and the integration of learning.

To support the priority of diversity with dignity, we propose, as a third goal, *standards without standardization*. Such standards would affirm the rich diversity among architecture programs, establish a more coherent set of expectations at all schools that would support professional preparation, and bring into closer harmony the scholarly activities of students and faculty.

Fourth, the architecture curriculum at all programs should be better connected. A connected curriculum would encourage the integration, application, and discovery of knowledge within and outside the architecture discipline, while effectively making the connections between architectural knowledge and the changing needs of the profession, clients, communities, and society as a whole.

Fifth, each school of architecture should actively seek to establish a supportive *climate for learning*—where faculty, administrators, and students understand and share common learning goals in a school environment that is open, just, communicative, celebrative, and caring.

Sixth, educators and practitioners should establish *a more unified profession* based on a new, more productive partnership between schools and the profession. The priorities for sustained action between the academy and the profession should include strengthening the educational experience of students during school, creating a more satisfying system of internship after graduation, and extending learning throughout professional life.

Finally, we urge schools to prepare future architects for lives of civic engagement, of *service to the nation*. To realize this last goal for renewal, schools should help increase the storehouse of new knowledge to build spaces that enrich communities, prepare architects to communicate more effectively the value of their knowledge and their craft to society, and practice their profession at all times with the highest ethical standards.

Taken together, we propose an enriched educational climate in the academy and the profession—dedicated, with equal intensity, to promoting professional competence, and to placing architecture more firmly behind the goal of building not only great buildings but more wholesome communities.

DESIGNS FOR RENEWAL:
SEVEN ESSENTIAL GOALS

An Enriched Mission

THE FIRST PRIORITY for renewing architecture education and practice focuses where it must: on the larger purposes of architecture itself.

Membership in any profession, whether law, medicine, teaching, journalism, accounting, or architecture, entails not only the mastery of a body of knowledge and skills but at its best the honoring of a social contract to advance basic human values. Lawyers train to serve clients competently, but as a *profession*, law must be judged, in the end, by how well its practitioners collectively promote the goal of a more just society. In the case of architecture, the larger purposes relate not only to building competently and fulfilling the wishes of clients, but to helping foster, through design, more wholesome neighborhoods, safer streets, more productive workplaces, a cleaner environment, and more cohesive communities.

As one teacher told us, architecture "must serve the interests of client and society, uplift human spirit, preserve the environment, address social issues, and expand aesthetic frontiers. If it does only one or several of these things, it is quite meaningless."

At the core of *all* professions, then, is a public trust. Indeed, any profession that drifts too far in its daily routines and priorities from that mandate risks, in the end, the loss of public confidence and support. Similarly, if professional education is to be successful, its methods and content must flow from goals that include but also *transcend* credentialing students, tenuring faculty, and learning how to practice profitably.

Perhaps the single most heartening discovery we made in our contacts with hundreds of faculty, administrators, students, and alumni was the depth, range, and ambitiousness of social convictions they express for their profession.

An intern at a small firm in Lancaster, Pennsylvania, had this to say: "Some of the major issues on the horizon are the social problems inherent in suburban sprawl, changing our car-crazed lifestyles, creating meaningful communities that people can care about, and educating the American public about this decline in the man-made environment."

A faculty member at the Southern California Institute of Architecture wrote us: "I see architects and architecture as having a wide impact on society in terms of seeing global issues as well as local ones, fostering a sense of connection and responsibility between people, between people and places, in big ways and in small ways, such as bringing in enough light to a room where a child will study and houseplants grow."

A twenty-four-year-old student at North Carolina State University told us: "I perceive architecture as being the pursuit of balance in man's presence on earth. Working for justice to the environment, to the different groups of our society. Working for the good of all but respecting all individuals."

"Architects should cast themselves in the role of 'stewards,' caretakers of the environment," said a Texas A&M faculty member. "Buildings should mediate between, or become a part of, the human-environment connection. Human well-being in places requires knowledge and understanding. These qualities can only be gained when we allow ourselves time and place to think, to probe, to question, to gain insight."

Clark T. Baurer, a partner of the Chicago firm Ware Associates, stated: "In a society that continues to splinter, fragment and stratify, architecture is one of the few professions or disciplines that can speak to the whole of an individual and a society. I would like to think that architecture could appropriately respect and poetically integrate the material and the spiritual, art and technology, the emotional and the intellectual."

We found, in short, a powerful wellspring of social idealism among architecture students, practitioners, and educators that is encouraging and remarkable at a time of general public skepticism toward all professions. The critical challenge, however, is to give greater clarity to the many goals we heard expressed so that the ideals can become more concretely connected to the daily routines of architecture education and practice.

What has society, in fact, entrusted to the profession of architecture, and might that historic and legal mandate be enriched so that architects could be more effectively engaged in society's most consequential problems? Most essentially, how might schools themselves add knowledge and clarity to that mission?

In the first third of this century, as reformers and legislators grew alarmed at the housing shortages, tenements, primitive plumbing, tuberculosis, immigrant poverty, and city fires that accompanied the growth of cities, states introduced language into their licensure laws aimed at ensuring the competence of architects and others in the building fields. That language remains part of the licensure laws of most states, and centers on safeguarding three broad public needs—health, safety, and welfare.

Maryland's code of business occupations, for example, says that the purpose behind regulating architects is "to safeguard life, health, public safety and property, and to promote the public welfare. . . ."[1] In similar language, Minnesota's registration codes say: "In order to safeguard life, health, and property, and to promote the public welfare, any person in either public or private capacity practicing, or offering to practice, architecture . . . in this state, shall be licensed or certified as hereinafter provided."[2]

The fundamental obligation to teach students to build safe, sturdy, healthy structures in which the air is wholesome, fire exits accessible, and windows secure in high winds is no trivial matter. It is, in fact, vital to maintaining public trust in the profession, and we concluded that many design studios could be doing more to make those basic concerns a more integral part of students' work. Still, the terms "health, safety, and welfare" have, for most educators and practitioners, been stripped of their larger meanings, summoning mainly regulatory mandates rather than the nobler ideals of the profession.

It now occurs to us that those terms, if freed from the amber of regulatory language, might point the way to a promising framework for defining, more boldly, the mission of architecture education and practice.

Might "health" include ecological and social health? Could "safety" include not only the proper number of fire exits and staircases, but the promotion, through architectural design, of safe streets and neighborhoods? Might "welfare" include social and economic opportunity and sustainable communities?[3]

We are not proposing that these broader possibilities and meanings be codified in formal regulations. We *are* suggesting that the conceptual framework for an enriched purpose for the profession may, in fact, already exist, camouflaged in the dry language of regulations. As Nicholas J. Mangravit, an independent architect and city planner in Columbia, Maryland, stated to us: "The architect is still licensed to protect health and welfare. While this may have originally been meant in a very narrow sense, it remains a good springboard for broader environmental and social responsibilities which are critical to current practice."

The point is that architects have special knowledge and skills, and a unique way of looking at problems and the world, that could contribute powerfully to meeting the broad challenges confronting individuals, neighborhoods, and nations. A more generous reading of the familiar mandate of "health, safety, and welfare" by architects and the organizations representing them would, we believe, be a promising first step toward opening the way to a broader consciousness, both within the profession and among the public, of the larger possibilities of architecture.

Still, enriching the content of those terms is just the beginning. What remains is to identify specific goals on behalf of society toward which the unique talents of architects could be beneficially employed. After weighing the many concerns and ideals expressed to us by hundreds of faculty, students, architecture school administrators, and practitioners around the country, we propose four specific priorities where the efforts of the profession might be creatively channelled to enrich the mission: *building to beautify; building for human needs; building for urban spaces; and preserving the planet.*

BUILDING TO BEAUTIFY

The commitment to beauty in public works seems at risk in this country. During the Depression years of the 1930s, every bridge, tunnel, and public building had unique, distinguishing aesthetic features. Indeed, since ancient times, the artistic dimension of the profession, the ability to uplift the spirit as well as create shelter, has been architecture's distinguishing hallmark. Ellen Dunham-Jones, a faculty member at MIT, has commented that public architecture in this country needs to recapture this sense of the uniqueness of each project, the "particularity of place."

Creating and preserving what is inspiring in the built environment has always been at the heart of the public trust of architecture. Schools of architecture and the profession must support each other in sustaining that commitment. To enrich the educational mission, we therefore urge schools to prepare future practitioners to be effective public "advocates for beauty," so that architecture's value is felt by *all* communities and individuals, not just those who can afford the fees. And further, architecture programs should, through readings and curricular offerings, actively foster a broad appreciation of the many different ways that beauty is manifested and valued in cultures around the world.

"I have this belief," said the late international architect Edward Durell Stone, "that great architecture will give everyone, the man in the street, the uneducated man, the uninformed man, an exhilaration. . . . The idea that architecture is something that can only be appreciated by a minuscule minority of precious initiates is all wrong."[4]

The supremacy of aesthetics in architecture education and practice has powerful moorings. As previously described, design in most architecture curricula remains an enduring Beaux Arts legacy. The Architect Registration Exam, administered by the National Council of Architectural Registration Boards, requires candidates to demonstrate an ability to use design to satisfy life, health, and safety standards. Engineers and other building professionals have been prevented by courts from engaging in activities that infringe on the aesthetic expertise that is the professional preserve of trained architects.[5] State and federal courts have even held that ensuring more attractive communities legitimately falls within the "police power" of states. A 1954 U.S. Supreme Court ruling states:

> The concept of public welfare is broad and inclusive. . . . The values it represents are spiritual as well as physical, aesthetic as well as monetary. It is within the power of the legislature to determine that the community should be beautiful as well as healthy, spacious as well as clean, well-balanced as well as carefully patrolled. . . .[6]

Still, the public's long-held association of architecture with the aesthetic has proven both an asset and a challenge. Architecture's ability to inspire has always been at the heart of its appeal. The architectural press, and the profession itself, have nurtured this mystique. As a consequence, however, the profession has acquired a rarefied image that makes it seem, at times, less vital to broad community concerns than other professions. During our campus visits and in our surveys, we heard repeatedly that one of the great challenges facing architecture is to convince the public—and those in government and the private sector who fund public works—of the value of beauty in the built environment.

"As irrelevant as this may seem, all people, and especially architects, should be taught to create and understand beauty. It is not valued in our streets," said William D. Stanton, a project manager with Nancy Coslee Power & Associates, a Los Angeles garden design firm with an architectural emphasis.

"The issue for me," said James Chaffers, a professor at the University of Michigan's College of Architecture & Planning, "is how do we define a quality environment, quality spaces, quality relationships? If we can't clarify and translate quality, and get it into the air and into the public's mind, I see us becoming dinosaurs, less and less necessary. We've become convinced that quality is too expensive. But in the long run, quality is less expensive. If we could really get at quality, we'd have to turn work away."

What, then, are the possible implications for architecture education? Over the years, schools of architecture have been criticized for overemphasizing the artistic dimensions of design, often at the expense of technical and practical knowledge. Some have suggested that architecture schools are at times more preoccupied with the merits of modernism *vs.* post-modernism than in serving the more essential goals of the profession and society. "In three years," one department chair told us, "Habitat for Humanity is going to be the largest builder of affordable housing, while architects sit around debating the merits of deconstructionism. What's wrong with this picture?"

As justified as such comments might be at some schools, we nonetheless believe that architecture education should reaffirm, unapologetically, its historic commitment to design excellence, *in the context of a balanced curriculum.* Surely, the nature and value of beauty itself is an apt subject for any aspiring architect, and further, it can readily invite collaboration with other disciplines from art history to anthropology. At Princeton University, for example, architecture students can take a course taught by an anthropologist, "Art, Society, and Culture," which investigates "the creation, interpretation, and evaluation of visual art forms as cultural and social acts. . . ."

But it is also essential that the commitment to the aesthetic have a *civic* dimension, and we urge schools of architecture to prepare future practitioners capable not only of creating beauty, but also able to communicate, clearly and convincingly, its value to the public. And we call upon national architecture organizations to continue to add their voices to elevating the importance of aesthetics in elementary and secondary education.

In an age of strip malls, sterile government buildings, fast food commercialism, and suburban sameness, we believe that educators and the profession should become more effective advocates for beauty in the structures that touch so permanently the lives of everyone in society.

Building for Human Needs

The task of architecture is to create environments that are not only beautiful, but comfortable and practical, relating designs to the psychological, economic, and spiritual needs of clients, inhabitants, passers-by, and future users. "The two primary purposes of architecture have always been to ennoble someone or something and to make someone's life more livable," said Kenneth E. Carpenter, dean of Louisiana State University's School of Architecture.

The intended and unintended impacts of architectural structures on human behavior are profound. Colors and materials can affect everything from emotions and blood pressure to worker productivity and corporate profits. To a child, a school can feel welcoming or forbidding. To an elderly or handicapped person, a structure can be inviting or insurmountable. The design of a house or an office can promote conversation and community, or separation. The nation has seen many instances of how architecture can promote well-being, or at the other extreme, how urban housing projects such as Pruitt Igoe in St. Louis can foster violence and social fragmentation.

"Architecture will not survive if separated from human use and need," a twenty-one-year-old North Dakota State University student told us. "I think it's unrealistic for people to think they can do far-reaching social reform through a building, but they can design an effective building that can provide for better human response."

"Space affects the way people behave," says Walter Wendler, dean of Texas A&M's College of Architecture, which operates a psychophysiology laboratory that conducts research into how environments affect such matters as productivity and emotions. "If you do not believe it, you should not practice architecture. If as architects, we do not find the means to understand this effect, there are probably too many of us. If as architects we can understand and demonstrate this and carry it into the spaces we make, there are far too few of us."

To enrich their mission, we urge that architecture practitioners and educators assume greater leadership in studying how environments affect human well-being, productivity, and happiness. The curricula and design sequences at architecture schools should foster a climate of caring for human needs by including more frequent contact with clients and communities and by placing more emphasis on "environment-behavior" studies. "Our schools are greatly in need of more classes which address how the built environment affects human behavior," a twenty-four-year-old student at the University of Michigan wrote us. "We cannot continue to teach as it was done decades ago. Society is changing. The values and the needs of the people we design for are changing. In order to meet their needs, we must change with them. . . ."

Further, the profession and the academy should collaborate on producing new knowledge aimed at clarifying how architects can create more effectively environments that enhance these goals. Put another way, "caring" design education should be as preoccupied with the *interior* of buildings as the *exteriors*. "It is vital for the architect to understand that he is a designer for humans. Accordingly, more knowledge should be gained about the physical needs, emotional needs, and broader social requirements to ultimately make the architecture successful for the users," a faculty member at a midwestern public program wrote us.

We found, in fact, classes and studios where such issues are being effectively explored. Norwich University offers a course called "Human Issues in Design" which explores "the range of human needs and desires . . . that must be fulfilled through the built form."[7] At Andrews University, students take a lecture course in human behavior for one academic quarter, followed by a design studio dealing with that topic. Faculty at the University of Michigan, the University of Wisconsin–Milwaukee, Roger Williams University, Kansas State University, and the University of Washington have developed rich offerings in environment and behavior.

In a graduate-level course titled "Design Methods and Programming," North Carolina State University professor Henry Sanoff translates the notion of caring design into practical terms. He teaches students the importance of group dynamics, interviewing, and listening skills in developing designs that respond to human needs. And he is critical of what he calls "introspective architecture" that makes little reference to anything but the creative impulses of the architect. Such reasoning, Sanoff says, is contrary to caring architecture, and tends to go like this: "I'm human, I'm designing for other humans, so why can't I be the model for what all other human beings need in the built environment?"

Sanoff spent much of a three-hour class sharing his recent experience of helping develop plans for a new elementary school in Davidson, North Carolina. "The problem at the core is that clients can express their needs, but they do not understand the spatial consequences of their ideas."

This is what Sanoff accomplished at the Davidson school. He had teachers create architectural models and drawings of their ideal rooms. He asked teachers and administrators to describe the teaching strategies and learning goals of the school. And he asked parents, teachers, and grade-school students to write a "Wish Poem," with each stanza beginning, "I wish my school. . . ." As a result of what he learned from children, for example, door handles in the school are placed just thirty inches from the floor—a bit low for adults, but just right for young children. From the suggestions of teachers, classrooms are arranged along one side of the halls only, to reduce noise. And there are large open spaces in the halls to allow students to gather for special projects.

"This is the difference between 'good' design and 'right' design," Sanoff says. "The rightness or wrongness isn't easily photographed. It has to do with whether the building works for those who will use it."

In the end, building to meet human needs means helping architecture students become effective teachers and listeners, able to translate the concerns of clients and communities into *caring design.*

BUILDING FOR URBAN SPACES

The architecture profession is rooted in the urban setting. Nowhere are the social impulses of architecture clearer than in its forays into city and community planning, from Frank Lloyd Wright's Broadacre to Le Corbusier's urban designs. The influence of architects, of course, has always been limited in shaping the urban landscape, and this is true today, with nonarchitect developers and others in the building field increasingly taking on the architect's role. Still, cities always have been, and will likely remain, the greatest canvas of the profession.

In the last half century, however, urban, suburban, and even rural America have moved relentlessly from an era of growth and seemingly limitless space to life in increasingly cramped quarters, and society can ill afford architecture that ignores context. British architectural writer A. Trystan Edwards, FRIBA, made the point elegantly fifty years ago:

> Can a haphazard assemblage of buildings, each conceived in isolation, and expressing nothing but its own immediate purpose, really be described as a city? What attribute is it which makes a building urban? My answer . . . may seem simple and tautological, but . . . in order that a building may become 'urban,' it must have been *urbane*. Now, urbanity . . . is nothing more nor less than good manners, and the lack of it is bad manners. I think I shall have little difficulty in showing that there can be both good and bad manners in architecture.[8]

A significant number of schools are, in fact, applying architectural expertise to urban problems and preservation. Twenty-three schools offer specializations in preservation within the architectural degree program, and twenty-four offer degree specialties in urban design. Yale University, with its Center for Urban Design Research, is among the leaders in studying the problems of urbanism and applying architectural knowledge to those issues. Students get first-hand experience in New Haven, and have participated in design "charettes" in Norfolk, Connecticut, among other places.

Still, the future requires, we believe, a more holistic architecture education which stresses, *at its core*, not only the creation of newness but also the skills and knowledge needed to value and enrich what already exists through preservation and restoration. Schools of architecture should also take a greater lead in developing new knowledge to make life in shrinking spaces safer, more wholesome and satisfying.

"The contemporary American city should be our focus," Michael Underhill, director of Arizona State University's school of architecture, wrote us. The challenge, he said, is "to find ways to grow cities with social systems that work, with environmental sensitivity and cultural meaning."

Design problems should *routinely* include such matters as safer streets, promoting mass transit instead of cars, and creating inviting gathering places that foster community instead of social isolation. James F. Barker, FAIA, dean of Clemson University's college of architecture, put it this way: "The challenge will be for architecture to directly address the real problems of our time: homelessness, urban decay and crime, and the destruction of 'community.'"

We found considerable interest on campuses in studying design in the urban context. Students at the New Jersey Institute of Technology are engaged in projects connected with its setting in Newark. The architecture school is participating in the redesign of a science park, the replanning and redevelopment of the downtown area, and in the development of prototype housing. "We use Newark as a lab," said Urs Peter Gauchat, dean of the college of architecture.

Similarly, University of Illinois architecture students have helped carry out urban projects in East St. Louis, and have learned valuable lessons not only in urban ecology and design, but also in the realities of white flight and disinvestment. In one studio, students met low-income residents to discuss what might be done to provide better housing. Students worked in teams, competing to come up with an acceptable project. They presented each idea to a panel of residents to select one proposal, then the entire class went to work on the preferred project.

For more than a decade, Ball State architecture school's Muncie Urban Design Studio has collaborated with community groups, the city government of Muncie, Indiana, and private industry to help revitalize neighborhoods. The "MUDS" studio meets in a restored storefront in downtown Muncie, and students learn how to confront and solve practical architectural problems on actual construction sites. In particular, the MUDS studios have set out to prove that it's possible to design and construct a high-quality, low-maintenance home for $49,500.

At North Carolina State, an architecture course called "Anatomy of a City," taught by Peter Batchelor, has won national honors. At the University of Virginia, the second year in the studio sequence is now devoted to gaining an understanding of the city and its natural setting.

Taught by architects, planners, and historians, it places priority on urban issues rather than single buildings.

A number of rural architecture schools, meanwhile, including Mississippi State, Auburn, and Clemson, have developed programs in which all students spend their final year in an urban-based studio which exposes them to the problems and culture of cities.

In an increasingly congested world—with more people, less space, and scarcer resources—architecture schools should prepare graduates to apply their design knowledge to preservation and renovation as much as the creation of "newness." The goal must be to help cities and smaller communities become safer, healthier, and more wholesome. A student at MIT said it best: "If architecture does less than make great cities, it fails."

PRESERVING THE PLANET

As a final priority, we propose that architects and architecture educators assume a leadership role in preserving the environment and the planet's resources. It is this priority, we are convinced, that could have the most far-reaching implications about the way schools, and the profession itself, conduct themselves in the next century. "In the next ten to twenty years architecture should become synonymous with stewardship of the built environment," Harry F. Kaufman, dean of the Southern College of Technology, wrote us.

Mary Ellen Winborn, a construction administrator with Lindberg Architects in Port Angeles, Washington, had this to say: "I believe preserving the environment is quickly becoming the larger purpose of architects in the next ten or twenty years. We have a responsibility to use less and get more. We need to be economical and rethink some of the extravagant ways of our modern predecessors."

"Architects today too frequently exist to serve their 'own-ness,'" a fifth-year student at Mississippi State said. "Architecture should be about creating a better environment—for everyone."

In our campus visits and in hundreds of conversations with educators, students, and practicing architects, we found widespread enthusiasm, particularly among students, for more environmentally responsible, or as it's come to be called "sustainable," architecture. But we found less certainty about how to incorporate that goal into the methods and content of architecture education.

A number of schools we visited or learned about are making impressive beginnings. California State Polytechnic University–Pomona's new Center for Regenerative Studies is an interdisciplinary enterprise involving the whole university, dedicated to preserving the environment. Students can earn credit for time spent at the solar-powered center, located right on campus. Eventually, housing for up to ninety students will be available at the center. Activities include growing crops, raising fish, processing sewage, and building structures with nontoxic renewable materials. The center is a hands-on laboratory of sustainability.

The University of Virginia is establishing the Institute for Sustainable Design to enhance research and instruction. Ball State's architecture program has been conducting a comprehensive review of its curriculum and teaching methods to incorporate the principles of sustainable design.

The educational implications of sustainable design for architecture programs, as we heard them repeatedly described, go well beyond joining a "green" movement, or adding a course or two to an already crowded curriculum. Alan J. Plattus, associate dean of Yale's School of Architecture, emphasizes that for sustainability to succeed, it must be part of "the mainstream of architectural education and practice, and not become identified with a special interest group or time-bound set of ideas." The goal, he continued, "cannot be to win converts to a movement . . . but rather to develop models that demonstrate the centrality of these issues within the core of our discipline."[9]

Intriguingly, many of the ideas being advanced in support of sustainable architecture echo reform proposals advocated by others for decades. For example, a program called Project EASE (Educating Architects for a Sustainable Environment), directed by Marvin E. Rosenman, chair of the Department of Architecture at Ball State, sponsored several national meetings of educators in 1994 and 1995 to consider what sustainability might mean to the education of architects. Participants came up with a long list of suggestions that have been advanced by reformers for years. Among them:

- tying studios to more "real life problems";
- developing "a fabric of many voices in studio instruction";
- replacing the "architect as hero" model with "architect as team player";
- "acknowledging the curriculum as one phase in life-long learning";
- "promoting an interdisciplinary/collaborative approach among designers, sociologists, ecologists, etc."; and
- "developing a solid theoretical and research base."

As we view it, the enriched mission of preserving the planet adds powerfully, first of all, to the goal of making learning in the studio context more *integrative*. Sustainable architecture requires, by its nature, looking at the connections in design problems, studying the community context of buildings, and seeking comprehensive solutions. If schools were to teach sustainable architecture, for example, design problems would be more likely routinely to include variables such as present and future energy and materials costs.

Such questions also point to the need for architecture schools to join with the profession and other disciplines to develop more new knowledge about sustainable design. The profession, schools, and students should expand their knowledge, for example, of energy, the use of renewable resources, the recycling process, the use of carcinogenic materials, and the safe disposal of waste.

Sustainable architecture also suggests a curriculum built around collaboration and teamwork, not only with other architects but with other disciplines. Simply stated, it is hard to imagine that architects, if conditioned in school to design alone, or only with fellow architects, could ever have much impact on preserving the world's resources.

Finally, sustainable architecture requires an educational program aimed at fostering social engagement and more effective communication with clients and the public. Schools must prepare graduates, in other words, to be effective teachers and learners, able to guide clients and communities to make sound design choices that balance both private and public concerns.

Susan Maxman, a practitioner from Philadelphia and recent past president of the American Institute of Architects, described her experiences in persuading clients of the benefits of sustainability this way: "My clients don't start out knowing that they want a 'green' building, but it takes a lot of talking and doing to convince them to spend more money and to convince them there is a payoff down the road. You have to convince people to spend upfront money to save long-term. You need to go to the legislature and convince them to spend more upfront on public works."

EACH OF THE FOUR PRIORITIES we have described to enrich the mission of architecture education and practice—building to beautify, building to meet human needs, building for urban spaces, and preserving the planet—connects to the larger goal of stewardship for the physical environment that we, as a nation, will pass on to our children. At the same time, enriching the mission of architecture must also mean strengthening connections with other professions and disciplines, and the college campus offers virtually endless possibilities for meaningful collaboration. We imagine, for example, faculty from architecture working together with schools of education and social scientists to develop elementary and secondary school buildings that further the goals of education reform. And we imagine architects teaming with medical schools or schools of public health to design better health facilities.

To those who worry that architects have neither the "clout" nor expertise to undertake such an imposing civic and social agenda, the answer is that they are right—if architects and architecture educators see the profession as acting in isolation, or locked in competition for turf. Architects, wrote Eric A. MacDonald, a manager at the Michigan State Historic Preservation Office, must "break down barriers of expertise and improve communication between architects and the public, and other professionals. . . ."

"Instead of fighting with allied professions," a student from the University of Illinois told us, "we need to work with other professionals in the future to ensure the health, safety and welfare of people, and also to leave a lasting contribution to future generations."

Enriching the mission of architecture, in the end, means embracing the professional responsibility to build not only for the moment, but for the ages.

Diversity with Dignity

THE DIVERSITY of the types and philosophies of architectural programs and the richly varied backgrounds and talents among the nation's architecture faculty are strengths that must be preserved. The landscape of architecture education should be dynamic and varied, serving the changing needs of students, the profession, and society through excellence in teaching, research, and the application and integration of knowledge. There should be, in other words, multiple peaks of excellence, enabling architecture education as a whole to respond more creatively to emerging opportunities and to serve effectively the wide-ranging career goals of future architects. Architecture scholarship, with its distinctive features and traditions, has much to gain from, and much to offer, the college campus.

There has emerged over the last century, first of all, an extraordinary level of *diversity among architecture programs*. The architecture field and its students have been, we believe, well served by the variety of settings in which education takes place: in freestanding academies, schools of technology, land-grant institutions, liberal arts colleges, research-oriented campuses, and community colleges. Some programs are housed in self-contained academic departments, while others are in colleges or units with related disciplines such as construction science, landscape architecture, interior design, or urban planning.

Further, a broad range of program *types* has helped open the doors of the profession to a more diverse group of students. There are five-year undergraduate bachelor's degree programs that accept students directly from high school, six-year master's degrees, and three- to four-year graduate programs. Over the years, schools have developed variations of those basic program types, and quite a few offer students a choice of all three. Some schools have selective admissions policies, while others admit students with a wider range of capabilities who must then survive a rigorous "winnowing" process later on.

WE ALSO FOUND, among architecture programs, a rich *diversity of missions*. That variety mirrors, in key respects, the scholarly emphases and strengths of the campuses they are part of, whether research, teaching, or service. Beyond those overarching priorities, many architecture programs have developed their own distinctive personas and specialties. Prospective architects can steep themselves, for example, in Miesian modernism at the Illinois Institute of Technology, or travel about ninety miles east to the University of Notre Dame and be part of a neoclassical counterrevolt. The University of California at Berkeley, Yale University, and Pratt Institute are leaders in urban architecture. Boston Architectural Center, Rice University, the University of Cincinnati, and Drexel University offer distinctive work-study programs that leave graduates well versed in office life. Mississippi State has been a leader in exploring the potential of computer technology in design education, and Texas A&M operates half a dozen architectural research centers studying a range of matters from the impact of natural hazards to improving life for impoverished communities along the Mexican border. Ball State has a growing reputation as a center for environmental issues.

Finally, along with diversity of programs and scholarly missions, there exists, among the nation's nearly 3,800 full- and part-time architecture faculty, a rich *mosaic of talent*.

Students at Princeton can study with eminent faculty including historian Anthony Vidler, sociologist Robert Gutman, and the world-renowned designer Michael Graves. Pratt Institute has more than one hundred adjuncts, many in active practice. At North Carolina State, senior faculty such as Peter Batchelor, Georgia Bizios, Robert Burns, Roger Clark, and Henry Sanoff are genuine practitioner-teachers, combining scholarly productivity and teaching excellence with professional accomplishment.

Remarkably, fewer than half of the nation's full- and part-time architecture faculty—48 percent—are licensed architects, according to data from the National Architectural Accrediting Board. And we found that many architectural faculty hold their highest academic degrees in fields other than architecture, from engineering to philosophy. At the fifteen campuses we visited, fewer than one-quarter of architectural faculty held doctoral degrees in *any* field—placing them outside the usual mold of academia.

Surveying this varied landscape of faculty talents, credentials, and backgrounds, Gregory S. Palermo, of the University of Iowa, has suggested that architectural faculty might be grouped under five broad headings:

- those whose backgrounds are mainly academic;
- those who combine teaching with limited practice, representing the core of full-time design teachers;
- those with well-balanced careers in teaching and practice—an increasingly rare breed;
- practitioners who teach, mostly on a part-time or adjunct basis; and
- practitioners who work outside academia but devote time to mentoring graduates and providing internship opportunities.[1]

During campus visits, we met faculty whose scholarly work reflected commitment to civic engagement and the application of learning to consequential problems. Sharon E. Sutton, of the University of Michigan's College of Architecture and Urban Planning, operates programs as part of the school's Urban Network, which centers on how young people can participate in shaping the built environment. At Ball State's Housing Futures Institute, Stan Mendelsohn has been working to develop a "therapeutic urban village" for the elderly in Bloomington, Indiana.

Some faculty are also deeply engaged in published or funded research activities. A 1992 survey by the Architectural Research Centers Consortium identified $21 million in funded research projects by architecture faculty in approximately thirty leading research campuses, exploring areas from acoustics and lighting to history and theory, and even tropical planning and design. Teachers at North Carolina State have received outside funding to explore areas such as software development and masonry. Faculty at Princeton have engaged in extensive consulting work, research, and scholarly publication on a vast range of subjects from Italian Renaissance architecture to structural design issues and city planning.

In our campus visits, we were also impressed, time and again, by the dedication of faculty *to teaching*, even at the expense of their own advancement on campus. "Programs in architecture often find themselves in vulnerable positions within their institutions," wrote William G. McMinn, FAIA, dean of Cornell's architecture school. "Design faculty are strongly committed to teaching rather than sponsored research, or to their practice rather than academic scholarship."[2]

In some schools, the teaching loads of faculty, especially junior instructors, are formidable. At one midwestern school we visited, the teaching load for the program's sixteen faculty members was said to be 160 percent above accreditation guidelines, while salaries were 26 percent below the campus average. In our survey, we found that nearly two-thirds of faculty agreed that teaching should, in fact, be the *primary* criterion for evaluating their performance.

If anything is clear, then, there is no single "right" approach to architecture education, nor any single form of scholarship that ought to predominate. The need for diversity in programs and faculty talent is, in fact, more urgent than ever as a changing job climate has made it essential that graduates leave school prepared to move among careers within, and beyond, "traditional" practice. The American Institute of Architecture Students, in a recent review of possible career options, listed no fewer than 107 "alternative" job paths for those studying architecture, from jewelry design to facilities planning, yacht building, and computer software design. Strikingly, traditional practice in architecture firms accounted for only five of the possible options listed.[3]

"There is a relationship between diversity in training and diversity in practice," said Ralph Lerner, dean of Princeton University's School of Architecture. "We ought to celebrate the differences."

Almost unanimously, practitioners, students, and educators believe that the diversity of philosophy and content of the nation's schools of architecture is a strength that ought to be preserved (table 9).

Table 9

THE DIVERSITY OF CONTENT AND PHILOSOPHY
AMONG U.S. SCHOOLS OF ARCHITECTURE IS A
STRENGTH AND SHOULD BE PRESERVED

	AGREE	DISAGREE	DON'T KNOW
Administrators	91%	8%	1%
Faculty	93	5	2
Students	87	7	6
Alumni	90	8	2

SOURCE   The Carnegie Foundation for the Advancement of Teaching,
Survey on the Education of Architects, 1994.

Still, in considering all the evidence, we were concerned that the
diversity of architecture education and scholarship is increasingly threat-
ened. Looking at higher education as a whole, diversity has been placed
at risk by a tendency to imitate the most prestigious and wealthiest
research institutions. Several years ago, in a report by The Carnegie
Foundation entitled *Scholarship Reconsidered: Priorities of the Professoriate,*
we warned that a single, research-driven model of scholarship has come
to dominate higher education, and many institutions have become "more
imitative than *distinctive.*"[4]

"Rather than defining their *own* roles and confidently shaping their
own distinctive missions," we wrote, "campuses increasingly seek to gain
status by emulating research centers."[5]

In a similar way, architecture programs and faculty find themselves
under growing pressure from their own campus leadership, and even from
their own ranks, to conform to more traditional models of scholarly
activities and rewards. "Whatever we do," said the dean of one public
institution, "it's no longer going to be possible for someone to get tenure
without a serious scholarly component."

And we were concerned that the methods and yardsticks used to
assess scholarship throughout higher education do not adequately affirm

the talents and interests that are specific to architecture faculty. More than two-thirds of faculty we surveyed felt that their schools should give more recognition to professional service and the application of knowledge, which together constitute much of the scholarly output of architecture teachers.

At a prestigious New England program, not a single studio instructor has won tenure in the last sixteen years. Younger faculty, especially, find it nearly impossible to establish their academic *bona fides* at this school where exceptions are hardly ever made to campuswide criteria for tenure and promotion developed with traditional scientists in mind. The architecture department awards long-term contracts to studio instructors to ease the worst effects, but many simply throw up their hands and leave after the requisite seven years, convinced they'll never satisfy the tenure criteria.

At another research-oriented institution we visited, an architecture faculty member with six years' teaching experience told us he was "very likely" to quit because, "I do not expect to be given tenure. How does one bridge design to research, within the context of university standards?"

A faculty member at a prestigious private institution described the difficulties in gaining recognition on his campus: "The way I, as a designer, am evaluated and what constitutes 'research' is not in line with academic tradition. The problem for a young designer-academic is that there are few opportunities to build buildings. We are stuck between the humanities and the professions. When I sit down with the dean each year, the messages are mixed and the requirements for promotion and tenure are blurry."

The realities are often difficult for senior faculty as well. A widely published teacher at an East Coast public institution told us that the university administration has been pressing architecture faculty to publish more, but has not provided additional time, resources, or administrative support to compete with wealthier schools in the chase for scarce research funds. Library and laboratory facilities at many schools are limited, and release time from teaching duties is hard to come by, particularly at schools with small teaching staffs.

A senior scholar we met put it this way: "If funds are sufficient, you buy yourself an assistant to give you time. But research is rarely funded at that level. So it's really very hard to teach, conduct architectural practice, and do research."

When we asked architecture faculty if they agreed or disagreed with the statement that architecture schools should count published research more heavily in making faculty promotion and tenure decisions, 58 percent disagreed.

In raising these concerns about the pressures to conform to a more research-driven model of scholarship, we are not suggesting that the architecture field wouldn't benefit from more and better research. To be sure, many architecture scholars argue, with considerable justification, that the collective research effort in the field should increase for the sake of the profession and the credibility of the discipline. We heard repeatedly, in fact, that architecture faculty have had difficulty convincing government and private funding sources of their skills as researchers.

Still, our deeper worries center on those architecture schools and faculty who are being pressed to "change their stripes" more by the external imperatives of prestige than by their *own* planning or self-defined objectives. Such change for the sake of conformity will, we believe, ultimately diminish the quality of research and, in all probability, widen the gulf between architecture schools and the profession. All too often, schools with few resources for research have sought to imitate ranking research centers. The result has been that research standards are diluted, the commitment to teaching and learning is compromised, and diversity among programs is diminished. The last thing schools of architecture need to do is to go down the same path of conformity that much of higher education has taken in recent years.

In this context, we found, unsurprisingly, serious morale problems among teachers at the fifteen schools we visited. On the positive side, more than half of the faculty we surveyed—52.4 percent—said they were "more satisfied" with their academic careers than they were five years ago. But 40 percent also said they were likely to leave their school of architecture within five years, for reasons other than retirement. Revealingly, nearly 44 percent said their jobs were "the source of considerable personal strain."

Salaries were a frequent complaint. Only 4 percent of faculty we surveyed called their salaries "excellent," 24 percent considered them "good," 45 percent rated their pay "fair" and about 24 percent said "poor." Interestingly, deans and other administrators took an even grimmer view: only 22 percent called faculty salaries at their schools "excellent" or "good," and 78 percent considered them either "fair" or "poor."

Faculty dissatisfaction over salaries relates not just to the amount, but to where those salaries seem to place architecture in the academic scheme. According to a 1994–95 survey of faculty salaries at over eight hundred institutions, average architecture faculty salaries are somewhat ahead of education, social work, communications, sociology, speech pathology, psychology, art and drama, English or foreign language, but well below the "harder" professional fields such as accounting, business, and engineering.[6]

We sensed, then, that some architecture programs and faculty are at a difficult crossroads. Universities are pressing faculty in all disciplines to produce more research, and architecture teachers who lack research expertise or are carrying heavy teaching loads and active practices have found it increasingly difficult to thrive or even survive in the campus climate. Further, we found that the architecture community is itself quite divided over the wisdom of pressing for more research, and if so, who should set the priorities. Finally, few campuses truly recognize or support faculty engaged in the *scholarship of integration*—making connections between architecture and other disciplines.

How, then, should we proceed?

In *Scholarship Reconsidered: Priorities of the Professoriate*, we suggest that the success of the drive for better education will be determined, in large measure, "by the way scholarship is defined and, ultimately, rewarded." We argue that the tired old "teaching versus research" debate should be replaced by giving the term "scholarship" a more capacious meaning, "one that brings legitimacy to the full scope of academic work."[7] We proposed that the work of the professoriate in all fields—including architecture—be thought of as having four separate yet overlapping functions, each richly deserving affirmation by all institutions of higher education:

- *the scholarship of discovery*, research that increases the storehouse of new knowledge within the discipline;

• *the scholarship of integration*, including efforts by faculty to explore the connectedness of knowledge within and across disciplines, and thereby bring new insight to original research;

• *the scholarship of application*, which leads faculty to explore how knowledge can be applied to consequential problems in service to the community and society; and

• *the scholarship of teaching*, which views teaching not as a routine task, but as perhaps the highest form of scholarly enterprise, involving the constant interplay of teaching and learning.[8]

The goal of widening the scope of scholarship beyond the old dichotomies of teaching and research relates to the need to affirm and sustain multiple missions among schools and faculty. Regarding architecture education nationally, we envision a landscape of programs that rewards scholarship in all forms, but in which each program is free to develop fully its own particular scholarly strengths and make its own unique contribution to the collective effort of architecture education. At no school should any facet of scholarship be neglected, nor should individual faculty be expected to lead unbalanced scholarly lives.

It is also clear that each program will assign different weight to each aspect of scholarship depending on tradition, resources, the nature of the faculty, and the culture of the institution as a whole. The scholarship of discovery would undoubtedly be a stronger priority at schools where the resources and expertise for research were adequate. Elsewhere, the scholarships of application and the integration of knowledge would be more vigorously pursued. At *all* programs, teaching and learning would be at the core. Based on our classroom observations, we concluded that many faculty, both beginners and veterans, could use help with teaching skills. Nearly 60 percent of the architecture faculty we surveyed agreed that their school or department would benefit from more sustained teaching training.

However the mission is defined, each program should develop a system of faculty recognition consistent with what the school and the campus as a whole are trying to accomplish. We found, in fact, campuses where warm relations exist between architecture educators and university administrators, and one of the key factors seems to be reasonable agreement about how scholarship should be defined, supported, and rewarded. Mississippi State University, for example, fought and won the right to open the state's only architecture program twenty-three years ago, and remains highly esteemed by the university's leaders. The program has further boosted its campus status with its national leadership in computerized design. There is also apparent agreement between university administrators and architecture faculty about the value of applied scholarship to serve the needs of communities and the state.

"The architecture school fits in beautifully here," Mississippi State's president, Donald Zacharias, told us. "Much of this has to do with personalities, and the calibre of students the school attracts. But this school is also continuing to build and strengthen its reputation nationally."

Urs Peter Gauchat, of the New Jersey Institute of Technology's School of Architecture, told us that "within our university administration, the program in architecture is held in high esteem. The School of Architecture is treated on an equal footing as other schools and colleges, even though with its 650 students it is the smallest academic college or school on campus."

No one should underestimate the challenges of changing a century-old university culture that has, by and large, valued one form of scholarship above the rest. Nonetheless, we find reason to hope that a more balanced view of scholarly work may be within reach. In a 1994 Carnegie survey, we asked chief academic officers of all 1,380 of the nation's four-year institutions whether their schools had reexamined faculty roles and rewards in the last five years. Sixty-six percent of those responding said such a review had been completed or was underway, and another 17 percent said that a review was planned (table 10).[9]

As relates to architecture education, we are especially eager for universities to give greater recognition to professional and civic service— *the scholarship of application*. We found it encouraging, in our national survey of chief academic officers, that 38 percent have already developed new methods of evaluating faculty in the area of applied scholarship.[10] And 41 percent told us that, for purposes of faculty advancement, applied scholarship counts more at their institution today than five years ago.[11]

Table 10

IN THE PAST FIVE YEARS, HAS YOUR COLLEGE OR
UNIVERSITY REEXAMINED FACULTY ROLES AND REWARDS?

	PERCENTAGE
Yes, the review has been completed	21%
Yes, the review is still underway	45
No, but we plan to initiate a review soon	17
No, we do not plan to initiate a review soon	18

SOURCE   The Carnegie Foundation for the Advancement of Teaching, Survey on the Reexamination of Faculty Roles and Rewards, 1994 (survey of provosts).

In recommending greater recognition of applied knowledge, we recognize that not all community or professional activity should be considered "scholarship," in the formal sense. Building a building or performing community or professional service does not constitute scholarship unless that activity relates directly to the intellectual work of the scholar and leads to results that are broadly relevant and susceptible to scholarly review. As Harvard historian Oscar Handlin observed, our troubled planet "can no longer afford the luxury of pursuits confined to an ivory tower. . . . Scholarship has to prove its worth not on its own terms but by service to the nation and the world."[12]

We suggest that elevating the importance of *applied* knowledge, so central to the nature of architecture, will enrich the *discovery* of knowledge, not diminish it. As we wrote in *Scholarship Reconsidered,* "The *scholarship of application* . . . is not a one-way street. Indeed, the term itself may be misleading if it suggests that knowledge is first 'discovered' and then 'applied.' The process we have in mind is far more dynamic. New intellectual understandings can arise out of the very act of application—whether in medical diagnosis, serving clients in psychotherapy, shaping public policy, creating an architectural design, or working with the public schools. In activities such as these, theory and practice vitally interact, and one renews the other."[13]

Beyond defining architecture scholarship more generously, how might a more expansive view of scholarly work be more fairly assessed and supported? To begin with, we believe there are certain basic principles of scholarly assessment that should be universally embraced.

First, the focus of evaluation should center on the work that faculty are assigned. Surely, if young faculty are ordered to spend most or all of their time and energy on teaching, it is unfair to penalize them for failing to do other work for which they are given neither time nor resources to accomplish.

Second, broad scholarly expectations for assessing faculty work should be clear from the beginning, and those expectations ought to be arrived at collegially, not imposed from above. The overall framework for assessing scholarly work must be decided campuswide, but each college or department ought to have more freedom and autonomy to determine how those broad principles of scholarly assessment might be appropriately applied.

Third, in evaluating the productivity of faculty, universities should recognize that no professor can be expected to engage continuously, and with unflagging energy, in a single form of scholarly activity year after year. In other words, the reward system of higher education should be more individualized and flexible to promote renewal and avoid burnout. Scholarly assessment should recognize career ebbs and flows, and certainly no faculty member should be branded as "dead wood" if the focus of his or her work shifts for a time from published research to teaching, or the application of knowledge, or the integration of learning.

In *Scholarship Reconsidered*, we propose that colleges and universities develop "creativity contracts," in which individual faculty would be asked to define their professional goals for a three- to five-year period, allowing for shifts along the way among the different kinds of scholarly pursuits. The goals are to increase and sustain productivity *across a lifetime* and to encourage the diversity of scholarship within institutions.[14]

In the context of architecture education, we urge schools to establish a genuine climate of scholarly integration. In practice, the degree to which such integration succeeds in any program appears to depend to a considerable extent on the willingness of faculty and administrators to support and reward such collaboration. More than a dozen architecture programs, for example, are housed in multidisciplinary colleges that include construction management, landscape architecture, art or engineering, but proximity alone is no guarantee of interdisciplinary activity.

Efforts at integration often face institutional obstacles such as scheduling and giving faculty credit for the time and energy they expend in developing courses in collaboration with others outside their own discipline. Even at supportive institutions, the degree to which integration occurs seems to rest mainly on the receptivity of the design instructor in charge, and also depends on the time and willingness of nondesign faculty who often get little credit for appearing in studio.

An interdisciplinary seminar formed in 1995 by Clemson University's College of Architecture, Arts, and Humanities illustrates the challenges. The seminar brought together students and faculty from architecture, landscape architecture, construction engineering, finance, and engineering with the aim of helping the Catawba Indian nation in North Carolina with its housing and development problems. James Barker, dean of the college, told us that making such integrative scholarship succeed involved not only structural changes in departments, but changes in the culture of the university and its rewards.

"One of the first challenges was to come up with a common language among students and faculty," said Bob Bainbridge, an architecture faculty member who helped plan the seminar. "I'd say, let's do a 'charette.' Someone from another discipline would throw out a term like a 'cost proforma.' Are we dealing with 'customers' or 'clients'? So our first assignment was to identify five jargonistic terms from each discipline, provide definitions, and develop a glossary."

Each school had to figure out where classes would meet, and how to award faculty credit toward their departmental teaching load. Once the deans assured faculty that such issues were not going to be obstacles to success, the administrators stepped aside and let faculty work out the details. The departments gave faculty release time to plan the course, and the university provided a $22,000 planning grant.

In a critical sense, then, these changes in scholarly climate must come from faculty themselves. Faculty are, collectively and individually, the gatekeepers of the academic culture and its values, and little change of any consequence will occur without their full participation. Our recommendations, however, are also directed at university administrators and trustees whose leadership is key to changing the reward systems which faculty of all disciplines on campus must live under. The essential point is that the university culture as a whole must change so that all forms of scholarship are recognized and rewarded.

At the same time, we agree with William McMinn, of Cornell University, that the profession and national organizations have a critical role to play in advocating more effectively the cause of architecture education with university administrators. Specifically, McMinn has urged that the national architecture groups and their local affiliates promote conferences that improve the dialogue and understanding among practicing architects, teachers, and university administrators about the special goals and strengths of architecture education.[15]

To summarize, the challenges to achieving or sustaining diversity with dignity—meaning respect for scholarship in all its forms—are surely not unique to architecture education, nor can they be addressed in a vacuum, without reference to the larger forces of conformity bearing down on all of higher education. Still, the pressures to conform to a single mode of scholarly activity and rewards have special poignancy for architecture programs, whose great strengths—dedication to teaching, and the application and integration of knowledge—are often among the least affirmed on campus. We agree that the gravity of problems confronting communities and the physical environment demand more architectural knowledge. At its best, the scholarship of discovery contributes in a vital way to the store of architectural knowledge while renewing the intellectual climate of the university as a whole. Still, it is simply unacceptable to judge all architecture faculty by a single, narrow set of criteria—as wrong as it would be to judge all practicing architects solely by their aesthetic talents.

Schools of architecture should take pride in their diversity and hold fast to it. And practitioners, national architecture groups, and university administrators should renew their commitment not only to confront architecture education's flaws, but also to help sustain the variety of programs and faculty talents that contribute so richly to the profession and to campus life.

Standards Without Standardization

So far, we have proposed that schools of architecture enrich their mission so that future architects can contribute to beautifying and preserving communities and the environment while creating sound, economical structures that fill the needs of users. We then suggested that to realize those priorities, there ought to be diversity with dignity, a scholarly environment in which the many strengths of architecture programs and faculty are preserved and rewarded.

We now propose, as a third goal, *standards without standardization*—a framework of common expectations for students in all programs, but one which also affirms the diversity so vital to the well-being of the profession. Certainly it is reasonable to expect that all students should graduate with a grasp of a widely agreed-upon set of skills and knowledge needed to contribute as a beginning member of the profession.

Specifically, we propose new language and categories for the architectural accreditation standards, language driven by the conviction that the standards used to evaluate student work and program performance should be organized not so much around blocks of knowledge as around *modes of thinking*: the discovery, application, integration, and the sharing of knowledge. Using those new categories, the work of students and the goals of architecture programs would thus be brought into closer harmony with the principles of scholarship for faculty work already described. Further, by shifting the principle focus of standard-setting from content to modes of thought, the resulting accreditation standards might be less prescriptive, more flexible, and more supportive of the rich diversity among programs.

What we envision, then, is a clearly defined middle ground between no agreement whatever among architecture programs concerning the content of a professional degree program—"letting a thousand flowers bloom"—and at the other extreme, a rigid, unitary set of standards which stifles diversity or creative approaches.

For much of this century, higher education has accepted voluntary self-regulation in standard-setting and accreditation because it feared the alternative—outside control by public agencies. The accreditation organizations that emerged, in architecture and scores of other disciplines, serve a number of essential ends. They establish a framework for ensuring a certain level of quality of educational programs on campus—all the while preserving diversity and leaving intact the authority of the campus. They provide various regulatory and governmental bodies with the yardsticks they need to determine which schools are eligible to participate in federal and other programs—while keeping government itself clear of the role of gatekeeper. Last and hardly least, the establishment of standards by accrediting bodies helps define and protect the boundaries of the disciplines and professions.[1]

Since its founding in 1940, the National Architectural Accrediting Board, Inc., has fulfilled these functions for architecture education, and the fifty-three "performance criteria" the board uses in accrediting programs are, beyond question, the closest thing in existence to "national standards" in architecture education. Those criteria outline, in broad strokes, what all students pursuing a professional architectural degree should know, understand, and be able to do upon graduation, regardless of the school they attend. In a setting, however, where architecture educators fiercely value their diversity and many bridle at the mention of "standards," or worse yet, a "core curriculum," the accrediting board has always had to step gingerly, prescribing what all graduating students should be able to demonstrate, while leaving it up to each school to accomplish that goal as they see fit. What the accrediting board's criteria seek to do is not to prescribe a core curriculum, but to identify a core body of knowledge.

John M. Maudlin-Jeronimo, executive director of the board, described the role of accreditation this way: "NAAB is *not* concerned with telling schools how to teach or with *prescriptive* curricula—whether a student has to complete two rather than one course in mechanics, or three rather than two courses in statics. . . . Rather, NAAB evaluates 'outcomes': entry-level qualifications should broadly state that a student is able to design a simple architectural system. . . ."[2]

In 1983, the criteria underwent a landmark change. The end verdict of the process remained the accreditation or nonaccreditation of programs, but the focus of the criteria shifted from programs per se to student performance. We concluded that in many respects the system of architecture accreditation that has evolved is exemplary, among the best in higher education. The fifty-three accreditation standards and the process of applying them are focused where they ought to be: more on student outcomes than on counting library volumes or the number of PH.D.'s held by faculty. Recently, Robert Geddes, co-author of the 1967 "Princeton Study," praised the accrediting board's criteria as "a brilliant statement about what architecture is."[3]

As presently written, the fifty-three standards are grouped under four broad headings:

Fundamental Knowledge—the social, environmental, aesthetic and technical skills and knowledge areas that provide the broad context for the study and practice of architecture. Among the twenty-seven criteria under this heading, students are expected to be aware of methods of historical inquiry, demonstrate an understanding of "the ecological impact of buildings and their occupants," understand basic principles underlying two- and three-dimensional design, and understand basic theories of lighting, acoustics, environmental control, and other building systems, to name a few examples.

Design—criteria aimed at ensuring that graduates can apply and synthesize fundamental knowledge into an actual architectural form. Eleven separate areas of knowledge fall under this heading—among them, gathering and analyzing information about human needs and behavior to inform the design process, designing for persons with differing physical abilities, and integrating natural and imposed site constraints into project planning and designing.

Communication—aimed at ensuring that students can express design concepts to others using words, technology, and other design media. Under this heading, students are expected to demonstrate, among other things, an ability to communicate architectural ideas in written and oral form, to use computer technology to display and use information, and to communicate with those who review or construct projects.

Practice—which calls upon students to demonstrate awareness and understanding of the business side of architecture and the ethical and legal responsibilities of the architect. Among the ten criteria under this last category, graduates are expected to show that they are aware of factors affecting the project process in various types of practice, understand contract negotiations, and understand the architect's legal and ethical responsibilities for public health, safety and welfare, property rights, zoning, and other factors affecting building design.[4]

Along with evaluating *what* is being taught, accrediting teams are also supposed to determine whether students at particular schools are getting a well-balanced education. According to the accrediting board, programs with an undergraduate component can devote a maximum of 60 percent of coursework to core architectural knowledge and skills, and all performance criteria must be covered in these required courses. A minimum of 20 percent of the curriculum must be occupied with liberal studies and other general university requirements. At least 20 percent must be the "students' part," electives that allow all students time to pursue their own goals and interests.

How effectively are schools of architecture, as a whole, achieving the many objectives of the professional degree programs described in the NAAB standards?

We found, in our campus visits and research, a mixed report card. There is, overall, strong satisfaction among students, alumni, faculty, and administrators with how schools are covering many individual subjects, particularly design. We were concerned, however, that the programs that have evolved at many schools seem to be either-or propositions—emphasizing either the technical and practical aspects of architecture or its more theoretical sides, but often failing to accommodate both adequately. The results, say some long-time observers, are critical gaps in the professional preparation of many students.

George Iwon, executive secretary of the Minnesota licensing board for architects, who also served as executive secretary of that state's design selection board from 1976 to 1995, told us that, in his experience, recent graduates are "ill-prepared in a number of areas: marketing, accounting, presenting themselves, and writing. In most architecture schools, when they are asked to present their projects to juries, it is mostly to architects. They only learn to speak architect talk."

Further, many programs lack integration and leave inadequate time for electives or liberal studies. At a number of programs we learned about, students are not getting anything approaching 20 percent of their credits as electives. Most disappointingly, many design studios seem not to be living up to their vast potential as settings where integration of knowledge might be fostered.

"In the fifteen years that the performance-based standards have existed," John Jeronimo, of the accrediting board, told us, "lack of integration of technical and practical knowledge into design work is probably the single most widespread area of program weakness."

A faculty member at Boston Architectural Center agreed: "Schools of architecture should find ways of identifying and developing the three aspects of professional practice into their curriculum: design, management, and technical mastery. Architectural programs should encourage these aspects so that the primacy of 'design' as the 'core' of the profession is recognized, but also recognizing the necessity of the other aspects. Students should be encouraged to pursue non-design roles if they are indeed better suited to these pursuits."

Still, from what we could determine, most schools of architecture do a good job covering many of the architectural "basics"—architectural history, architectural theory, drawing, technical courses such as structures

Table 11

PLEASE INDICATE HOW STRONG YOU CONSIDER THE
CURRICULUM AT THIS SCHOOL IN THE AREAS OF DESIGN,
STRUCTURES, DRAWING SKILLS, AND ARCHITECTURE HISTORY
(PERCENTAGE RESPONDING "STRONG")

	ADMINISTRATORS	FACULTY	STUDENTS	ALUMNI
Design	90%	76%	93%	92%
Structures	92	73	72	80
Drawing skills	85	54	76	75
Architecture history	82	78	81	86

SOURCE   The Carnegie Foundation for the Advancement of Teaching, Survey on the Education of Architects, 1994.

and environmental systems, and, above all, design. In our surveys, each of those areas was praised by majorities of alumni, students, faculty, and administrators (table 11).

Beyond those basic subjects, however, opinion seemed less favorable. Administrators and faculty were more positive than students or alumni concerning how effectively computer-aided design is covered at their schools (table 12). Writing skills were cited as a weakness by majorities of administrators, faculty, and alumni (table 13), and a majority of faculty, students, administrators, and alumni disagreed that their schools were effectively preparing students for opportunities involving non-Western and developing nations. Finally, we heard repeatedly that students were not getting enough experience in school actually building things with real construction materials.

We found that the curricular area of greatest dissatisfaction was "professional practice." Nearly eight out of ten alumni we surveyed disagreed that the curriculum at their alma mater dealt effectively with contract negotiation and financial management, and nearly 60 percent disagreed that architecture students leave school "well prepared as entrepreneurs."

When recent alumni of the University of Washington were asked

Table 12

PLEASE INDICATE HOW STRONG OR WEAK YOU CONSIDER
COMPUTER-AIDED COURSES AT THIS SCHOOL

	STRONG	WEAK	DON'T KNOW/ NOT AVAILABLE
Administrators	71%	29%	0%
Faculty	60	35	5
Students	44	47	9
Alumni	34	53	13

SOURCE   The Carnegie Foundation for the Advancement of Teaching, Survey on the Education of Architects, 1994.

what elements were missing from their education, they cited liberal studies; construction experience; legal and real estate knowledge; how to operate a business; how to market oneself; research, writing and presentation skills; and communication and negotiation skills.[5]

In a 1993 survey, 655 architectural professionals in California were asked what curricular areas needed to be improved at the schools they attended. The four most frequently mentioned were business practice, practice realities, liberal studies, and marketing.[6]

A 1995 report sponsored by the National Research Council concluded that architecture students generally have a good grasp of design but too often leave school with "little knowledge of business, economics, and management and that this adversely affects their ability to serve their clients, to understand the concerns of their employers, to manage projects effectively, . . . and to qualify for more responsible positions."[7]

Architects abroad express similar concerns. A recent study of British architects found that fewer than one in four felt "adequately trained" in practical matters such as client relations, office and budget management, computer-aided design, property development, accounting, facility management, marketing, or computerization.[8]

"While we continue the core areas of design, technology, history, and theory and practice, an adjustment of focus and content is necessary. The

Table 13

PLEASE INDICATE HOW STRONG OR WEAK YOU CONSIDER
TEACHING OF WRITING SKILLS AT THIS SCHOOL

	STRONG	WEAK	DON'T KNOW
Administrators	33%	66%	1%
Faculty	27	65	8
Students	52	42	6
Alumni	39	59	2

SOURCE   The Carnegie Foundation for the Advancement of Teaching, Survey on the Education of Architects, 1994.

professional area is one that truly needs our attention," said William C. Miller, dean of the University of Utah Graduate School of Architecture. "Every practicing architect I speak to says it's a problem that we spend so much time on theoretical analysis," agreed a recent Ivy League graduate.

To be sure, we discovered from reading several dozen accreditation reports that accrediting teams routinely call these problems to the attention of school administrators, often in blunt terms, and many programs take seriously those recommendations. Still, it also appears that at a number of architecture schools there seems to be little relationship between the standards and the decisions shaping curricula. We concluded that this apparent disconnection has at least three causes: lack of widespread understanding of the standards among faculty and students, doubts and concerns about the usefulness of the accreditation process, and lack of coherence in the standards themselves.

The standards are, in many cases, not well enough known or understood to have a meaningful impact on curriculum development. Schools are required to keep copies of the latest accreditation reports and self-assessment in their library, but we encountered few faculty or students familiar with the fifty-three criteria.

Second, some educators seem to regard the standards and the

accreditation process as neither valid nor useful. During our school visits, faculty and administrators complained that the language of the standards is often vague, making compliance difficult to measure and inviting overly subjective judgments by accreditation teams. In some places, we heard criticism of the competence of accreditation teams. On some campuses, the rite of accreditation is viewed as a costly waste of time that plays little or no strategic role for improvement. Some we spoke with dismissed the standards as mere minimums. The harshest critics considered the accreditation process itself as worse than useless, fostering a "checklist mentality," trampling diversity and failing to weigh fairly the strengths and unique characteristics of each program.

We did not hear such criticisms everywhere. At some campuses we visited, accreditation was praised as a useful opportunity for the school community as a whole to consider its strengths, shortcomings, and plans for the future. In 1993, the accrediting board surveyed eighty-eight architecture deans and department heads and found, in fact, a wide range of opinion on accreditation: 66 percent agreed that "the standards established by NAAB insure an adequate architectural education and training for students," and 52 percent agreed that "NAAB promotes excellence in architectural education." But 47 percent disagreed that the NAAB "represents my concerns about architectural education"; 40 percent disagreed that schools have enough autonomy within the accreditation process; and 39 percent disagreed that NAAB is "flexible" in evaluating each school's needs and circumstances.[9]

Most essentially, we concluded that the accreditation criteria seem to have had only a limited impact on curricula at some schools because the language, the organization, and in some respects even the intentions behind the standards could be clearer. The standards are supposed to describe what students should know and be able to do upon graduation from a professional degree program. Yet, as we read them, the fifty-three criteria are an eclectic assortment of learning goals ranging from the minimal to the breathtakingly grandiose, many of which are achievable, if at all, only as *lifetime* goals.

Is it really conceivable, for example, that most students will graduate from school with genuine understanding of "the values, needs, and ethics that guide human behavior," or "the basic principles governing the formation of diverse cultures and human behavior," while busily engaged with learning the intricacies of environmental systems, building codes, and computer-aided design?

We believe, then, that the content of the standards and the procedures for applying them could be improved. In terms of content, the goal ought to be a new framework for the standards that puts clearer emphasis on the connections within knowledge, and joins students and faculty together in a common quest for learning. Specifically, we suggest that the same four categories we have proposed to define the scholarly work of *faculty*—the scholarships of teaching, discovery, integration, and application—might, with some modification, also be used as new headings under which to group the criteria for evaluating student performance as well.

In terms of *procedures*, we suggest that a more open process, in which the NAAB standards would become more universally understood, might breathe fresh life into the standards, empower students, and set the stage for a wider dialogue about the goals of architecture education.

We therefore recommend that the four existing headings under which the NAAB standards are currently organized be revised.

Instead of "fundamental knowledge," we suggest that the first major heading be changed to *discovery of knowledge*. The education of students about the scientific, social, aesthetic, political, and environmental foundations of architecture should not be about "teaching" disembodied skills and facts. The standards should stress active inquiry and learning by doing, rather than the accumulation of facts from texts, required lectures, or design problems handed ready-made to students. Further, students should be partners in extending the knowledge base of the profession through reflective practice. Learning to define problems, asking the right questions, and weighing alternative approaches must be at the heart of architecture study. The language of accreditation standards ought to reflect more forcefully that students, through active learning, should be partners with faculty, not only in learning the established traditions and knowledge of the field but also in the search for new knowledge.

The second new heading we propose, replacing "design," is the *integration of knowledge*. At virtually all schools, design is quite rightly considered the heart of the curriculum. Still, the term "design," as commonly used by architects and architecture educators, has taken on limited connotations, focusing more on the aesthetic and theoretical dimensions of design than on the integrative nature of the process itself. We agree, in fact, with the accreditation board statement that "design education develops the ability of the student to synthesize social, environmental, technical and aesthetic considerations into a cohesive and unified architectural entity and includes an understanding of process and product."[10] As Alan Plattus, associate dean of Yale University's architecture program, aptly put it, design should be about "making knowledge, not just consuming it."

We suggest that the term "integration of knowledge" is a more capacious and accurate description for this kind of learning. By emphasizing more forthrightly the term "integration," this new language might also prompt consideration of new standards calling more directly for students to make connections between the discipline of architecture and other related fields such as interior design, landscape architecture, and engineering.

The third category of standards we propose is the *application of knowledge*, replacing the current category called "practice." Scholarship in architecture, in a profound way, has to do with applied knowledge, and we believe that such activities ought to be more vigorously affirmed and assessed. What we have in mind is a set of criteria that encourages students to apply what is learned in school to office life and to the larger social context. Specifically, it is our hope that our proposed category would encourage educators to place the economic, management, legal, and ethical issues facing architects in broader context by stressing the connections of ethical issues not only to practice but to issues of public and social policy as well.

Finally, instead of "communication," we propose a fourth new category, the *sharing of knowledge*. The existing accreditation criteria already recognize that graduates should be able to communicate their architectural design ideas using a variety of media. With our new heading, we would hope to elevate to a higher priority the goal of graduating students capable of communicating the concepts and the value of architecture to everyone involved—fellow architects, others in the building process, clients, users of buildings, and entire communities. As we have already stressed, graduating students able to communicate clearly to everyone affected by the built environment is at the core of enriching the mission of architecture.

WE RECOGNIZE that changing the categories of the NAAB standards is only a first step. Creating a new framework within which to think about scholarly work by students would, we expect, open the way to broader discussion and reconsideration of the fifty-three criteria themselves. And we further hope that such a review might result in fewer, less prescriptive and more parallel criteria.

Moving from the matter of content to procedure, we recommend that the written products of the accreditation process be more broadly and publicly distributed. In particular, the accrediting board should make available, in compact booklet form, a list of the fifty-three criteria for distribution to every student during freshman orientation at all schools of architecture. If these criteria represent the broad expectations of the profession about what graduates should know, shouldn't they be far better known and understood by the students who must fulfill them?

To summarize, the essential principles of a new framework of *standards without standardization* are these: The diversity of programs should be preserved, while establishing a clear set of expectations uniting all professional programs. The language of the standards should stress, with greater clarity, the interconnectedness of architectural knowledge within the discipline and beyond. The standards themselves should be far better known within the architecture community, and especially among faculty and students. Finally, a new framework of standards should be constructed that helps expand the meaning of scholarship itself and create a scholarly partnership between students and faculty.

A Connected Curriculum

THE CURRICULA at the nation's 103 accredited architecture programs must respond to a multitude of seemingly conflicting demands. In line with the standards just described, they must, first of all, arm students with the skills and knowledge needed to enter the profession as reasonably competent beginners. Graduates should also leave school liberally educated—able to communicate clearly in written language and speech, and aware of the ethical and moral issues needed to guide practitioners through a lifetime of civically responsible practice in a multicultural and interdependent world.

And with all of that, curricula must be flexible enough to allow students time to pursue their own dreams and specialties—more important than ever in a job environment where many graduates are finding careers in fields having less and less to do with the traditional task of building buildings. Curricula must also be responsive to the changes and demands of the profession, state regulators, accrediting bodies, and certainly not least, the academic culture of the campuses in which they are located.

Taken together, architecture curricula should develop in graduates the habits of mind that MIT's Donald Schön calls "reflective practice"—the ability to be constantly conscious of each step of the design act even as it is being committed.

"A good architect," said the late Italian master builder Pier Luigi Nervi, "is someone capable of seeing the main problems of a design, capable of examining with serenity the various possible solutions, and who finally has a thorough grasp of the technical means necessary to accomplish his project."[1]

The long-standing challenge of balancing liberal studies with practice skills is, of course, one that has historically faced *all* of professional education, not just architecture, and if anything is clear from our travels, it's that no school, in the space of five or six years, can hope to achieve more than a solid beginning in meeting both these goals. Architecture school should be understood as the first step in a lifelong learning process that extends through internship and passage of the licensing exam and continues throughout professional life. The American Institute of Architects clearly embraced the concept of extended learning by adopting, in 1992, a continuing education requirement for membership.

Kenneth Carpenter, dean of Louisiana State University's school of architecture, put it this way: "The role of the schools in educating architects needs to be more clearly defined as the first few years of a much longer education. It is foolish to pretend that in eight years immediately following graduating from high school one can become a knowledgeable architect."

Patricia K. VanderBeke, a solo practitioner in Chicago, made a similar point: "Architecture schools should accept that they are but one facet in the creation of architects and concentrate on opening doors, illuminating possibilities, and providing the groundwork for a lifetime of creative thought and the application of the results."

The demands on architecture curricula, then, are truly daunting. They are supposed to guide relatively unsophisticated eighteen- and nineteen-year-olds, many of whom arrive with shaky academic, communication, and study skills, to a position where they can discover their voices as designers and express those ideas with clarity to architects and nonarchitects. They must initiate students into the traditions of architecture, its great personalities, its language, its history and theories. Students are expected to gain the technical know-how to design and build structures that actually stand, and become familiar with a rapidly growing body of technical knowledge. They are supposed to understand the rudiments of business. They must be adept at drafting, and increasingly, they need knowledge of computer design tools. Students are expected to have the beginnings of the worldliness that comes from travel, diverse experiences, and most especially, a broad-based liberal education. And finally, the curriculum should reflect the reality that many, if not most, students may pursue multiple careers.

A faculty member at a West Coast school stated the challenge this way: "Architectural curricula should not be designed to 'keep up' with the profession and society. There is no single way of practicing architecture and there should not be a homogeneous architectural curriculum. Schools should not expect that all students will become licensed professionals, although they must prepare students to be so if they choose. Architecture in the coming century will be shaped by a variety of practices. An architectural education *educates* a student in a way of thinking and *trains* a student in particular techniques and practices."

There is no single "right" curriculum to accomplish these many difficult tasks. Still, we are convinced that there are, in fact, several broad strategies which, taken together, might help all schools create curricula that are more satisfying to students, that meet the changing needs of the profession, and that connect learning more effectively to the needs of clients, communities, and society as a whole.

The broad strategies we recommend are these: a more *liberal* curriculum; a more *flexible* curriculum; and a more *connected* curriculum.

A More Liberal Curriculum

The continued vitality of the architecture field depends on the ability of practitioners to communicate clearly and convincingly to architects and the lay public. In a rapidly changing world, students need to be able to look beyond the confines of a single discipline and view problems in their totality. To understand the ethical choices entailed in *any* profession, students should be exposed to how the great figures in history, literature, philosophy, and art have struggled with life's moral dilemmas. These are the needs that make a true liberal education so essential to the future of the architecture profession.

Many schools of architecture have long recognized, of course, that liberal education doesn't take place only in classrooms. Architecture schools, on the whole, have done well in providing students with opportunities for study and travel to architectural meccas around the globe, from Rome to China to Columbus, Indiana. Most programs also sponsor lecture series, bringing on campus not just practitioners but poets and scholars as well. Students are clearly happy with these aspects of their program. Sixty-nine percent of students we surveyed believe their schools are effective in providing field trips, 74 percent believe the same about study abroad, more than 90 percent about lecture programs featuring practicing architects, and 77 percent about visiting critics and juries. Such enrichment opportunities are not frills but are, in fact, the very experiences that help provide future architects with the perspective needed in a global, multicultural society.

Still, visiting a variety of architecture programs, we found questionable the oft-stated claim that architecture education is, per se, a "liberal education." One of the most sobering moments of our campus visits occurred when we asked a fourth-year student to describe his school's humanities courses. He replied: "What are humanities?"

"Most of our students know little history, philosophy, literature, Western and non-Western traditions, and see general education courses as necessary evils to be forgotten as soon as completed," a faculty member at a West Coast public program told us. "It is the rare inspired student who can draw on a philosophy course to develop a design project, who can synthesize and inform the design process by reference to and understanding of our culture. We as faculty must do more to help students make these connections. We burn our students out and they have no time or energy for intellectual pursuits."

Diane Ghirardo, a faculty member at the University of Southern California and recent past president of the Association of Collegiate Schools of Architecture, had this to say: "I do not believe we can continue to treat these disciplines as trivial addenda to the grand project of Design. They are instead the necessary foundation for any engagement in the built environment."[2]

We recognize, of course, that tipping the scales too far in the direction of liberal studies, if done at the expense of professional knowledge and design experience, runs the risk of educating dilettantes. Still, it was striking that some of the most eloquent architects and interns we met also happened to be the most ardent advocates of a more liberal architecture education.

"Getting your foot in the door, being technically prepared for an entry-level job, is not what it means to educate an architect," a Massachusetts architect who hires interns told us. "In the end, people with only that kind of knowledge have very limited capacities. We look for people who can play every aspect of the game, and *want* to play it. We want people who have the knowledge of where the screws go, and yet they see that as a *design decision.*"

"The creative process is complicated and demanding," said Harold L. Adams, FAIA, RIBA, chairman of the Baltimore-based multinational design firm RTKL Associates, Inc., in his 1993 Tau Sigma Delta Gold Medallist address. "Without a broader base of knowledge—quite apart from design— our reputation as leaders will continue to erode. . . . We need to go beyond the basic design, construction methods, and architectural history courses and require students to study broadly in the liberal arts and sciences."[3]

A 1994 survey of employers by the Kansas State University's Department of Architecture found that recent graduates received their lowest ratings in the areas of "written communication," "verbal communication," "leadership," and "social maturity/polish."[4]

The need for a liberal architecture curriculum is particularly urgent for students who begin their professional programs directly from high school. Such students display a wide range of academic capabilities, study habits, and maturity. At some schools, architecture students are among the cream on campus. A recent study at the University of Tennessee, for example, found that architecture students academically outperformed most other undergraduates except engineering students.

Nationwide, college entrance examination scores indicate that high school seniors with an interest in architecture study fall somewhere in the middle of all college-bound students. In 1995, according to The College Board, such students scored slightly above the average of 495 in the math section of the SAT, but averaged only 413 in verbal skills—fifteen points below the national average. And in our survey, nearly 58 percent of faculty disagreed that the quality of students had declined in the past five years.

But six out of ten also agreed with the statement that "the undergraduates with whom I have close contact are seriously underprepared in basic skills."

"In this class," one design instructor told us, gesturing around his studio, "out of twenty students, maybe seven or eight have what it takes to go on to be architects. And this is a good class."

In common with many undergraduates, architecture students often display weak oral and written skills. At the start of a fifth-year lecture class we attended, the instructor felt obliged to give a five-minute talk on the difference between "its" and "it's."

"The writing skills of 50 percent of my students are just atrocious," he told us.

"It's not just grammar and spelling," said a faculty member at an East Coast private institution. "They can't construct an argument with a beginning and an end and something in the middle."

Assessing the communication abilities of recently hired interns, a partner of a Washington, D.C., firm told us: "You see students who are well organized and good designers, but can't put two sentences together."

The ability to speak and write with clarity is essential if architects are to assume leadership in the social, political, and economic arenas where key decisions about the built environment are being made, and some administrators and faculty we met are taking those skills seriously. Ball State's architecture program, for example, has a "Writing in the Design Curriculum" program that brings English faculty directly into the studio and prompts both students and faculty to be more mindful of the importance to architects of writing well.

Overall, however, it was disturbing to find how undervalued clear communication is at some programs. At one prestigious school we visited, we read two final theses: one clearly and grammatically written, and a second ungrammatical and nearly incomprehensible. Both papers, however, received identical grades of "A" from the same design instructor. Whatever criteria counted in arriving at those grades, writing ability was surely not high among them.

At another school, an adjunct professor told us that when he tried to count grammar and other writing skills into student grades, he was "hammered" by students in their faculty evaluations, and the school administration offered little support.

Architecture students are certainly not alone in having problems with communication skills, study habits, or maturity levels. Nor is it reasonable to expect architecture schools to spend scarce time and resources remedying skills that should have been mastered in high school or earlier. Still, given the urgent need of the architecture profession to communicate clearly not only to architects but to the public, those shortcomings deserve higher priority from architecture educators.

ARCHITECTURE SCHOOLS, then, have throughout their history faced the formidable challenge of providing students both professional knowledge *and* a liberal education. That seems more difficult than ever, given the growing complexities of the profession and society, especially within the confines of most five-year bachelor's programs. Still, we are concerned that simply lengthening degree programs or, as some propose, making architecture education a purely graduate study similar to law or medicine, might address the problem of liberal education but might also undermine the goal of student diversity by making it even harder for less affluent candidates to attend.

One promising approach is being tried at the University of Minnesota, which has recently phased out its five-year bachelor's program and replaced it with a "one-plus-three" graduate degree. Harrison Fraker, recent past dean of Minnesota's College of Architecture and Landscape Architecture, now dean of the College of Environmental Design at Berkeley, said students earn a four-year undergraduate bachelor of arts in architecture degree. This program is designed to provide a broad liberal education but includes one highly concentrated year of architecture education and three academic quarters of design studio. Students then move on to a three-year graduate program for a professional master's degree.

Fraker says the new curriculum aims, first, to produce more mature, liberally grounded architecture graduates. It also addresses a criticism often leveled at typical three-year graduate programs, that students leave with too little studio experience. Finally, Fraker said the new program will dispel what he calls the "myth" of the five-year professional bachelor's degree: "In my ten years here, maybe six students have actually graduated in five years," he said.

To enlarge the perspective of the profession, then, there is no substitute for a genuine liberal education. The breadth and depth of this need was stated eloquently by the late British philosopher Michael Oakeshott:

> This, then, is what we are concerned with: adventures in human self-understanding. Not the bare protestation that a human being is a self-conscious, reflective intelligence and that he does not live by bread alone, but the actual enquiries, utterances and actions in which human beings have expressed their understanding of the human condition. This is the stuff of what has come to be called a "liberal" education—"liberal" because it is liberated from the distracting business of satisfying contingent wants.[5]

A More Flexible Curriculum

More flexibility within the curriculum relates, in a crucial way, to the lifelong needs of students whose education should prepare them for career changes. It relates to the need to make architecture education a less harried experience for students, to allow time to explore specialties within the architecture field. Finally, flexibility is the necessary precondition for discovering the connectedness of knowledge—within the discipline, to other fields, and to life.

At many programs, we found that the curriculum is jam-packed to the breaking point. Students are marched in lockstep through five or six years so laden with required courses that little time is left to pursue a specialty or simply explore their own intellectual interests. The National Architectural Accrediting Board, in its guidelines, prescribes that schools should allocate at least 20 percent of the credit hours required for graduation to elective courses. In our examination of about fifty curricula, however, we found that many schools are falling well short of that guideline. At one five-year program, only six credits out of 180 required for graduation are electives. At a six-year master's program we visited, only nine credit hours out of 128 were designated as free electives for undergraduate architecture students.

With the exception of students entering architecture studies from other disciplines later in their academic careers, we found that only the most determined students can find time to concentrate on an architectural specialty, or foreign language, business management, art history, or philosophy. "I wanted to minor in Italian," a student in her junior year told us. "Every semester there is an architecture class or a studio that conflicts. There's just no chance."

As one administrator put it: "It seems important to design an educational system for architects that is not a lockstep program in which each credit hour is accounted for. Instead, a program should have enough flexibility to accommodate, to some extent, the specific interests of a particular student. This will encourage a higher degree of engagement by the student and ensure a certain amount of breadth."

We were encouraged to find widespread recognition among architecture educators of the need for more flexibility, and a number of schools we learned about are actively reviewing their curricula to give students more options.

The University of Hawaii at Manoa is planning to replace its professional bachelor's and master's programs with a single, six- to seven-year doctor of architecture program. The new program aims to provide more integration of academic and professional content, allow students to take the licensing exam sooner and with better prospects of passing it, and give students more opportunities to interact with other disciplines on campus.

In 1993, the University of Arkansas dramatically restructured its curriculum to create more flexibility and integration. Previously, according to Arkansas' architecture dean, Daniel D. Bennett, the curriculum had grown so overburdened and rigid that only six of 163 credits required for graduation were free electives, and students typically carried eighteen to twenty-one credit hours per term. By consolidating and integrating courses and by introducing more practice and technical content into design studios, students now take a more manageable sixteen hours per term. Students have twenty-four credit hours of free electives, and nine fewer required courses.

At Norwich University, a five-year program in Vermont, a relatively generous thirty-six of 167 required credits are free electives and students have wide latitude to concentrate in nonarchitectural courses. Thematic integration of subjects and the "bridging" of lecture and design subjects are among the keys to flexibility. Faculty who teach lecture courses also teach studios, further setting the stage for integration and flexibility.

Louisiana State University's architecture program is streamlining its curriculum and reducing course requirements to increase flexibility and give students more time to pursue specialties or liberal studies. And at Mississippi State, a new curriculum which took effect in 1995 eliminated and consolidated a number of required courses, lightened the required semester courseload to four instead of five or six, and increased the number of electives.

More flexibility would also help schools offer more specialized programs. At the University of Illinois, students entering the two-year graduate phase of the degree program must choose an area of specialization from among four options: design; history and preservation; practice, technology, and management; or structures and research. "Selecting an option forced me to think about why I like architecture and where I wanted to go with it in my life," one student told us.

Some schools, of course, are in a better position than others to support specialization and new graduate programs on their own. Still, we learned of a highly promising new alternative strategy that would pool the resources of schools and practitioners to promote more widespread curricular flexibility and specialization. Five programs in the Washington, D.C., area—Catholic University, the University of the District of Columbia, Virginia Tech, Howard University, and the University of Maryland—have recently pooled their resources and established an American Academy for Architectural Conservation Studies. Its mission is to provide interested students at their schools and elsewhere with credit-granting courses in historic preservation, preservation architecture, architectural conservation, and preservation technology—specialized areas that few schools can provide their students adequately on their own.

We can imagine a national network of such specialized architectural academies around the country, jointly sustained by area schools and practitioners, each offering architecture students specialized programs in emerging or neglected areas such as environmental sustainability, environment and behavior, or computer visualization science which many schools, on their own, might not be able to support.

More flexibility in the curriculum would contribute powerfully, then, to the goal of a more coherent, broad-based and caring architecture education—one which enables students to pursue their individual interests and dreams, and at the same time responds more fully to the changing needs of the profession and the society.

A MORE INTEGRATED CURRICULUM

Making the connections, both *within* the architecture curriculum and *between* architecture and other disciplines on campus, is, we believe, the single most important challenge confronting architectural programs. Allan R. Cooper, director of California Polytechnic State University–San Luis Obispo, stated it this way: "Students can no longer afford to work in sublime isolation from others, nor can faculty continue to ignore the essential interdisciplinary nature of architectural decision making."

The good news is that architecture, by nature and tradition, holds vast potential as a model for the integration and application of learning, largely because of its most distinctive feature—*the design studio*. The integrative possibilities of studio extend far beyond architecture. MIT's architecture school, for example, recently began an experimental course that applies the studio methods of architecture to teaching software design. Beyond question, the design studio is a model that many other disciplines on campus, as well as elementary and secondary schools, could well profit from.

For much of this century, design has dominated the architecture curriculum at nearly all schools. It is a *place*—the design studio—where students spend as much as 90 percent of their time and energy. It is a *product*—the tangible result of thinking about and making architecture.

And it is a *process*—a way of thinking during which the many elements, possibilities, and constraints of architectural knowledge are integrated.

At its best, the design studio sequence provides the connective tissue that brings together, progressively, the many elements of architecture education. Encouragingly, we found numerous examples of "integrative" studios where the discovery, the application, and the integration of knowledge are creatively pursued.

The American Institute of Architects, in its annual Education Honors program, cited a required course in "Environmental Choices" at the University of Virginia, taught by dean William McDonough, that also involves more than a dozen faculty and visiting lecturers from a variety of disciplines. Southern California Institute of Architecture received a grant from the National Endowment for the Humanities to help support and develop a humanities faculty that could collaborate in the teaching of design. Lawrence Technological University recently revamped its entire studio sequence to integrate lecture class content such as lighting, structures, and computer-aided design into studio projects.

A stated goal at MIT is to bring other disciplines into studios and juries. Ecologists, civil engineers, historians, real estate experts, and landscape architects have participated in studio and workshop projects. Jan Wampler, a studio instructor, called this approach "a great advantage. Our faculty, in one way or another, is always interested in the process of getting a building built. In my studio this year, we are trying to find new forms of architecture for serving the needs of the elderly. Geriatric experts have joined me in lectures, in desk crits and on juries."

Students at California State Polytechnic University–Pomona take eight interdisciplinary general education courses emphasizing broad, integrative themes such as "consciousness and community," "authority and faith," and "individualism and collectivism."

Another promising integrative approach is the design-build studio, which involves students in authentic projects from the program stage through construction, and often includes real clients. At Ball State, students in Bruce Meyer's design-build studio negotiated with a farmer in Albany, Indiana, to convert a barn into a livable dwelling for an agreed-upon price of $80,000. Students broke into teams, some dealing with interior modelling, others with exterior features, others with mechanical systems. Each team had to deal directly with contractors, arrive at budgets, make sure designs complied with codes, and get the client to agree to the plan and budget.

The design-build studio, said Meyer, is an excellent setting for students to be "total architects," to learn to work with others in teams, communicate with clients, reach compromises, and shepherd a project through the complete building process, from conception and design, to negotiation, to construction.

Alongside such encouraging examples, however, we also found design studios that are not living up to their integrative potential. At some schools, we heard students complain that studio was more about personalities than principles, a place where the ideology of the instructor became the true curriculum. Repeatedly, we heard that many design instructors resist the idea of coordinating design projects with what students are learning in support courses. Fifty-five percent of faculty surveyed disagreed, for example, that design projects at their schools effectively integrate principles of environmental sustainability. Fewer than one in three students felt that computers were being effectively integrated into design studio. And 57 percent of faculty and 71 percent of alumni disagreed that design studio projects effectively integrate key aspects of practice such as building codes.

William C. Miller, dean of the University of Utah Graduate School of Architecture, had this to say: "The studio as an environment of synthetic activity should expand its focus beyond the formal and aesthetic. Schools should treat the entire curriculum holistically and integratively, ensuring that core areas (design, technology, history, practice) work in tandem with each other. The separation of these core content areas is artificial and needs reconsideration."

We are not proposing that all design studios be turned into intellectual cuisinarts. There must always be occasions, especially at the beginning of the professional program, when students can simply discover and dwell on the *art* of architecture, freed from the constraints of budgets, codes, or clients. Nor do we suggest that *all* design projects and juries should incorporate practice or technology issues. We are suggesting that schools should offer students creative ways to bring professional practice and technical issues such as codes, budgets, client relations, and ethics much more frequently into the design process.

To the perennial question, then, "does design studio take up too much student time?" our answer is this: at schools which offer studio sequences that allow students to leave school with a narrow base of architecture knowledge, there is too much studio. At schools which use the studio to guide students through a gradually more complex and integrated exploration of architecture in its many dimensions—aesthetic, cultural, historic, practical, and technical—there can hardly be too much.

Computers also hold enormous promise as an integrative tool for architecture education. To realize that potential, however, schools must do far more than simply add a course or two in computer-aided design.

At the University of Colorado, Boulder, Mark D. Gross says the school of architecture has developed a three-pronged, highly-integrated "computing curriculum" that has students and faculty working together to explore computers as tools for representation, for analysis and decision making, for gathering information, and for exploring design theories and methods.

A few schools are pioneering ways to stretch the boundaries of design with technology. Columbia University recently established a "paperless studio" using computers, and Texas A&M developed an "Electronic Design Studio" for upper-level and graduate students. The idea, said Vallie Miranda, who runs the Texas A&M studio, is not to replace traditional studios or other forms of graphic representation, but to integrate more technical and practical issues into design and to offer students a means to test their concepts far more quickly. Using software developed at Texas A&M called "Energraph," students can instantly select a part of the world, time of year, and time of day, and test how much heat and light their building design will gain or lose depending on which direction it is facing.

Since 1994, Mississippi State has taken part in an interdisciplinary project with the National Science Foundation which allows students to create lifelike, three-dimensional computer simulations of their design ideas. Using advanced visualization tools, students can invite their classmates and design juries to take a computerized stroll, or "fly-through," around and inside their designs.

The key to the Mississippi State program, however, is its exploration of the computer's potential as an integrative tool in the design process. Sophomores are required to purchase laptop computers, and each year, they introduce more sophisticated elements into their design projects. And further, computers do not replace traditional design methods such as model building.

Finally, a number of schools require students to "pull it all together" in a final design thesis. In our surveys, more than eight out of ten administrators, faculty, students, and alumni agreed that such an integrative project should be a requirement for a professional degree. Yet a recent study by the University of Puerto Rico indicated that only about one-third of accredited five-year bachelor's degree programs *require* a thesis for graduation. We agree with the overwhelming majority of those we surveyed: all graduates should be required to pull together, in a single piece of design work, what they have learned in the professional degree program and express their design concepts clearly—orally, in writing, and in two- and three-dimensional representations.

TO SUMMARIZE, THESE THREE PRINCIPLES—more liberal content, more flexibility, and more integration—are, we believe, the necessary ingredients for sustaining a connected curriculum that would help schools realize the learning goals set forth in the NAAB standards, and the enriched purposes of the profession. The seeds of curricular renewal already exist in promising practices we found in classrooms and studios, and our visits to campuses convinced us that there is no single correct way to realize that goal. Still, we also conclude that it is the programs that have veered to the greatest extremes—that stress, for example, the theoretical or artistic at the expense of the practical and technical, or are so practical in orientation that they seem more like trade schools than professional programs—that are serving least well the students, the profession, and society.

What matters most, in our view, is not the amount of time spent in school, but how well the time is spent. The focus must be on changing the prevailing values within schools of architecture and higher education generally which foster fragmentation and inflexibility to those that encourage making the connections across the disciplines and to the vital concerns of communities.

We were impressed by the number of schools hard at work doing exactly that. Still, administrators at several programs around the country also told us that there is a need for better information about exemplary curricular programs among schools. Some deans and department heads, particularly in parts of the South, have begun on their own to hold regular exchanges of ideas. The Association of Collegiate Schools of Architecture and the American Institute of Architects, through their annual awards and honors, have published and distributed examples of promising practices and course descriptions. We urge the national architectural organizations to continue to promote promising practices so they can become even more widespread and better understood.

GOAL FIVE

A Climate for Learning

HEALTHY LEARNING COMMUNITIES share certain unmistakable characteristics—openness, fair play, clarity of communication, inclusiveness, tolerance, caring, joyfulness, and commonly held purposes. Whether the focus is kindergarten, college, or architecture school, any talk of realizing enriched educational missions or higher levels of student and faculty scholarship is hollow unless the climate for learning in the school community itself is supportive, not corrosive.

In the context of higher education, the learning community begins with respect for the rich resources each member brings to it. It rejects prejudicial or dogmatic judgments, honors diversity, and seeks to serve effectively and empower the full range of people in our society. Clear and civil language in all forms—written, oral, and three-dimensional representations—is also at the heart of a well-functioning community of learning. A healthy learning climate must be caring, where every individual feels affirmed and where all activities, inside and outside the classroom, are humane. Finally, learning communities are places of celebration, where the traditions, purposes, and accomplishments of the institution are regularly recalled and rituals are shared in a spirit of joy and common cause by all members of the community.

These basic building blocks of a healthy learning climate raise special challenges for architecture programs. Do schools of architecture promote clear, open communication within classrooms, studios, and design juries? Do they foster effective communication with nonarchitects in other schools and disciplines on campuses? How successfully have they opened doors to diverse groups of student and faculty? Are students given the time and support to partake fully in the social and intellectual benefits of campus life?

We propose, for all architecture programs, that these essential issues be regularly assessed:

First, is the school a communicative place?

Effective learning requires a free exchange of ideas. In the end, the value of any architecture program must be measured by the clarity and civility of language in all its forms. True learning is undermined in a climate of excessive competition, coercion, intolerance, or isolation from others on campus.

We were, first of all, concerned by the sense of social, physical, and intellectual isolation of architecture schools on their own campuses. Because of their exceptionally heavy workload, most students find it difficult to participate in activities that make for a well-rounded college experience. In our survey, we asked students how much time, in a given week, they spend on extracurricular activities. Twenty-eight percent said they spent none at all, and only one in four said they spent three hours or more per week.

"My problem," one student told us, "is that they tell you they want you to be 'well-rounded,' but we have eighteen hours of classes a semester and they are all right here in this building."

Further, more than 73 percent of students surveyed agreed that they "often feel isolated from others outside the architecture school." One wheelchair-bound student told us that he had hoped to become involved in a campus group for disabled students, but a crushing load of required courses left no time for mingling with nonarchitecture students.

"There are architecture students and normal students. I don't consider myself a real college student," one particularly unhappy student told us. "I've come this close to transferring out of here I was so frustrated. My whole life is in architecture."

"I used to have lots of friends outside architecture school in my freshman year, now I don't have any," a third-year student told us. She said a guest lecturer recently added salt to the wound by asking her class, "Do you want a family, or do you want a career?"

At the same time, however, we discovered a number of programs trying to make connections on campus and beyond. Some have course offerings open to nonarchitecture students. For four decades at Yale, Vincent J. Scully's lecture course "Modern Architecture" has packed that school's largest lecture hall with as many as five hundred students and has influenced generations of both architects and liberal arts students. Texas Tech replaced its standard freshman-year "Introduction to Architecture" course with a new, more interdisciplinary course that aims to attract nonarchitecture students as well. Since 1983, Ball State's award-winning "LUNCHLINE" elective uses conference calls to allow students and faculty to talk and exchange design ideas with the world's most prominent practitioners.

The University of Minnesota's architecture program is opening its doors to students attending liberal arts colleges in Minnesota such as Carleton College and St. Olaf College, allowing them to gain first-hand appreciation of the challenges of the built environment and those who shape it. Minnesota's college of architecture recently introduced a series of interdisciplinary courses for both architects and nonarchitects, titled "Introduction to the Built Environment," "The Design Professional and Society," and "Ecology, Design, and People."

Next, in evaluating the quality of communication within schools of architecture, we urge all programs to consider whether design juries and other forms of student evaluation encourage free-flowing dialogue, or one-way conversations in which faculty and invited critics talk while the student audience looks on silently. In our visits to campuses and architecture firms, we found ambivalence toward design juries. Many praise juries as a toughening experience and a lesson in poise and communication. As one West Coast faculty member told us: "Juries can sometimes get nasty, but I say, welcome to the profession!"

Others contend that juries foster excessive egotism and an adversarial approach to clients. And others recall their history of arbitrariness and outright cruelty to students. A senior member of a large firm had this to say: "There is a history of viciousness and this gets institutionalized. I think the real cause is that most architects don't have the verbal skills to elucidate and to keep the focus of the jury on the content of the work. However, when the jury system works well, nothing is better."

After observing several dozen juries at fifteen schools of architecture, we concluded that juries at many programs seem less hurtful and more constructive than they were in years past. In our surveys, students, both male and female, overwhelmingly support the jury system. Roughly nine out of ten agreed that "critiques of my work by design juries are generally respectful and constructive."

In recent years, some programs have replaced the label "jury," with its trial-like connotations, with more benign-sounding terms like "review." At the Southern California Institute of Architecture, students and faculty have explored ways to encourage students to be more active in discussions during "pinups"—when student drawings are pinned up and displayed for critique—and to give students choices on the manner in which they would like to be reviewed. Even rearranging the physical setting can help—placing critics and students around a table to promote a greater sense of parity. At California State Polytechnic University–Pomona, one professor makes videos of her students as they make their presentations and lets them look at the videos to evaluate themselves as effective communicators. The University of Oregon has one of the oldest traditions of noncompetitive studio evaluation. Since the 1920s, the term "jury" has never been used and, in fact, studio work is not graded except for "pass" or "no pass."

At Miami University in Ohio, professor Thomas Dutton has used a "round robin" format. Four critics talk with students individually about their projects. When their work isn't being reviewed, students participate in reviews of other students. The round robin format seems to foster more open, back-and-forth dialogue, and more participation by students themselves.[1]

Despite the apparent progress in ridding most schools of egregious abuses, communication problems remain in many juries. Too often, the proceedings seem almost Kafkaesque—a sleep-deprived student facing a panel of inquisitors, with the "right" answers so subjective as to be unknowable. As one upper-level student at an East Coast private program told us: "We don't have dialogues. You speak, and the professor responds."

Indeed, it was disturbing to witness, at many schools, an unwritten code of silence among students themselves. In the competitive, high-pressured and high-stakes environment of juries, the agreement seems to be, "I won't ask any questions about your projects if you'll do the same."

Our survey found that 58 percent of administrators agreed that their school should offer alternative ways to evaluate design projects, and two out of three students also felt that way. But from what we could determine, only a minority of design faculty are seriously experimenting with changes in or alternatives to the jury method.

"It is indeed ironic," wrote Kathryn Anthony in her acclaimed 1991 study, _Design Juries on Trial_, "that throughout the term, design instructors encourage their students to be creative, go out on a limb, take a risk—and then when it's all over most of those same instructors rely on the same technique they've been using for years."[2]

We recommend that alternative approaches to evaluation of design projects be more vigorously explored. Kathryn Anthony wrote that the focus of reform ought to be on increasing student participation, emphasizing the process as well as the product of design, making the criteria for evaluation less mysterious and subjective, and lessening the tension and public humiliation.[3]

Instead of separate evaluations of each student's work, for example, the entire studio's collective work might be displayed, almost like an opening night at a museum or gallery. Students would be required to prepare a brief written statement explaining clearly their goals and how they were achieved. Jurors and fellow students would be asked to write their reactions and suggestions concerning each project.[4]

In other words, schools must do more than simply change the name of juries to a more benign-sounding label. The real challenge is to modify the format of evaluation to make it into a more communicative experience in which respectful dialogue contributes to student learning.

To summarize, schools of architecture should afford students more opportunities to join in the social and cultural discourse of campus life. Within the programs themselves, we urge more openness, tolerance, and collegiality between students and faculty, in design juries and other settings. The quality of any school, in the end, must be measured by the quality of its communication.

Second, is the school a just community?

A just learning community is a place where the diversity of peoples, cultures, and ideas is affirmed in the curriculum and the classroom, a place where differences in background and perspective among students and faculty are celebrated.

A just setting begins with *inclusion*. Schools must, first of all, actively acknowledge the contributions of all groups to the history and body of knowledge that make up the architecture curriculum, and at the same time, aggressively seek to diversify their student body and faculty.

The need for inclusiveness is more urgent than ever. Repeatedly, we were told by practitioners and educators that much of the future of the profession lies beyond U.S. borders, in developing nations and in non-Western cultures. Major immigration and demographic changes in this country will undoubtedly result in significant changes for the profession, especially in the larger "gate cities." An aging population is creating new architectural needs. People with Alzheimer's disease, or AIDS, or poor vision, for example, require new building types. There are increased needs for hospices and for transitional housing for the homeless. In response to more ethnically diverse clients, architects will need to gain knowledge of a greater variety of tastes and traditions. Their education must prepare them for all of these emerging challenges.

"To use a cliche, the world is coming closer together," a thirty-two-year-old Illinois student said. "The faces of architects are slowly changing. The larger purpose of architecture for the future will be to respond to economic, environmental, human diversity and the social needs of the changing population."

The need for diversity, then, goes to the heart of the vitality of the discipline and profession of architecture. In 1990, the Association of Collegiate Schools of Architecture's "Code of Conduct for Diversity in Architectural Education" offered six reasons to diversify both the faculty and the student body: promoting social justice, improving the climate of architectural education for all, recruiting the best talent from the widest possible pool, increasing sensitivity to the full range of future clients, teaching students to work in a global marketplace, and fostering diversity within the profession.[5]

Beyond those overarching reasons, there is the sheer excitement and ferment that comes from genuine diversity. At Pratt Institute's school of architecture, you can walk into an architecture studio and find a Hispanic mother of three, four Israelis, three African-Americans, a German, a Ukranian national, and students aged twenty to forty. The curriculum includes a *required* architecture history course in non-Western cultures in Africa, pre–Columbian America, India, Tibet, Southeast Asia, China and Japan, along with Western cultures. At the time of our visit, both the dean and the chair of the department of undergraduate studies were women.

The inclusive climate we found at Pratt remains, however, the exception. Thirty years after the dawn of the civil rights era, architecture remains among the less successful professions in diversifying its ranks — trailing, for example, such formerly male-dominated fields as business, computer science, accounting, law, pharmacology and medicine.[6] As a recent headline in *Progressive Architecture* magazine suggested, it remains, in key respects, a "white gentleman's profession."[7]

Progress has, in fact, been made in bringing more female students into architecture schools. In 1972, approximately 6 percent of students in architecture schools were women.[8] By 1994–95, women accounted for 30 percent of total architectural enrollment and graduates, according to data from the National Architectural Accrediting Board (table 14).

Still, fewer than 19 percent of all full- and part-time faculty are female, and fewer than 10 percent of all tenured faculty are women, according to NAAB statistics (table 15). Michael Kaplan, of the University of Tennessee–Knoxville, who has studied the issue, found that forty out of 108 architecture schools in the United States and Canada had no tenured women at all in the fall of 1992, and twenty-seven schools had one. In a 1990 survey sponsored by the Association of Collegiate Schools of Architecture, professors Linda Groat, of Michigan, and Sherry Ahrentzen, of the University of Wisconsin–Milwaukee, found that fully two-thirds of faculty women believed that sexism is inherent in architectural education.[9]

Nine of the fifteen schools we visited had fewer than 4 percent African-American students, and eight had no African-Americans on their faculty.

Table 14

FEMALE AND MINORITY STUDENTS IN
ACCREDITED ARCHITECTURE PROGRAMS
(ACADEMIC YEARS 1990–94)

	1990–91	1991–92	1992–93	1993–94	1994–95
Total enrollment					
(full/part-time)	38,233	39,626	37,127	36,096	35,524
Female	10,925	10,892	10,585	10,281	10,885
(% total)	(28.6%)	(27.5%)	(28.5%)	(28.5%)	(30.6%)
Female graduates	1,309	1,304	1,346	1,373	1,355
(% total)	(25.8%)	(29.4%)	(26.6%)	(28.2%)	(30.3%)
African-American	1,969	2,045	2,172	2,004	2,018
(% total)	(5.2%)	(5.2%)	(5.9%)	(5.6%)	(5.7%)
African-American					
graduates	214	200	168	180	172
(% total)	(3.4%)	(4.5%)	(3.3%)	(3.7%)	(3.8%)
Hispanic	2,535	2,188	2,496	2,877	2,952
(% total)	(6.6%)	(5.5%)	(6.7%)	(7.9%)	(8.3%)
Hispanic graduates	266	214	242	263	272
(% total)	(5.2%)	(4.8%)	(4.7%)	(5.4%)	(6.1%)

SOURCE   National Architectural Accrediting Board Annual Statistics.

Women, furthermore, remain a small minority within the profession itself. In 1972, only one percent of the members of the American Institute of Architects were women. As of June, 1995, roughly 10 percent were women: 5,622 out of 53,706.

The rate of progress has been even more discouraging for ethnic and racial minority groups, and if anything, minority practitioners told us that the outlook today may be bleaker than ever. A prominent African-American architect in Washington, D.C., told us that selection committees at the General Services Administration and at Fortune 500

Table 15

FEMALE AND MINORITY FACULTY IN
ACCREDITED ARCHITECTURE PROGRAMS
(ACADEMIC YEARS 1990–94)

	1990–91	1991–92	1992–93	1993–94	1994–95
Total faculty (full/part-time)	3,666	3,456	3,542	3,758	3,789
Total female	622	568	631	713	712
(% total faculty)	(16.9%)	(16.4%)	(17.8%)	(18.9%)	(18.8%)
Female tenured	NA	91	98	113	125
(% total faculty)	NA	(7.6%)	(8.1%)	(9.2%)	(9.7%)
Total African-American	92	110	102	126	123
(% total faculty)	(2.5%)	(3.2%)	(2.9%)	(3.4%)	(3.2%)
African-American tenured	NA	44	40	44	40
(% total faculty)	NA	(3.7%)	(3.3%)	(3.6%)	(3.1%)
Total Hispanic	93	95	110	164	149
(% total faculty)	(2.5%)	(2.7%)	(3.1%)	(4.4%)	(3.9%)
Hispanic tenured	NA	22	25	38	42
(% total faculty)	NA	(1.8%)	(2.1%)	(3.1%)	(3.3%)

SOURCE   National Architectural Accrediting Board Annual Statistics.

companies simply don't take blacks seriously as potential lead designers. Instead, firms like his typically get the "crumbs" from major projects.

"We have held joint conferences with the AIA to bring these issues to light. The AIA has not spoken up as it should. They have thrown up their hands. My answer to them is, if people who looked like you were being denied work, I wouldn't have to tell you what to do. You would do it," he said. Harry Robinson, FAIA, dean of historically black Howard University's School of Architecture and Planning, told us bluntly that "most of the architecture world is very hostile to our students."

One fifth-year African-American student told us her recent decision to transfer from a predominantly white public university was prompted by a faculty member who told her that she would never succeed as an architect, [because she was] a woman and a black. "I don't think anyone should tell you that. For that reason, I plan to leave the country. I don't want to practice architecture in the United States."

Architecture schools cannot be held solely responsible for solving long-standing problems of race and gender within the profession. We recognize that progress in minority and female recruitment and advancement depends on a stalwart commitment on the part of university administrators and the profession, and on public policies affecting access to higher education in general. Further, the harsh truth is that the architecture profession has difficulty competing with law or medicine in terms of salary prospects—a serious concern for less affluent students who must weigh their career options against the need to pay back crushing student loans.

Still, we were concerned by the sense of fatalism we found at some schools, and within the profession generally, on this issue. A top official of one of the five national architecture groups told us that he believed the diversity issue will simply "solve itself," given time.

Revealingly, our surveys found that male faculty, as a group, take a considerably more sanguine view than female faculty concerning diversity. Two-thirds of female faculty members disagreed that their institutions were making good progress in promoting gender diversity, and more than 70 percent disagreed that good progress is being made in promoting racial diversity among faculty.

By striking contrast, six out of ten male faculty thought their schools were making good progress in achieving ethnic diversity among faculty (table 16).

Clearly, there are no simple solutions, but we believe a more flexible and more inclusive architecture curriculum would help. Professors Sherry Ahrentzen and Linda Groat recently described some of the issues schools ought to consider in order to foster a more just atmosphere where the values of *all* students are affirmed.

There is, to begin with, a "comfort factor" or "critical mass" of

Table 16

THIS SCHOOL IS MAKING GOOD PROGRESS IN FOSTERING
GENDER, RACIAL, AND ETHNIC DIVERSITY AMONG FACULTY
(SURVEY OF FACULTY)

	AGREE	DISAGREE	DON'T KNOW
Gender diversity:			
All faculty	71%	27%	2%
Male	79	18	3
Female	33	67	0
Racial and ethnic diversity:			
All faculty	53%	42%	5%
Male	60	36	4
Female	21	72	7

SOURCE   The Carnegie Foundation for the Advancement of Teaching,
Survey on the Education of Architects, 1994.

women and minority students needed to lessen the likelihood of
harassment. Ahrentzen and Groat's research suggests that schools with the
highest proportion of women and minorities had the most hospitable
environments. Second, they concluded that minority and female students
tend to find traditional, highly competitive learning settings like studios
and juries inimical, and they recommend that schools expand their
teaching repertoires. Third and most important, Ahrentzen and Groat
said that curricula should include more of the social and human aspects
of architecture—issues often more in line with the career goals and
interests of female and minority students.[10]

An African-American architect we spoke with put it this way: "We
don't have any 'jazz,' any 'hip-hop,' any 'blues' in the way students learn
to design buildings. We have African-American architects who leave
school well trained only as 'classical musicians.'"

A common problem also seems to be the often fleeting nature of
faculty expertise on less traditional subjects within architecture depart-

ments. We found, for example, course offerings listed in catalogs in such areas as Chinese architecture, but such courses were not regularly offered because the single faculty member able to teach it was on leave, or otherwise occupied. And when we asked faculty at schools we visited why there were few or no courses with a non-Western perspective, the frequent reply was that no one at the school was equipped to teach them.

As one possible solution, we suggest that architecture faculty collaborate with other departments on campus such as Near Eastern, Far Eastern, African, and women's studies to develop courses—possibly team-taught—combining the expertise of those disciplines with architectural expertise. Such interdisciplinary partnerships might not only put those courses on a more permanent footing, they would also provide rich opportunities for collaboration between architectural and nonarchitectural students and faculty on campus.

Some schools have taken steps to make their curricula more inclusive. Stanley Hallet, dean of Catholic University's architecture program and an expert on Moslem architecture, teaches an advanced "Cultural Intervention Studio" that relates modern design to the traditional forms and rituals of a diverse range of cultures from English rural villages to underground communities in southern Tunisia. One student took the lessons he learned from the ancient cliff dwellings of the Anasazi Indians in the American Southwest and designed two modern, below-the-earth rooms.

Students at MIT have gone to Minsk, Belarus, to help create new settlements for villages contaminated by Chernobyl; to Kingston, Jamaica, to help replace derelict downtown communities; to New Delhi, India, to help with designs to revitalize slums; and to São Paulo, Brazil, to create more housing for underserved populations.

At California State Polytechnic University–Pomona, a weekly lecture series in the atrium of the college serves as a vehicle for promoting diversity. The series, devoted mainly to Latin America, brings in authorities whose lectures help make students aware of work rarely covered in regular coursework. The Southern California Institute of Architecture has set a goal of making sure that 50 percent of those who sit on review panels are women.

In what may be a promising model, professors Linda Groat and Sharon Sutton recently began a thorough self-assessment of their architecture program to identify how the curriculum, styles of teaching, and the general climate for learning might be improved to better reflect our changing society.

We recognize that schools of architecture are not on an even playing field in establishing more just and diverse learning communities. Schools in more urban, multicultural regions have a profound advantage in attracting students from many backgrounds. Wealthier campuses can attract more international students, develop more multicultural curricula, and are likeliest to have "need-blind" admissions policies to support disadvantaged students.

Still, all schools of architecture should place higher priority on achieving a more just and diverse learning climate, and the profession should be a supportive partner in helping schools achieve that goal.

Third, is the school a caring place?

To promote a caring climate for learning, schools of architecture must be places where students feel supported rather than hazed. An overly competitive, intimidating atmosphere takes away from this purpose. The point is that the education of architects must develop in students a sensitivity to the needs and concerns of *others*—from individual clients to whole communities.

These are the questions, then, that should be addressed: Do programs provide enough counseling, or are students left to fend largely for themselves? Are schools supportive of the changing needs of students, many of whom are working or raising families? Are policies aimed primarily at helping students realize their dreams, or are they calculated to winnow down each class to a select few?

Architecture students, particularly those admitted directly from high school, have among the most demanding schedules of anyone on campus. Conservatively estimating a work week of fifty hours, each student will devote, on average, about eight thousand hours to earn a professional degree. In contrast to the twelve to sixteen credit hours per academic term taken by most undergraduates on campus, architecture students commonly carry eighteen, twenty, or even more credit hours per term. And increasingly, students feel pressed to find outside jobs during their school years both to help pay for their education and to improve their chances of getting a foothold in the profession after graduation.

We found wide variation in the degree to which schools are inclined to nurture students, or test their endurance. The dean at one of the more empathetic schools we visited explained his program's philosophy this way: "We take the view that if young kids make their way to our school, we ought to keep them. They are now our children. That's what we think about education. We are here to heal and empower, not to disembowel."

At other schools, especially those that have relatively open admissions, we found policies that sometimes amount to academic hazing—or as one senior design instructor told us, "to separate the wheat from the chaff." Such schools tend to see themselves as "gatekeepers" of the profession, charged with ensuring that only the most capable and determined students survive to graduation.

Interestingly, a number of architecture students and graduates we met agreed that architecture education *should* be a hardening experience. As one student at a West Coast campus told us: "A value of it being so rigorous and taking so much commitment is that you leave with a commitment to the field."

Nonetheless, we are concerned that life for many architecture students is socially isolated and exhausting, and leaves little time for any but the most determined students to explore the connections between architecture and other fields of study. More than eight out of ten students we surveyed said they found the work load at architecture school "overwhelming" (table 17).

An architect from Connecticut viewed the matter this way: "Architects need to get out of the centuries-old philosophy that one must earn a 'red-eye' badge of courage by 'surviving' school, in order to be a 'real' architect."

Students repeatedly complained to us about inadequate career support and academic counseling. Fewer than 15 percent of students surveyed "strongly agreed" that they could get good counseling at their school; 56 percent said they are "not well informed" about state licensure

Table 17

I OFTEN FEEL THAT THE WORK LOAD AT
ARCHITECTURE SCHOOL IS OVERWHELMING
(SURVEY OF STUDENTS)

	AGREE	DISAGREE	DON'T KNOW
All Students	81%	18%	1%
Male	79	19	2
Female	84	16	0

SOURCE   The Carnegie Foundation for the Advancement of Teaching, Survey on the Education of Architects, 1994.

requirements; and 45 percent felt uninformed about the Intern Development Program, the required program of internship in some forty states.

Students told us, furthermore, that lack of administrative staff at their schools increased their sense that "the system" is chaotic and indifferent to their needs. The job of counseling often falls to faculty, adding strain on their time while yielding less than satisfying results for many students. At one private institution, lack of administrative support has crippled a recent initiative to offer a pregraduate internship program.

The counseling picture is considerably better at MIT. The architecture department is served by MIT's Career Services Office, which operates an Architecture Internship Program. During January vacation break, the program provides about forty students with jobs in Boston-area architecture offices. The school also has Operation Firm Openhouse. For a week, architecture students visit nearly two dozen firms to get a taste of office life, and the visits often end with informal conversations about the relevance of education to practice. The Career Services Office sponsors lecture series and workshops by alumni and others on topics such as the Intern Development Program, how to start a firm, and alternatives to traditional practice.

Some schools aren't always accommodating of the schedules of older, "nontraditional" students trying to balance school demands with family needs. As one thirty-five-year-old student with children put it: "The program doesn't allow for the fact that if my kid gets chicken pox, I can't be here. Faculty don't understand this."

When we asked students whether they had ever *seriously* considered dropping out of architecture school, 29 percent said they had. In fact, many never make it to graduation. At one public institution we visited, the freshman class of 213 shrank to 50 by the sophomore year. Some don't make the grade academically; others either are overly stressed by the long hours and social isolation, find that architecture isn't for them, have conflicting family demands, or have to leave for financial reasons.

High attrition rates are, of course, not unusual in professional schools, including law and medical schools. Still, the irony is that, rather than weeding out only the least talented from entering the profession, some deans told us that it is often the most capable students who end up questioning whether the stresses of architecture school are worth the likely rewards.

We believe, in short, that the relatively good morale we discovered among many students is encouraging, but tells only part of the story. Most of those contented students are, after all, the *survivors* of architecture education, not the many who decide to leave school for one reason or another.

The choice between a caring climate or a "boot camp" strategy, then, is not just a sentimental issue. Historically, the value of having architecture education take place on a college campus has rested on the belief that students can gain a breadth of knowledge and intellectual growth that would be difficult to achieve in an office apprenticeship system. We are concerned, however, that the pressure-packed nature of student life at some campuses is undermining that advantage. Are schools of architecture truly producing "problem solvers," able to integrate knowledge from many disciplines, and ready to relate empathetically to a variety of clients, if students have little time to pause and reflect on the deeper meanings of what they are learning?

"The chief complaint," said a faculty member of a midwestern program, "is that teachers and students are absolutely overwhelmed. That's the biggest obstacle to a more integrative, holistic approach to architecture education." A practicing architect in Minneapolis agreed: "To have to occasionally put in long hours is fine. But to have to practically kill yourself to do well, to finish the project, is ridiculous. I think it's the result of a male-dominated profession—a competitive, prove-your-manhood, 'I'm tough, I can take it' sort of thing. That mentality is counterproductive to the nature of our profession. We need to learn to cooperate, not compete."

Finally, is the school a celebrative place?

Architecture education is, for many students, an all-consuming way of life. We suggest that at all schools, there should be occasions when the community as a whole can come together to celebrate its traditions, its purposes, and its individual and collective accomplishments. Given the special creativity inherent in architecture education, we believe architecture schools could be campus models for combining celebration with genuine learning.

Several programs offer examples of what we mean. Every spring for the last ninety-three years, architecture freshmen at Cornell take part in "Dragon Day." The class builds an enormous dragon on wheels and parades it around campus—while fending off the pranks and taunts of engineering students. As architecture chairman Kent Hubbell told us, it is a moment of celebration, but also a serious design problem that demands creativity and cooperation.

A number of architecture programs, such as North Dakota State, hold annual Beaux Arts weeks, featuring gala balls, guest speakers, and awards and scholarships ceremonies. The University of Washington has an annual Recognition Day, which brings together the school community and alumni benefactors to honor student and teacher accomplishments and to award scholarships.

Freshman orientation can be a time of tedium and long lines, but at some programs, it is an occasion to pass on traditions. At the University of Oregon, all students admitted to the graduate program in architecture are invited in the spring to "Graduate Day," which pairs new students with existing ones and features lectures, field trips, and discussions about the school. In the fall, when the new students enter, there's a further orientation that extends and enriches the personal contacts.

Friendly gathering places can also foster a sense of belonging and can help break down the barriers that divide faculty, administrators, and students. At California State Polytechnic University–Pomona, there's the "Cafe," a coffee and sandwich shop in a lovely courtyard behind the college building. Students and faculty can be found there at all times, seated together on benches, finding ready companionship. North Carolina State has an inviting courtyard where students can gather.

THESE, THEN, ARE THE ESSENTIAL BUILDING BLOCKS for creating, at every architecture school, a healthy climate for learning. Within each category, and from school to school, different questions and issues will, of course, be raised. We suggest that schools of architecture should regularly evaluate themselves as *learning* communities—places where students are supported, not put on trial, where communication is clear and mutually respectful, where all groups are actively sought out, and where the community regularly celebrates itself.

We recommend that each school of architecture set aside one day a year to evaluate itself as a community for learning. Written forms might be distributed to students and faculty on how the climate, curriculum, and methods of the program might be made more *communicative, just, caring,* and *celebrative.* The results could become the basis for a community-wide meeting, to which university administrators, alumni, and area practitioners might also be invited.

Architecture education programs have an opportunity to join forces with every other professional program on campus by considering, periodically, whether they have healthy climates for learning. Beyond question, the attitudes and behavior fostered in school will profoundly influence professional behavior later on.

A Unified Profession

T HE WORLDS OF ARCHITECTURE PRACTICE AND EDUCATION depend on each other for their purpose and vitality. Both bear responsibility for preparing students for gainful employment and for continuing the lifelong professional education of architects. In the end, the academy and the profession also share an obligation to serve the needs of communities, the built environment, and society as a whole. It is inconceivable that these goals can ever be effectively realized in an atmosphere clouded by miscommunication, mistrust, or lack of respect.

As Joseph Bilello, former education director at the American Institute of Architects, has written, better connections between the academy and the profession hold the promise of improved access to vital information for students, educators, and practitioners. Better relations would also help eliminate misunderstanding about the needs of the profession and the context of architectural education. Finally, they would improve greatly the chances for constructive change within the profession.[1]

Historically, however, relations between the academy and the profession have been marked more by tension than cooperation. Those tensions worsened during the difficult economic times in the late 1980s and early 1990s, but there is no question that the underlying problems run deeper than a passing economic phase. Some have even suggested that relations between the worlds of architecture education and practice have sunk to an all-time low.

Practitioners we met or heard from tended to blame schools for producing graduates with insufficient knowledge of the technical and practical side of architecture and little experience in team work. In our survey of architecture school alumni, fewer than half—46 percent—

thought their alma maters had done a good job fostering the ability "to work cooperatively in interdisciplinary teams."

"What this means," a partner in a large North Carolina firm told us, "is schools are only dealing with the first 15 percent of architecture. The attitude is, you'll learn the rest in an office. Schools don't teach students how a wall is assembled, there's practically nothing about the constructability of buildings or knowledge of CAD. They dump it on the profession to teach it later on."

Such critics add that schools, as a whole, have failed to grasp fully the economic and technological changes transforming the profession, and have instead continued to educate students for jobs which, for the majority of graduates, don't exist. "The model that schools offer their students is having your practice and being a designer," said a New York architect who has taught at a number of architecture schools over the years. "The model is one that nine out of ten will fail to emulate."

Further, educators have been accused of keeping the profession at arm's length, seeking money or political support but being less willing to heed the advice of practitioners about what students should be learning. Several leading practitioners, for example, told us they'd be eager to work with architecture schools, but they find educators frequently disinterested. "I'm not invited to schools," said Wilson Pollock, FAIA, principal partner of ADD Inc., a Cambridge, Massachusetts, firm with a national reputation as a progressive and successful model. "It amazes me, frankly, that they haven't done so. About a year and a half ago, the AIA College of Fellows sent out a questionnaire, and said that it would be great if we could share our knowledge with students. I said 'great,' and I've never heard a word since."

Finally, critics blame schools for helping perpetuate false images of practice. "Architecture schools should stop perpetrating the myth of the architect as visionary genius and encourage, along with design and theory, training in management, technology, and do everything in their power to discourage future generations of prima donna architects," an Indiana practitioner told us.

"We must get beyond our self-created image as world saviors and artistic geniuses and realize that architecture means building something," said a twenty-one-year-old student at a private, East Coast program. "The field is turning too introspective and closing itself to clients and the other professions. That's why contractors are doing all of the building—no one wants to hear about the selfish interests of the architect. Schools should address this instead of perpetuating it."

Conversely, those schools that have tried perhaps too hard to respond dutifully to every proposal from the profession may, in the end, be doing a disservice to the profession and to students. The chairman of the department of architecture at a large southern university wrote us: "One of our faculty recently proclaimed that employers value our graduates because they work hard and don't question things much. Personally, I don't see this as a strength, or a desirable characteristic. It is not surprising in the context of a profession that seems more interested in having students learn computer-aided design than anything else."

Many educators, for their part, claim that their harshest critics are ignorant of, and unsympathetic about, the realities of academic life. Many practitioners, they believe, have never fully grasped that professional schools are supposed to educate students, not just train them. "The treating of architectural education as the presentation of routines and skills that are necessary for the practice of architecture will surely prepare the student for a profession that has ceased to exist," said David Cronrath, chairman of the College of Architecture at the University of Nebraska–Lincoln.

Educators also accuse practicing architects of failing to fulfill their historic responsibility to provide educationally meaningful internships. "One of the major issues will be recognition, by the professional community, of their role in the education of the architect," Robert Zwirn, director of Louisiana State University's School of Architecture, told us. "The minimum years of education and internship are currently eight. Students take longer preparing for the Architecture Registration Exam. Thus, 40 percent of the educational process is assumed by the office. I have found few architects who will acknowledge this or fully understand its implications."

The consequence of this tit-for-tat is that neither educators nor practitioners seem truly to understand what the other wants or needs.

And surely some responsibility rests with the five national architectural organizations themselves. As Dee Christy Briggs, past president of the American Institute of Architectural Students told us, "The academy hasn't changed to meet the needs of the profession, and many practitioners have no idea what faculty must deal with to survive in the campus setting. There's a real need for much better understanding on both sides."

Still, in weighing all we learned, we concluded that the gloomier assessments of current relationships may be overblown, and ignore the close, daily relationships many schools enjoy with the profession and the genuine efforts at improved relations that are underway at some. Indeed, any fair assessment of the current state of relations must note that a great deal of cooperative interaction takes place *routinely* at practically all schools of architecture. Virtually all schools maintain at least some contact with alumni through committees and fund-raising efforts, and invite practicing architects and other professionals to participate in juries and lecture series. Educators at some schools we visited are, in fact, well connected to the profession, through their own initiatives and practices. The associate dean at Mississippi State's architecture school, for example, has served as president of the state chapter of the AIA.

Further, in recent years, state, local, and national chapters of the architecture organizations, as well as some schools, have moved to foster better understanding between educators and area practices. Kansas State University, for example, lists no fewer than forty ways that its faculty and students routinely interact with the profession—including alumni programs, student mentoring, exhibits and lectures, faculty presentations at local AIA conferences, participation by area architects in design juries, tours and field trips to professional offices, scholarships and gifts from practitioners, design charettes, career fairs, and professional practice fellowships for faculty.

The challenge, however, is to unite the nation's practitioners and educators, and the organizations representing them, around an explicit agenda for sustained collaboration. In considering all that we learned, we concluded that many of the principles and prerequisites we have suggested for renewing architecture education might apply, with equal force, to helping create a more unified profession. Those principles include more open and respectful communication between schools and practicing architects; a mutual appreciation of, and support for, the rich diversity of the worlds of education and practice; and more caring and flexible approaches to the lifelong education of architects, particularly during internship.

Specifically, we propose that this new, more unified partnership should be based on at least three priorities that architecture educators and practitioners have a vital stake in addressing together: *partnerships to enrich schools, partnerships to support experience,* and *partnerships to sustain learning.*

Partnerships to Enrich Schools

Practicing architects routinely participate in school life—as visiting critics, guest lecturers, adjunct instructors, or mentors. We propose that this rich base of experience and ideas be made an even greater part of classroom and studio life, and in discussions about the priorities of the curriculum itself. And further, we recommend that firms regularly invite faculty and administrators to spend time in offices to exchange ideas and to help educators and practitioners keep abreast of the realities of practice and academic life. Scholars and practitioners have much to learn from each other.

Two of the nation's leading examples of close, continuing relations between practitioners and schools are the Boston Architectural Center and The Frank Lloyd Wright School of Architecture, the latter located in Scottsdale, Arizona, and Spring Green, Wisconsin. At BAC, virtually all faculty are practicing architects who teach as unpaid volunteers. At The Frank Lloyd Wright School, enrollments are limited to thirty-five students, who work as "apprentices" with practicing architects. Much of their time is spent working at the firm connected to the school, Talieson Architects, and the learning-by-doing philosophy seeks to give students first-hand experience in all facets of architecture, from design to construction.

Some schools invite design teams from architectural firms to join students in studios to analyze designs using case studies. At California

State Polytechnic University–Pomona, students working in studios with professionals get exposed to ways of thinking about the entire architectural process, from design through construction. At Oklahoma State University, faculty recently set a goal of visiting every architectural office in the state, and the school's AIAS chapter is helping place students in area offices during spring breaks.

Next, we urge schools to use alumni committees as forums for candid, sustained dialogue between practitioners and educators on the educational mission of programs and the curriculum. The University of Kansas recently made it a priority to transform the school's professional advisory committee from a fund-raising group to one that is truly engaged in school affairs.

A number of schools such as Mississippi State and North Carolina State routinely survey exiting graduates and alumni. Such surveying provides valuable information about the latest changes and needs in the building industry, the types of careers graduates are pursuing, and the appropriateness and adequacy of the curriculum.

Student mentoring is another excellent way for practitioners to enrich schools. At Washington University–St. Louis, area practitioners are teaming up with faculty to advise students. The Universities of Nebraska and Kansas each have extensive mentorship programs that pair students with active architects. Effective partnerships between educators and practitioners may also hold the key to addressing curricular shortcomings in professional practice. The University of Illinois at Champaign–Urbana offers students a "practice lab," integrating technology, design, and business, with the help of area architects. Students periodically visit a Chicago architecture firm and work on projects the firm has completed, analyzing how they might have handled it differently. Members of the firm act as clients, providing students with reactions to proposed changes.

The new partnership between architects and educators should begin, then, with enriching the daily activities and the curriculum of schools themselves. By involving practicing architects more often in studios and classrooms, and on committees that deliberate about curricula and school missions, the academy and the profession can reach a fuller understanding and respect for their missions, while developing a sense of shared purpose based on strengthening the profession and meeting the needs of future architects.

PARTNERSHIPS TO SUPPORT EXPERIENCE

Internships, before and after graduation, are the most essential link connecting students to the world of practice. Yet, by all accounts, internship is perhaps the most troubled phase of the continuing education of architects.

During this century, as architectural knowledge grew more complex, the apprenticeship system withered away and schools assumed much of the responsibility for preparing architects for practice. However, schools cannot do the whole job. It is widely acknowledged that certain kinds of technical and practical knowledge are best learned in the workplace itself, under the guidance of experienced professionals (table 18).

All state accrediting boards require a minimum period of internship—usually about three years—before a person is eligible to take the licensing exam. The National Council of Architectural Registration Boards (NCARB) allows students to earn up to two years of work credit prior to acquisition of an accredited degree.

The Intern Development Program (IDP), launched by NCARB and the American Institute of Architects in 1979, provides the framework for internship in some forty states. The program was designed to assure that interns receive adequate mentoring, that experiences are well documented, and that employers and interns allocate enough time to a range of educational and vocational experiences to prepare students for eventual licensure.

As the IDP guidelines state, "The shift from school to office is not a transition from theory to pragmatism. It is a period when theory *merges* with pragmatism. . . . It's a time when you: apply your formal education to the daily realities of architectural practice; acquire comprehensive experience in basic practice areas; explore specialized areas of practice; develop professional judgment; continue your formal education in architecture; and refine your career goals."[2]

Table 18

POSTGRADUATE INTERNSHIP, RATHER THAN SCHOOL,
IS THE BEST TIME FOR STUDENTS TO ACQUIRE
MOST OF THE TECHNICAL KNOWLEDGE REQUIRED FOR
LICENSURE AND EVENTUAL PRACTICE

	AGREE	DISAGREE	DON'T KNOW
Administrators	63%	36%	1%
Faculty	52	40	8
Students	65	29	6
Alumni	60	39	1

SOURCE   The Carnegie Foundation for the Advancement of Teaching, Survey on the Education of Architects, 1994.

Whatever its accomplishments, however, we found broad consensus that the Intern Development Program has not, by itself, solved the problems of internship. Though we found mutually satisfying internship programs at several of the firms we visited or heard about around the country, at many others interns told us they were not receiving the continuing education and experience they needed.

The truth is that architecture has serious, unsolved problems compared with other fields when it comes to supplying on-the-job learning experiences to induct students into the profession on a massive scale. Medicine has teaching hospitals. Beginning teachers work in actual classrooms, supported by school taxes. Law offices are, for the most part, in a better financial position to support young lawyers and pay them living wages. The architecture profession, by contrast, must support a required system of internship prior to licensure in an industry that has neither the financial resources of law or medicine, the stability and public support of teaching, nor a network of locations like hospitals or schools where education and practice can be seamlessly connected. And many employers acknowledged those problems.

"The profession has all but undermined the traditional relationship between the profession and the academy," said Neil Frankel, FAIA, executive vice president of Perkins & Will, a multinational firm with offices in New York, Chicago, Washington, and London. "Historically, until the advent of the computer, the profession said, 'Okay, go to school, then we in the profession will teach you what the real world is like.' With the coming of the computer, the profession needed a skill that students had, and has left behind the other responsibilities."

One intern told us she had been stuck for months doing relatively menial tasks such as toilet elevations. Another intern at a medium-sized firm told us he had been working sixty to seventy hours per week for a year and a half. "Then my wife had a baby and I 'slacked off' to fifty hours. The partner called me in and I got called on the carpet for not working hard enough."

"The whole process of internship is being outmoded by economics," one frustrated intern told us. "There's not the time or the money. There's no conception of people being groomed for careers. The younger staff are chosen for their value as productive workers."

"We just don't have the structure here to use an intern's abilities to their best," said a Mississippi architect. "The people who come out of school are really problems. I lost patience with one intern who was demanding that I switch him to another section so he could learn what he needed for his IDP. I told him, 'It's not my job to teach you. You are here to produce.'"

What steps might help students gain more satisfying work opportunities, both during and after graduation?

We recommend, first, that schools, practitioners, and local and national architectural organizations collaborate to increase the availability, information, and incentives for students to gain work experience *during school*. And we urge that the monitoring of those internships for their *educational* value be improved. A survey of 448 students attending 126 accredited and nonaccredited schools by the American Institute of Architecture Students found:

- 79 percent had worked in an architectural related job during their school years;

· 84 percent said their school did not assist in placing them in an internship; and

· 94 percent received money for their work, but only 21 percent got academic credit, and 69 percent received no IDP credit.

Despite their dissatisfactions, students overwhelmingly support pregraduate internship: 60 percent of those surveyed called their experiences "very valuable," and 30 percent said they were "valuable."[3]

Only about 15 percent of architecture schools *require* internship experience for graduation, according to the student survey. And from what we ourselves observed, pregraduate internship programs are struggling at many schools due to lack of administrative or faculty support, and students are often left to their own devices to find work and fit it into already crowded schedules.

"Our university has a very, very insignificant internship program that, if students have heard about at all, doesn't get used that much," a twenty-six-year-old student at a southern campus told us. "Job placement for undergraduates is very poor."

A large private institution we visited has been trying to launch an internship program in which students are prescreened by the school, and then apply to firms. Only about ten students have participated because of problems in getting faculty advisers, fitting the internship into the curriculum, and determining whether internship hours should count toward "design" credit or "elective" credit.

Another school we learned about requires students to work five hundred hours at firms or related activities. Faculty told us, however, that student interns get little monitoring. Further, the program is controversial because it carries no academic credit, and students often work without pay. "As matters now stand," said one professor, "students can go through the program without setting foot in an office, even with the five-hundred-hour requirement."

We suggest that schools collaborate with area and national employers, and through AIA affiliates and alumni networks, to create up-to-date job banks for students. Even among schools that require pregraduate work experience, job placement and counseling assistance appear weak. At minimum, all schools should keep students well informed about state licensing requirements, IDP membership, and the benefits of gaining IDP credit for undergraduate work experience. Area employers should cooperate in supporting students trying to gain IDP credit. And registration boards should continue to provide all schools with materials for students about the IDP program and the licensing program.

Further, the incentives for undertaking work experience during school might be improved. We suggest that schools consider awarding academic credit for such work wherever possible, and work with state and local AIA affiliates to help employers monitor work experiences so that students can gain academic or IDP credit. It may even be appropriate to consider modifying the National Council of Architectural Registration Boards' rule that forbids "double-dipping," or awarding both academic credit and IDP credit for work experience. Simply put, we are suggesting that anything that would create a more seamless transition between the classroom and the workplace and make IDP-eligible experience more attractive for students would be beneficial.

Several schools have effective programs that make work an integral part of the educational experience. For decades, the University of Cincinnati's Cooperative Education system has students alternately working and studying for three-month intervals. At Drexel University, most students work by day and attend classes by night.

At Boston Architectural Center, the majority of students work at full-time jobs in architecture during the day, often earning $12 to $15 an hour, and attend classes four nights a week. One-third of a student's academic credits must be earned in the "work curriculum," and credit is awarded based on increasing levels of responsibility. The school provides job search assistance, and students fill out forms each year designating the level of experience they expect to achieve. Forms are then checked by a work supervisor to confirm that the desired experience was obtained.

"Students here know what they are getting into when they are graduated and they are more immediately involved in the real world of architecture," said a practitioner who works as a volunteer instructor at BAC.

Among more traditionally structured programs, Ball State requires students to spend at least twenty weeks as interns after the third year. Most find jobs in or near Indiana, but some work as far away as Germany. Students also take a one-credit course called "Internship Preparation," during which they learn the requirements of the Intern Development Program. A faculty member serves as an IDP coordinator, assisting students in finding jobs and visiting offices in the area to monitor student progress. Most students earn IDP credit, reducing the amount of internship time required after graduation.

In 1992, Catholic University began offering an internship as a one-semester elective carrying three academic credits, but no salary. Students spend time in actual offices, and participating firms designate mentors to oversee the experience. Students devote about fourteen hours a week to the job, and are supposed to keep a diary of their experience. Once a year, all practitioner-mentors come to the campus for lunch with the faculty.

TO SUMMARIZE, we are not proposing that all schools *require* work experience for graduation. Such a requirement would pose insurmountable burdens for some students or schools, either because of location, administrative and financial difficulties, or the time involved. However, we do urge that schools, practicing architects, and state and national architecture organizations should collaborate to create more pregraduate work opportunities available for students who want them, and make those experiences more educationally meaningful. That means better counseling, more job information, more incentives and rewards for work experience, and better monitoring of those experiences for their educational value.

What about internships *after graduation?*

The success of internship for the forty-five hundred graduates entering the work force each year appears to rest on the good will and the resources of the employer and the assertiveness of the intern. And at a number of firms we visited, internship seemed frustrating for both employers and interns.

"In general, we probably don't do it as well as we should," confessed the senior partner of one of Boston's largest and oldest firms. "Ideally, someone should be exposed quickly to all phases of the process, starting with conceptual drawing, going to an interview and getting a commission and seeing a project through in the field. It's hard for an architectural firm to put people off and on teams. But that is what you would have to do because a project is only one phase at a time."

An intern at a New York City firm expressed her frustration: "Honestly, I've thought of getting out of the business so many times. I think this is a good firm. The people are wonderful. But I do feel that through school I've built some knowledge and talent and now it's like I'm just feeding them to someone without getting anything back."

At firms where internship seemed mutually satisfying to interns and employers, however, a key ingredient for success seemed to be a clear corporate commitment to regard interns not as burdens or expenses but as long-term investments. And further, such firms tended to regard continued training and education for *all* employees—young and old—as vital to their overall missions.

"We don't hire 'temporary people,'" said Wilson Pollock of ADD Inc., in Cambridge, Massachusetts. "We are investing in people and we want them to grow with us. We have one hundred hours in training for people. We spend $6,000 in hard costs per new employee. We train a lot more than other firms. We put our training programs on e-mail to our employees, and if they don't sign up by themselves, we encourage them."

"The attitude at this firm is that interns aren't just a bother, that it's a partnership," said Robert Workman, a founding partner of BSW International, a large, Tulsa, Oklahoma, firm which has gained national prominence building Wal-Mart stores. "I'm willing to bank that loyalty will keep people here. Our turnover is just a fraction."

Too few architects, said Workman, recognize that a commitment to the well-being and education of interns flows directly from a commitment to corporate growth and *profitability*. "Without profit, I can assure you that we can't do R&D, can't afford technology, can't afford education and training, can't find great support people. Without all these, your company is faking it."

In a recent policy change, BSW interns receive overtime during the first two years of employment. The firm sends a personal welcome letter to all newly hired interns, assists them in keeping up with IDP paperwork and recordkeeping, pays 100 percent of each intern's registration board fees, 50 percent of the cost of taking the licensing exam, and 50 percent of the cost of AIA membership. The firm also provides all employees, including interns, with a rich menu of continuing education courses and seminars, from contracts to drawing presentations, from team building and process management to seminars, to help prepare interns for the licensing exam. Bsw, furthermore, combines internship training with the notion of community service. Interns work on *pro bono* "teaching projects" and competitions, teaming with IDP sponsors on projects for area social service organizations such as Habitat for Humanity.

Much smaller firms have also found it possible to commit to the principle of treating interns as investments. At Yeater Hennings Ruff Shultz Rokke Welch, in Moorhead, Minnesota, interns told us that they are not subjected to "clockwatching," nor is there a sense of "rank" or "hierarchy." As partner Royce Yeater told us: "We make interns act like architects before they even think they can."

"They're really good here at providing us with diverse experiences," said one intern at Yeater Hennings. "It helps both us and the firm."

When all is said and done, internship, in order to be beneficial and effective, must be a partnership between the worlds of education and practice. It is unreasonable for schools simply to hand graduates off like footballs to firms which are then left to carry, largely on their own, the financial and pedagogical responsibilities of internship. At the same time, we concluded that what is urgently needed is a fundamental change in corporate behavior toward interns. It is time, we believe, for all firms that hire interns to take stock of their motivations—for the sake of both the profession and beginning architects.

At some firms, investing time and money in an intern seems to be regarded as a poor gamble. After all, doesn't a firm "lose" if, after investing time and money in training, an intern doesn't work out or decides to leave? In fact, at offices where internship seems to be working, the opposite is frequently true: loyalty breeds loyalty, and over time, firms that have the best intern policies are likeliest to attract and keep many, if not most, young architects. And even if a promising intern leaves, the profession as a whole will benefit in the long run by its collective investment in education.

In short, we feel that only firms prepared to treat interns as *investments* ought to hire them. Both interns and the profession as a whole would likely be better off if firms who regard interns as cheap labor or an expensive nuisance stopped hiring them at all.

Still, a serious problem remains. Many firms lack the corporate mindset, or profitability, to support internship as we envision it and, indeed, as some firms are already practicing it. To fill the unmet need for quality internship on a broad scale, we propose that state and national architectural groups, along with individual schools and firms, aggressively explore collaborative partnerships to support internship. The so-called "teaching office," developed by a task force of the National Institute of Architecture Education, is one possible approach. Details and financing of such offices still have to be worked out collectively by the industry and individual schools. But as described by the institute, office staff would continue to be the prime mentors for interns, while the teaching office would be affiliated with particular architecture schools, and faculty would work with practitioners to develop the curriculum for training programs.

Such offices might operate as community design centers or clinics, offering interns and practitioners alike the opportunity to apply their design skills to housing problems and other social service challenges. The resulting interaction between practitioners, teachers, and interns would surely strengthen the partnership between practice and education.

PARTNERSHIPS TO SUSTAIN LEARNING

Repeatedly, we heard architecture educators express pride in how effectively design education prepares graduates for lifelong learning. In our survey, an overwhelming 93 percent of architecture alumni felt they left school "well prepared as life-long learners." There remains, however, the challenge of assuring that *all* architects have sufficient avenues—within and outside their offices—to continue their education throughout their careers.

Continuing education for all practitioners can be, as Clemson University architecture dean James Barker has said, "the best bridge between education and practice we have yet established."

Underscoring that challenge, the American Institute of Architects reached a historic decision in 1992 to make continuing education a *required* condition for AIA membership. A gradual phase-in period began in January 1995, and starting in 1998, practitioners will need to pursue from twelve to thirty-six hours of learning experiences to meet the AIA membership requirement. The institute, since 1987, has also given a number of firms citations for outstanding education in practice.

Beyond such national recognition, local initiatives to encourage continuing education are gaining ground. The Kansas City AIA Chapter is setting itself up as a clearinghouse of information about what opportunities are available for continuing education, in schools or elsewhere. The chapter will provide calendars of courses and events and will also put the information on-line.

A few firms and individual architects may be able to sustain meaningful educational programs on their own, but beyond question, the academy and the profession will need to collaborate in new ways if the mandate for continuing education is to be effectively met on a broad scale.

Properly pursued, continuing education could be a rich vehicle for increased communication and understanding between academicians and practitioners, providing avenues for regular exchange of architectural knowledge and innovations. As faculty and practitioners meet more often to consider the future directions of architecture, the curricula of schools of architecture will be enriched by those discussions. And continuing education could help build support for expanding the knowledge base of architecture and lead to collaborative research partnerships.

"The building of a body of knowledge related to the profession is an important possibility for continuing education," wrote Marvin Malecha, FAIA, dean of North Carolina State University's College of Design. "Engaging practitioners in the creation of case histories, creating libraries with vital documentation regarding building performance and design process decision strategies not only complements the increasing demand upon office to document design decision strategies, but actually provides a greater base of information from which to act."[4]

"Through education there exists an opportunity to create a new breed of architect who can embrace the diverse changes that are taking place in contemporary world culture and across global market economies," a faculty member at a public, midwestern institution told us. "This requires the architectural community—students, educators, practitioners, governing agencies, to collectively *rethink* and *redefine* architecture and the *making* of architects."

If the possible rewards of continuing education are great, however, so are its pitfalls. At worst, the mandate could devolve into a routinized exercise of trade school courses and remediation. As noted in a 1993 report by the AIA's Architects in Education Steering Group, school resources could be spread too thin, schools could become embroiled in professional and competitive conflicts, and the most capable faculty might be recruited away by architectural firms.[5]

It is essential, then, that professional architecture associations collaborate with schools and other providers of continuing education to define and uphold goals for continuing education and assure accountability, quality control, and relevance of coursework. Each architecture school or other provider of continuing education should assess with care the training needs of faculty, the kinds of curricula schools are prepared to offer, and each school's motives for entering the continuing education field.

At the time the American Institute of Architects requirement was adopted, roughly one-third of accredited U.S. schools of architecture had established continuing education programs, but many other schools were considering such offerings, according to a 1993 AIA-sponsored survey.[6]

Among schools we visited, the Boston Architectural Center runs a continuing education program open to nonarchitects and licensed architects. Offerings include design, history and theory of architecture in Boston, visual studies, computer-assisted design, technology, and management. Continuing education has also proven to be an incubator for new programs that the school can pilot and offer on a regular basis.

California State Polytechnic University in Pomona offers one of the country's few weekend programs, designed primarily for those who began professional education but never completed it. The program maintains quality and credibility by ensuring that courses are the same as those offered on weekdays, and are taught by the same faculty.

We also found support for continued education at a number of firms we visited. At The Hillier Group, in Princeton, New Jersey, for example, everyone at the firm must attend regularly scheduled lectures given by members of the firm or outsiders. RTKL, the Baltimore-based firm, brings faculty in from universities to serve as continuing educators and design consultants.

The essential point is this: continuing education can only be meaningful if practitioners and educators aim high. As Donald Watson, FAIA, dean of Rensselaer Polytechnic Institute's School of Architecture, put it, lifelong learning must be based on a collaboration to improve both education *and* practice, "by nurturing the architectural discipline's unique capacity for design, reflection, and research through the notion of critical practices," and "by promoting lifelong learning that undertakes 'continuing education' as a leadership program—rather than one that seeks only to raise the minimum acceptable levels of practice . . ."[7]

IN THE END, THE GOAL for educators and practitioners must be a *unified profession*. The time has come for educators and practitioners to put their differences in perspective and join in common purpose around the three priorities we have identified—enriched learning during school, more satisfying internships, and sustained learning throughout professional life.

We are convinced that progress in meeting these challenges will only come when architecture educators and practitioners take steps to create avenues for more open, sustained dialogue, and fully acknowledge their shared goals and responsibilities. We are further convinced that these three priorities relate intimately to each other, and that progress and better dialogue in promoting continuing education, for example, will likely pave the way for progress in providing more educationally sound internships.

Schools have a responsibility to educate, not merely train, and to support as well as challenge existing practice. At the same time, no school can afford to neglect its continuing obligation to the profession. The profession, while understandably preoccupied with the daily imperatives of practice, must be more supportive of schools of architecture, understand better the realities of the academic setting, and demonstrate a widescale willingness to continue the education of graduates through supportive, educationally sound internships. Encouragingly, we heard from practitioners and educators who fully appreciate this.

"The school of architecture is the academy—a place to study the discipline," Henry V. Stout, an architecture and urban design consultant in Ruston, Louisiana, told us. "The school is not a vocational setting that prepares you to take an exam, or to be a practitioner, but rather a place in which to ponder, question and challenge the discipline's requisite knowledge; a place in which the discipline's knowledge base is being expounded through research and experimentation."

Service to the Nation

ERHAPS NEVER IN HISTORY have the talents, skills, the broad vision and the ideals of the architecture profession been more urgently needed. The profession could be powerfully beneficial at a time when the lives of families and entire communities have grown increasingly fragmented, when cities are in an era of decline and decay rather than limitless growth, and when the value of beauty in daily life is often belittled. Surely, architects and architecture educators, as well as the organizations that represent them, ought to be among the most vocal and knowledgeable leaders in preserving and beautifying a world whose resources are in jeopardy. Graduates should be knowledgeable *teachers* and *listeners*, prepared to talk with clarity and understanding to clients and communities about how architecture might contribute to creating not just better buildings, but a more wholesome and happy human condition for present and future generations. Schools of architecture, in other words, should educate students for both *competence* and *caring*—in service to the nation.

Over the years, Carnegie Foundation reports have expressed great concern that higher education as a whole has lost its direction, that it is no longer at the vital center of the nation's work. Many now view the campus as a place where professors get tenured and students get credentialed. This loss of direction reflects, to be sure, a larger confusion. We as a nation are uncertain about our priorities at home and abroad. We are shaken by an awareness of the limits on growth and resources that were once thought to be unending. For the first time in about half a century, America's colleges and universities are not collectively caught up in some urgent national endeavor.[1]

Reflecting on the most consequential policy matters of our time, we're struck that it was not tenured university scholars but intellectuals outside academe who most profoundly redirected public attitudes and education: Rachel Carson raised the nation's consciousness about pollution with her book *Silent Spring*; Ralph Nader challenged this country about product safety with *Unsafe at Any Speed*; Michael Harrington, in *The Other America*, shaped the debate about poverty; and Betty Friedan, in *The Feminine Mystique*, changed America's views on gender equity.

There is a growing feeling throughout this country that higher education is a private benefit, not a public good—and to a considerable degree, we felt a similar absence of civic engagement at some architecture schools as well. The department head of a large, public midwestern school of architecture told us that his institution had, in recent years, witnessed a "shift in purpose that was once inclusive of public issues and values to one that is increasingly private and individualized. Thus the most available 'common purpose' is business success."

This isn't to suggest that all, or even most, schools of architecture have been totally disengaged. To the contrary, we found outstanding examples of community-minded scholarship, and schools of architecture deserve huge credit for performing, collectively, millions of dollars worth of *pro bono* work every year through their involvement in a variety of housing and community projects in some of America's most depressed urban and rural communities.

Nonetheless, we concluded that schools of architecture could do more, in their own programs, to instill in students a commitment to lives of engagement and service.

Civic activism seemed more widespread during the era of modernism, and later in the 1960s. Architecture schools across the country opened "community design centers" which gave students opportunities to engage, first-hand, in the problems of depressed communities. Educators from architecture and other disciplines founded the Environmental Design Research Association, an organization that still survives. Courses in historic preservation education were introduced at the University of Virginia, Cornell University, and Columbia University, and soon spread to other campuses.[2]

As Robert Geddes, a former Princeton University architecture professor recalled recently, the 1960s were: "an optimistic, progressive era, the opening decade of the civil rights movement and the environmental movement. The borders of architecture seemed open, and architects worked in collaboration with urbanists, social and behavioral scientists, design methods and systems building technologists. Schools of architecture became 'Colleges of Environmental Design.' Urban renewal, social housing and new communities generated a new field, urban design, which joined architecture, landscape architecture and city planning."[3]

In the aftermath of those heady years, architecture entered a period of uncertain economics and increasing self-doubt. Many practitioners and educators began to question what influence, if any, architects could rightly expect in shaping the great social and political debates affecting national life. Such doubts continue to linger.

"It would be unrealistic to think that architecture alone, or perhaps even as a significant part of the whole, can solve social ills," a middle manager at a large, Somerville, Massachusetts, design firm said. An executive of a large Texas design firm undoubtedly spoke for many when he told us: "As an architect, what responsibility do we have to the community? When we build developments, we have often found ways to step a building back, build a pond, and be responsible to the community. But our primary interest has been serving the client. So here we are, we're architects and defenders of the environment while having to serve the client. We meet with the community. We agree to decrease the density of the development. We make a lot of traffic improvements. And then the development never gets built."

"The fact is," he continued, "my education exposed me in no way whatsoever to the realities and the politics of how a project gets completed. When we talk about architects and society, the society in which we live has no moral or ethical fiber. It's the dollar that motivates a lot of architects, zoners and planners. The best we can do is try to act as mediator between homeowners and developers."

Alongside such views, however, we also found encouraging evidence of a renaissance of interest in civic affairs at a number of campuses and firms. Among students especially, we discovered a growing thirst to connect professional goals to public service. A twenty-year-old student at

Catholic University had this to say: "Architecture schools should definitely stress social issues more. There is a need to move away from building monuments to ourselves and towards providing for the needs of the larger community."

This apparent reawakening seems most powerfully directed at protecting the environment, and addressing the social, physical, and economic needs of rural and inner-city communities. "I feel architecture has an ever-growing responsibility to try to improve upon current social conditions and problems," said a twenty-one-year-old at North Dakota State University. "Worldwide, we are desperately in need of better city planning to lessen crime and overcrowding while increasing efficiency and density for the expanding population. We also need more sustainable designs to put an end to the vicious cycle of destruction that often befalls buildings long before it should."

We also found a revival of civic consciousness at some firms we visited. Bsw International, the large Tulsa, Oklahoma, firm, has established an in-house "Green-Team," a research and development unit which views building costs as a long-term investment over the extended life of a building, and takes into account not just initial costs, but "sustainability" issues including energy efficiency, materials, and the construction process.[4]

How widespread this desire for civic engagement actually is among practitioners and educators is difficult to measure. According to Joseph Bilello, formerly of the AIA, only 5 percent of the AIA's membership identify the environment as a critical issue that they want to be involved in. Still, it was striking that a midsummer meeting of educators and practicing architects in 1994 on environmentalism in education at the Cranbrook Academy in Michigan drew 130 participants from across the country—roughly triple the usual turnout for such summer sessions. Thus, while issues like environmental sustainability may still be on the margins of day-to-day practice, we found reason to hope that a growing number of educators and practitioners are eager to see the profession assume greater leadership in solving these national and global problems.

The question is, how might architecture schools elevate the concept of service to the nation so that it is at the core of architectural education and practice? We recommend that schools pursue four broad strategies:

- establish a climate of engagement;
- clarify the public benefits of architecture;
- promote the creation of new knowledge; and
- stress the critical importance of ethical professional behavior.

STUDENTS AND FACULTY alike should regard civic activism as an essential part of scholarship. Students frequently complained to us, with considerable justification, that the demands of architecture education leave little time for service, and this is surely one more reason that the curricula at all schools should be made more flexible. In a climate of engagement, student empowerment and civic participation are not merely accepted but are considered essential preludes to *professional* engagement. Administrators and faculty actively seek and respect student views, and welcome student involvement in school affairs, including decisions affecting the curriculum.

At Ball State University's architecture program, for example, Marvin Rosenman, the department chair, routinely takes students to national architecture conferences, where they are sometimes the only students present. And students have actually initiated new, interdisciplinary courses and studios that became part of the curriculum. At North Dakota State, virtually every school committee has student members, and student evaluations of faculty members are taken seriously.

Open and communicative classrooms are also essential to establishing a climate of engagement. Design studios and juries should be opportunities for respectful, two-way exchanges. Classrooms and studios in which faculty critique, while exhausted students dutifully nod and listen, are hardly ideal breeding grounds for future leadership or civic responsibility. The point is that in order to have a climate of activism, schools of architecture must first establish a climate of participatory learning. "We cannot have a holistic, sustainable practice of architecture unless we have a comparable approach to education and life," a Ball State faculty member said. "We must practice what we preach."

Students, for their part, must also examine their own priorities. The American Institute of Architecture Students is a powerful presence on many campuses. The AIAS chapters at Roger Williams, Carnegie Mellon, Cooper Union, Kansas State, Michigan, and the University of Texas at Arlington are among the nation's strongest. At some schools, groups of students have formed their own organizations. At Pratt, African-American students have formed a Black Architecture Students Coalition to help ensure that non-Western perspectives are explored more extensively in the curriculum, and to encourage students to become more involved in issues affecting the minority communities that surround the school.

At some campuses, students seem shockingly detached from such activism. In 1994–95, fifty-four of the AIAS's 164 chapters had twenty or fewer members. Some, including several of the nation's most prestigious institutions, have no AIAS chapters at all.

For students to recognize the professional and ethical importance of civic engagement in their own lives, such behavior ought to govern the day-to-day conduct of each faculty member and the school as a whole. Such connections with communities can, of course, take many forms. The Southern California Institute of Architecture operates an early childhood program in the arts in its building. Young children attend in the afternoons and on Saturdays. Using architecture as a vehicle, the program introduces children to the built environment, asking them to imagine, for example, how Winnie the Pooh might have designed a house for Christopher Robin.

At Arizona State University, a member of the faculty is the architectural critic of the *Arizona Republic*, the state's largest newspaper. Several other faculty are involved in television programs about architecture, or serve on boards and commissions that influence major projects both on campus and in the metropolitan area. The school has a Joint Urban Design Program in collaboration with most of the cities in the metropolitan area.

"In our school," said John J. Meunier, RIBA, dean of Arizona State's architecture school, "we retain close contacts with the community around us, and attempt to provide some leadership for public opinion about architecture. Architects are too rarely admitted to the higher levels of decision making in our society. Their education needs to equip them to command respect because of their ability to add value in complex social, economic, political, and technical discourse. Their aesthetic abilities are respected, and that needs to be sustained, but they need to develop a high level understanding of the other domains.

"Our task," he continued, "is to educate our students to be able to conceive of responsible and beneficial change to the physical environment, and to provide leadership in the community to those same ends. In doing so, we believe that we are strengthening the field of architecture, and ultimately enriching the discipline and the profession."

To further promote a climate of engagement, all programs should provide students more opportunities to integrate civic affairs into the curriculum itself. A required fourth-year "Urban Design Studio" at North Dakota State illustrates the point. In 1994, teams of architecture and landscape students spent a week in Duluth, Minnesota, studying urban design problems first-hand. Groups of two to ten students prowled the city—some photographing buildings, streets, bridges, and waterways, while others studied the cultural, political, historical, and social attributes of the city. Some groups looked at vegetation and soil, while others examined demographics, transportation systems, sewage, politics, transit and pedestrian systems. They interviewed Port Authority officials, the Coast Guard, numerous city officials, ordinary residents, and even tourists on the street. They collected and analyzed detailed zoning and other technical documents to inform their design decisions. The next assignment was to pool their information, refine it into a single, arresting, well-communicated master plan for the city, and present it as if they were trying to "sell" their ideas to the city council. "That's what you have do with clients," said their professor, Rurik Ekstrom. "You have to turn them on."

ALONG WITH ESTABLISHING a climate of engagement, the second strategy we propose is for schools to help clarify the many contributions architects can offer the nation. Repeatedly, architects told us that the public continues to view the profession of architecture as removed from commonplace problems. A Boston Architectural Center graduate put it this way: "Our training in problem solving, organization, idea generation and expression provides us with almost infinitely adaptable skills. Yet the public perception of the field is one of abstraction and elitism."

We have already discussed our concern that schools of architecture, as a group, inadequately stress the importance of clear verbal communication. It is difficult to imagine that architects can ever assume leadership in community affairs if they cannot express themselves in jargon-free English in public forums, or to clients. In fact, we found that relatively few studio projects involve sustained contact with clients. In our survey, 58 percent of architecture school alumni disagreed that their schools effectively prepared them for "communication with clients."

"I do think we should revolutionize the audience we address and broaden our mission to educating the whole culture about architecture— not just architects," says Lawrence W. Speck, dean of the school of architecture at the University of Texas at Austin.

One way to promote better understanding is to begin at an early age. Administrators at Howard University make frequent trips to area high schools to give students who would not otherwise have thought about architecture as a career a taste of what it is about. The Universities of Michigan and Illinois, North Carolina State and Mississippi State are among roughly two dozen schools that operate summer programs for prospective students.

One of the most successful local outreach programs, the Center for Understanding the Built Environment (CUBE) in Prairie Village, Kansas, was founded thirteen years ago by architect Dean Graves and operated by his wife, Ginny. It uses a "teach the teachers" strategy, and provides curricula and has conducted workshops for more than six thousand teachers. The idea is to reach children early and bring the ideas of architecture alive in their own communities and lives.[5]

Schools, then, must place far greater priority in preparing graduates to be effective and empathetic communicators, able to advocate with clarity for the beauty, utility, and ecological soundness of the built environment. Educators and the profession simply must make better known what architecture can contribute to fulfilling human needs and promoting more wholesome communities.

As a THIRD STRATEGY, schools of architecture should place greater emphasis on generating new knowledge. America's schools of architecture must convince the public, and perhaps some of its own constituencies, of the need to strengthen and defend scholarly investigation in the search for solutions to society's most vexing problems—from homelessness to energy issues and building better schools and safer neighborhoods.

According to the Center for Construction Innovation, a research and development arm of M. A. Mortenson & Co., a Minneapolis-based construction firm, the building industry as a whole is "substantially behind the times," largely because of a lack of investment in research and development. In 1992, according to their data, research and development spending for the entire U.S. construction industry was $343 million, including product manufacturers, contractors, design professionals and others. That same year, the top five Japanese construction companies alone invested $700 million. If the trend isn't reversed, they warn, the American building industry will be increasingly vulnerable to foreign competition both domestically and abroad. In 1994, twenty-one of the top one hundred U.S. construction firms were foreign-owned. In 1980, only two were.

There are, then, both economic and social reasons for schools of architecture to affirm more vigorously a commitment to generating new knowledge, and surely the profession could provide important support for that effort. We find it encouraging that the American Institute of Architects and the Association of Collegiate Schools of Architecture jointly formed, in 1995, the American Institute for Architectural Research, whose stated goals are to advocate increased research, encourage collaborative research partnerships within the architecture field and with other parts of the building industry, and broadly disseminate the results.

As Donald Watson, FAIA, dean of Rensselaer Polytechnic Institute's School of Architecture, told us: "The discipline of architecture needs a rigorous knowledge base by which to support its premises and principles that define the relationship between human and community health, and between building and urban design. To date, the development of rigorous theory/knowledge building has been at the edges of the curriculum and frequently marginalized as something separate from architecture, that is: environment/behavior studies, building sciences, environment/technology studies, etc.

"As a result," Watson continued, "most architectural graduates are not prepared to understand the validity or value of their professional services. As a further result, the standing of the profession is marginalized in the public's eye. That is, without research, scholarship and a rigorous knowledge base, the profession cannot take stands on significant health, economic, social, political or ethical issues."

To have an impact on the critical issues affecting community life, then, there must be a seamless connection between learning, the generation of new knowledge, and community service. It is not enough, in other words, for schools to provide students with community experiences for their own sake. The goal should be to provide opportunities for students and faculty to work together in communities to produce genuine scholarship with broad applicability, and to disseminate those findings so that others can benefit from those experiences.

Pratt Institute's thirty-three-year-old Center for Community and Environmental Development (PICCED) is one of the nation's model multidisciplinary centers connecting architects to the physical, social and economic challenges of urban America. The center works closely with community groups to revitalize poor neighborhoods from the South Bronx to East New York. It has built centers for the homeless, housing for AIDS patients, self-help housing, and other projects throughout the city. "We think architects need to know what community planners know, just to be better architects. They need to understand how to deal with the community as the client," said PICCED's director, Ron Shiffman.

Architectural research can be of powerful help, of course, in developing affordable, environmentally friendly housing. Several years ago, BAC's Architectural Research Center designed and built the "Mac-Donald Starter House," a three-hundred-foot starter home, named for architect Donald MacDonald who designed it. The house costs only $28,000, putting it in reach of working class families with incomes as low as $25,000 a year. Bill Boehm, director of BAC's research center, said the purpose of the center is to make architects and students ask "the larger questions about community and environment, as well as about politics and economics."[6]

A number of architecture programs have combined, in a fascinating way, research and community service in rural settings. At Texas A&M, The Center for Housing and Urban Development, a state-funded outreach and community research center based at the architecture school, focuses on helping impoverished "colonias" along the Mexico-Texas border. It is not just a housing program, but a multidisciplinary attack on the social isolation and the economic and physical problems of these communities. Each year, Texas A&M architecture students help design and construct community centers to house services for Texas' most depressed communities. They gather information from residents and construction workers, take it back to the school to develop designs, then go back to the field to refine the designs with community input. "We've gotten a wealth of reality lessons," says Kermit Black, the center's director. "What we're really trying to do is help residents build communities."

Since 1979, Mississippi State University's Center for Small Town Research and Design has discovered that rural towns are excellent laboratories for studying community dynamics. Research coupled with community service is the key to the program, says Judith Van Cleve, a faculty member who has worked at the center for years. "Small towns are excellent for learning about the idea of community. You have to create a new package each time you go in. You can't just go in with a set program. You have to keep on with research."

Auburn University operates off-campus community centers in an urban setting—Birmingham, Alabama—and in 1993, it opened a rural center in Greensboro, Alabama, where students design, restore, and build homes in one of the state's most impoverished areas.

More recently, Mississippi State began applying its expertise in "virtual reality" technology to help community planners throughout the state. The small town of Madison, Mississippi, for example, wanted to build a better connection to a nearby interstate highway by turning an existing two-lane highway into a five-lane access road. The plan might have destroyed historic structures along the route. Using virtual technology, the architecture school was able to help town planners visualize the consequences of their plans and develop an alternative that solved the access problem while preserving the historic structures. "This is the service outreach of the school," said John M. McRae, dean of Mississippi State's School of Architecture. "And it can raise the public perception of the value of architectural services."

THE FOURTH AND FINAL STRATEGY WE PROPOSE for placing architecture education more firmly in the nation's service is to stress the importance of *ethical professional behavior*. In the short span of five or six years, schools can only begin to help students come to grips with the choices, the tradeoffs, and the priorities, involved in ethical professional behavior. Several years ago, in a Carnegie Foundation essay titled *Higher Learning in the Nation's Service,* we issued a challenge to *all* professional schools, including architecture programs. Students, we said:

> should confront the ethical implications of their chosen professions. In law, business, economics, and medicine, as well as in the sciences and arts, ethical issues arise every day. When does advertising edge over into deception and dishonesty? How does a physician determine the limits on his or her ability to save lives, and what does one do when that limit is reached? How should data be gathered, interpreted, and reported? What social concerns should influence the work of the architect, the geneticist, the industrial chemist, the newspaper reporter, the mining engineer?[7]

In the context of practice, the lines of ethical behavior are sometimes drawn narrowly or legalistically. But in the education of architects, ethics surely ought to encompass more than the minimum obligations not to cheat clients or skirt laws or regulations. Architects have an ethical responsibility to inform clients and communities of the full range of implications of their wishes and alternatives. As a long-time Pratt Institute faculty member told us, the goal of education should be "to establish an ethical stance on what *should* be built, to consider the built environment as an enabler, allowing people to fulfill their goals, aspirations, and dreams."

A Minneapolis architect stated the ethical challenge well: "Architects have several purposes. We must be client-oriented, for that is the only way to prosper in today's service-oriented economy. But we have to balance economic concerns with the need to create architecture that 'lifts the human spirit.' And we must do all this in an environmentally sensitive manner, using safe building materials and preserving the natural environment as much as possible. It is an increasingly difficult task, requiring continual development of our ethics, knowledge, sensitivity, and problem-solving skills."

Of all the reasons that this study has argued for better integration of liberal studies into the architecture curriculum, none is more important than the preparation of future architects to confront the ethical choices and tradeoffs of professional life. To have encountered and thought about Hamlet or Doctor Faustus in literature, to have read Hobbes and Rousseau in philosophy, and even to weigh the words of history's most notorious architect, Albert Speer, are experiences of incalculable value to anyone engaged in professional study.

In arguing for the centrality of ethical behavior, we are not suggesting that schools turn out architects who feel they must test their ethical compasses by "walking away" from any project of which they do not heartily approve. Aspiring architects should learn how to present forthrightly to clients the options, consequences and long-term costs of design ideas, with the goal of working together for a better environment. "Architects are most successful, in my mind, when they are able to view themselves as servants without relinquishing their responsibility for leadership," a faculty member at Boston Architectural Center told us.

A Minnesota architect told us he feels a responsibility to discuss the importance of environmentally sustainable design openly with his clients: "We have seen a few clients approach this on a moral level. But more often they see this as 'another set of regulations,' or 'another set of problems.' In every case, as an architect you should say to them, 'if it's worth doing, let's do it right. Let's look at the long-term effects, not because sustainability is a "hot button" issue, but to avoid short-cuts that will hurt you in the long run.' We have a professional responsibility to raise the issue with the client. But if they say no, or we don't have the bucks, then we will say, fine, but not before we have laid out honestly the implications."

IN THE END, EDUCATION'S MOST ESSENTIAL MISSION is to develop within each student the capacity not only to build with competence, but to judge wisely in matters of life and conduct. Harry Robinson, dean of Howard University's school of architecture, stated it perfectly: "Social responsibility is not a quality that can be deferred until [students] have completed their tenure . . . in the academy. It must be integral to education in the discipline of architecture."[8]

This imperative, then, does not replace, but rather enriches the need for rigorous professional study, and indeed, we have argued that schools of architecture should provide students and faculty with opportunities to combine scholarship with service to the nation. The aim is not only to prepare the young for productive careers, but to enable them to live lives of dignity and purpose; not only to generate new knowledge, but to channel that knowledge to humane ends.

BUILDING COMMUNITY

Building Community

WOVEN THROUGH THIS REPORT is the conviction that architecture education and practice must be tied to both private and public ends. Students come to school, first of all, to follow their own special aptitudes and interests. They are eager to get credentials and get a job. Armed with their new knowledge, they are prepared to extend learning over a lifetime. Serving the educational needs of every student, and the professions they will enter, must remain a top priority in architecture education, but such private concerns, while critically important, are not sufficient.

The academic and professional lives of architects must also be grounded in *public* purpose. Harking back to the three ancient principles of Vitruvius—firmness, commodity, and delight—the education of architects should prepare future practitioners dedicated to building technically sound, visibly pleasing, and useful structures for clients and users, and equally, to making life more comfortable, pleasurable, secure, and productive for all citizens, including the disenfranchised in our society. The scholarly activities of both faculty and students should relate not only to private goals and agendas, but to matters of consequence to the profession, and beyond that, to society as a whole. No less important than acquiring design skills, technical competence and business judgment, education must begin to help students develop the ethical grounding, the intellectual roundedness, and the maturity to weigh the impact of their work on present users and future generations.

The time has come, then, to demystify architecture, to elevate its place in the consciousness of the public and in the daily lives of communities. Schools of architecture, to begin with, can no longer afford to be strangers in their own settings. They should reach out to students in other disciplines, and form scholarly partnerships with other professional schools and academic departments. Such partnerships would surely enrich campus life, create new opportunities for integrated learning and research, and set the stage for lifelong collaborations with other professions. They would also help make a knowledge of architecture and design what it always *should* have been—an essential part of liberal education for *all* students, not just those in architecture schools.

None of this is likely in a campus climate where the intellectual community itself is fragmented, where discourse among disciplines is discouraged, or where one form of scholarly work is affirmed and rewarded above the rest. We urge university administrators to honor the distinctive scholarly contributions of design education, respect and preserve its diversity, and vigorously promote a campuswide climate of integration so that architecture and other disciplines are mutually enriched. Making the connections between architecture and other fields to strengthen communities must begin, then, on the campus itself.

Thirty years ago, the great modernist and Harvard educational leader Walter Gropius said that "responsible architects think very much in terms of the whole community."

"I have always told my students, 'I am not interested when you build a beautiful design in a gap of a street if you treated it only as a unit in itself, not considering the neighborhood which is already there,'" Gropius said. "'You have to blend in with the larger circumstance. This larger circumstance is the main thing and all limited objectives have to be subordinated to the whole.'"[1]

In describing architecture's capacity, then, to enrich the lives of its users and entire communities, utilitarian language alone falls woefully short. Eero Saarinen's architectural vision could construct a supremely functional suburban office complex that improved the bottom line for a tractor company, while also creating the St. Louis arch, whose "usefulness" lies only in its brilliant evocation of our nation's unity and grandeur.

Throughout history, architects like Gropius and Saarinen have struggled to discover the meaning of their work *beyond* its immediate time, place, and purpose. Such architects, through their farsightedness, have contributed not only to the pleasure and purposes of clients and users, but to the sense of commonality within the community itself.

We have seen cases where a skilled, imaginative, and caring architect can transform a building from one that discourages community to one that dramatically promotes it. The Woodruff Arts Center in Atlanta, as originally designed, housed symphonies and plays, but the Center's leaders discovered that patrons found the architecture forbidding, not warm, inviting, or uplifting. When the time came to renovate the structure, the Center's leaders took those concerns to their architect, Thomas Ventulett, FAIA, of the Atlanta firm Thompson, Ventulett, Stainback & Associates. Among other things, he moved the stairs to a more conspicuous space so that people became part of the scenery. He put canopies out to pull people in. He opened up the entryways to make them four stories high in glass, so that those doorways shone like beckoning lanterns. As a result, ticket sales are up, and Atlantans today look upon their center and its leadership with renewed pride. Through caring design, a community was enriched.

And yet, at times, even the most renowned architects seem not to have fully grasped the potential of their designs to shape, for better or worse, community life. The urban planner William H. Whyte tells the story of how Ludwig Mies van der Rohe was astonished when he saw people sitting, talking, sunbathing, and eating happily outside the Seagram Building which he and Philip Johnson built in midtown Manhattan. In designing Seagram, Whyte wrote, it apparently never occurred to Mies or Johnson that its steps and ledges would quickly become a "sittable public place" that promoted community in a small corner of New York City.[2]

We can't help but wonder: Couldn't the capacity of architectural design to enrich community life be less serendipitous, more carefully examined? It is here that we believe that architecture scholarship has so much to contribute. We have suggested that architecture education and the profession, as part of an enriched partnership, should collaborate to develop new knowledge aimed at ensuring that the impact of design decisions on the health, safety, and welfare of communities is better understood—the kind of research, for example, that William Whyte pursued in studying city pedestrian traffic and the "sittability" of urban places, or the "reflective practice" described so eloquently by Donald Schön of MIT. Such reflective, caring architects and scholars are, in the truest and richest sense, *lifelong learners*. They are the life blood of a profession that could be of incalculable value to a society in desperate need of both expertise and compassion.

Still, if excellent architecture has the potential to build community by creating shared spaces, careless architecture can as easily erode them, and the proof is on the very campuses where many architecture schools are located. The problem with much of academic architecture isn't just its ugliness or poor fit within its environment, writes David W. Orr in his recent book *Earth in Mind: Our Education, Environment, and the Human Prospect*. The haphazard, disconnected landscape of so many college campuses carries its own powerful, subliminal messages which undermine the learning community: that architecture is shaped by faceless authority, rather than those who teach or learn in it; that the environmental or energy costs of buildings don't really matter; that rich intellectual activity can be forced to occur in sterile places.

"My point," wrote Orr, "is that academic architecture is a kind of crystallized pedagogy and that buildings have their own hidden curriculum that teaches as effectively as any course taught in them."[3]

Cultivating the principle of building not just buildings but communities must begin, then, on college campuses themselves. We have proposed that all schools of architecture should evaluate themselves as *communities of learning*, places that foster connections rather than compartmentalization, cooperation along with productive competition. We are convinced, in fact, that architecture education, with its spirit of close engagement between students and teachers in design studio, could become a campus model in promoting greater community within higher education.

In the hard-edged, competitive world that architects and architecture schools inhabit, a call for community renewal and civic engagement as priorities might seem quixotic. Cultural and social connections have faded and the very notion of commonalities seems oddly out of place in the landscape of the campus or workplace. Too often, the academic and professional worlds are marked by vocationalism, the fragmentation of knowledge, and territoriality. The atmosphere seems to be more about self-preservation than renewal.

"There's so little involvement by architects in community organizations," an Indianapolis architect told us. "You just don't see it in our profession. We need to get the profession back to the status of community leaders."

It is naive to imagine, of course, that architecture alone could wipe away the deep divisions on our college campuses, or in our society. We don't pretend that architects, no matter how well educated, or a building, no matter how well crafted, could turn the tide on crime, the fragmentation of families, the problems of public education, health care, or the many other crises of modern life. Still, it is disturbing to hear so many architects professing powerlessness in civic affairs. Who better than the architect, after all, to remind society's philistines of Vitruvius's charge two thousand years ago to cherish firmness, commodity, and *delight* in our built environment?

And yet, our travels leave us hopeful that there is, among many architects and educators, a climate of growing readiness for more active community engagement. It is worth recalling that nearly 40 percent of students we surveyed said that their primary motive for entering the profession was not salary or prestige, but improving communities and the built environment.

The world needs more scholars and practitioners not only educated to prosper in their own careers but also prepared to fulfill social and civic obligations through the genius of design. There is much to be done. Federal studies calculate, for example, that $112 billion in work is needed to repair or replace increasingly squalid school buildings that our nation's children attend each day—at a time when neither government nor citizens seem much moved by the impact that collapsing ceilings, exposed wires, fire-code violations, or poor air quality might have on learning, or the lives of children. In these and countless other matters affecting the health, safety, welfare, and happiness of communities in this country, the voices of architects must be heard, and their talents meaningfully employed.

Perhaps more than any other time, it is during the college years that those qualities of competence, caring, and character should be cultivated. A twenty-year-old student from the Southern California Institute of Architecture expressed it perfectly: "The larger purpose of architects is not necessarily to become a practitioner and just build," she told us, "but to become part of the community that enriches society."

To that noble end, architects must unite with all professions, and all citizens, in common cause.

From october to december 1994, the Carnegie Foundation for the Advancement of Teaching conducted surveys of students, faculty, and alumni at fifteen accredited U.S. schools of architecture. In addition, a mailed questionnaire was sent to deans and department heads at all professional architecture programs in the United States.

The instruments used in the Survey on the Education of Architects were developed by the Foundation staff, in consultation with officials of the five national architecture organizations and other scholars in the field. The questionnaires asked for information concerning the background of individuals involved in architecture education and practice, their educational experiences, the academic climate at their schools, the future of the profession, and the role of architecture in society.

The number of questions per instrument ranged from 106 (for the survey of alumni) to 119 (for the survey of administrators). Seventy-six questions were asked on all four instruments; twelve identical questions were asked on three of the instruments; and thirty-five questions were asked on two of the instruments. Approximately ten items on each survey were asked of only one of the four groups (from eight on the survey of faculty to sixteen on the survey of students).

The majority of questions offers four response categories; a possible response of "don't know" or "not applicable" was provided when appropriate. Many of the tables included in this report combine response categories. For example, "strongly agree" and "agree somewhat" have been combined into "agree"; "strongly disagree" and "disagree somewhat" are combined and called "disagree."

All four surveys included open-ended questions regarding the purposes and priorities of architecture, the strengths and shortcomings of respondents' own educational experiences, and any changes they would like to see in architecture education and practice. Architecture school administrators were also invited to send additional information about promising practices at their institutions.

The survey of administrators was mailed to 168 deans, department heads, chairs, and directors of the 109 U.S. professional architecture programs in the United States, including 103 accredited programs and six candidate programs. Usable responses were received from 80 administrators—a 47.6 percent overall response rate. However, those replies included at least one response from 65 programs surveyed—a 59.6 percent institutional response rate. Follow-up telephone calls were later made to approximately 25 respondents seeking additional information about curricula, or amplification of views expressed in the survey.

The surveys of faculty, students, and alumni were distributed at the fifteen schools of architecture which were visited as part of this report. Those schools were selected, with advice from the five architectural groups and based on Carnegie's own research, to represent the geographic, programmatic, and philosophic mix of the nation's accredited professional degree programs. At each school, the dean or department head agreed to distribute questionnaires to ten faculty, thirty students, and twenty-five alumni.

In the survey of faculty, we requested that administrators select a mix of tenured, nontenured, and adjunct faculty. Responses were received from eighty-four faculty members, representing a completion rate of 56 percent.

In the student survey, we asked schools for a mix of respondents with a range of ability levels, freshmen through upper-class students, and a representative balance of males and females. Responses were received from 241 architecture students—a 53.6 percent response rate.

Twenty-five alumni from each institution were also selected to participate in the survey. We asked administrators to select alumni who had graduated within the past ten years, and we requested that persons who did or did not enter traditional architectural practice be included.

An explanatory cover letter for the instruments was prepared by the school itself on its own letterhead to increase the probability that alumni would complete and return the forms. Usable responses were received from 154 alumni—a 41.1 percent response rate.

Students and faculty returned completed instruments to the dean or department heads in charge of distribution in sealed envelopes to ensure confidentiality. They were then mailed in bulk to the Foundation. Alumni mailed their forms, in preaddressed and stamped envelopes, directly to the Foundation.

The surveys of faculty, students, and alumni were not designed to be probability samples. However, the fifteen schools from which our samples were drawn were, in our judgment, representative of the mix of accredited U.S. architecture programs, and the resulting pool of survey respondents closely approximated the ethnic and gender balance of the entire U.S. student and faculty universes.

Nationwide, 19 percent of architecture faculty were women in the 1994–95 academic year, according to data from the National Architectural Accrediting Board, Inc. In our sample, 18 percent were women. In the universe of architecture students, 28 percent were women in 1994–95, according to NAAB data, whereas in our sample, 38 percent were women. Finally, in our sample of architecture students, one-quarter were from various minority groups, compared with 23 percent in the entire architectural student population.

The Survey on the Reexamination of Faculty Roles and Rewards, also cited in this report, was administered by The Carnegie Foundation in the fall of 1994. Questionnaires were sent to provosts of all four-year institutions classified as Research Universities, Doctoral Universities, Master's Colleges and Universities, and Baccalaureate Colleges in 1994. A total of 1,380 survey instruments were mailed; completed questionnaires were received from 865 provosts. The overall response rate was 63 percent, and ranged from 54 percent for Doctoral Universities II to 78 percent for Doctoral Universities I. Detailed results from this study will be presented in a forthcoming report by the Foundation on the assessment of scholarly work.

To obtain a copy of the instruments used in the Survey on the Education of Architects, 1994, or for further information, please contact:

The Carnegie Foundation for the Advancement of Teaching
Data & Analysis Department
5 Ivy Lane
Princeton, New Jersey 08540

NOTES

INTRODUCTION   *A Profession in Perspective*

1. Joseph Bilello, "Interaction and Interdependence: University-Based Architectural Education and the Architecture Profession," in Special Education Task Force, American Institute of Architects, *Architectural Education Report* (Washington, D.C.: American Institute of Architects, 1991), 18.

2. W. Cecil Steward, "Influences from Within the Academy upon Architectural Education," July 1988 (white paper prepared for the NAAB Board and distributed at the National Architectural Accrediting Board Validation Conference, Woodstock, VT, Sept. 9–10, 1993), 14.

3. Henry N. Cobb, "Architectural Education: Architecture and the University," Walter Gropius Lecture, Harvard University, 1985; reprinted in *Architectural Record* (Sept. 1985), 47.

4. American Institute of Architects, *Architecture Fact Book* (Washington, D.C.: American Institute of Architects, 1994), 34.

5. Gregory Palermo, "An Architect's Architect(s): Liberating Possibilities in Architectural Education," July 1995; draft of paper for *Dichotomy*, the student journal of the University of Detroit Mercy School of Architecture (projected publication date August 1996).

6. Turpin C. Bannister, ed., *The Architect at Mid-Century: Evolution and Achievement*, volume one of the Report of the Commission for the Survey of Education and Registration of the American Institute of Architects (New York: Reinhold Publishing Corp., 1954), 94.

7. Dana Cuff, *Architecture: The Story of Practice* (Cambridge, MA: MIT Press, 1991), 23–26.

8. *Ibid.,* 26.

9. Arthur Clason Weatherhead, "The History of Collegiate Education in Architecture in the United States" (PH.D. diss., Columbia University, 1941), 27–29.

10. Kathryn H. Anthony, "ACSA Task Force on 'Schools of Thought in Architectural Education'" (paper prepared for the Association of Collegiate Schools of Architecture, 1 March 1992), 13.

11. William L. Porter and Maurice Kilbridge, *Architecture Education Study, Volume I: The Papers* (New York: Andrew W. Mellon Foundation, 1981), x.

12. Weatherhead, 63.

13. Bannister, 99.

14. F. H. Bosworth, Jr., and Roy Childs Jones, *A Study of Architectural Schools* (New York: Charles Scribner's Sons, 1932), 3.

15. Weatherhead, 164.

16. *Ibid.*, 173-74.

17. Bannister, 107.

18. Weatherhead, 193.

19. *Ibid.*, 195.

20. *Ibid.*, 244.

21. John M. Amundson, "An Historic Perspective of Accreditation in Architecture" (collected papers from the National Architectural Accrediting Board accreditation conference, New Orleans, 18 March 1976), 24.

22. Bannister, 103, 109.

23. National Council of Architectural Registration Boards, "1993 Circular of Information No. 2: Architect Registration Examination" (Washington, D.C.: NCARB, 1993), 2.

24. Palermo, 4.

25. Porter and Kilbridge, 842-45.

26. Bosworth and Jones, 30-31, 43.

27. *Ibid.*, 36.

28. *Ibid.*, 129.

29. *Ibid.*, 110.

30. *Ibid.*, 104.

31. Bannister, 442-49.

32. *Ibid.*, 109.

33. Robert L. Geddes and Bernard P. Spring, *A Study of Education for Environmental Design: The "Princeton Report"* (Washington, D.C.: American Institute of Architects, Dec. 1967; reprinted June 1981), 4.

34. Robert L. Geddes, remarks to the AIA/*Architectural Record* Walter Wagner Education Forum, 1995 AIA National Convention, Atlanta, GA, 6 May 1995.

35. Geddes and Spring, 22.

36. *Ibid.,* 22.

37. Geddes, remarks.

38. Porter and Kilbridge, 293–94, 326.

39. *Ibid.,* 300–307.

40. *Ibid.,* 683.

41. *Ibid.,* 733–34.

42. *Ibid.,* 737.

43. Robert Gutman, *Architectural Practice: A Critical View* (Princeton, NJ: Princeton Architectural Press, 1988), 97–108.

44. Dana Cuff, 43.

45. *Ibid.,* 118.

46. *Ibid.,* 121.

47. *Ibid.,* 108.

GOAL ONE   *An Enriched Mission*

1. Maryland Board of Architects, *Code of Maryland Regulations, Annotated Code of Maryland Business Occupations and Professions Article, Title 3,* Baltimore, MD, Oct. 1994, 2.

2. Minnesota Board of Architecture, Engineering, Land Surveying, Landscape Architecture & Interior Design, Rules Chapter 1800 & 1805, Statutes Chapter Parts 326.02–326.15, Chapter 326.02, Registration of Architects, Engineers, Surveyors, Landscape Architects, and Interior Designers, Subdivision 1.

3. Douglas Kelbaugh, letter to authors, 8 June 1995. We are indebted to Douglas Kelbaugh, of the University of Washington's College of Architecture, and his colleagues at Project EASE, for suggesting to us that broadening the definitions of "health, safety and welfare" could provide a basis for challenging architecture schools to more active engagement in community and national affairs.

4. John Peter, *The Oral History of Modern Architecture: Interviews with the Greatest Architects of the Twentieth Century* (New York: Harry N. Abrams, Inc., 1994), 68.

5. Carl M. Sapers, "The Scope of the Police Power—NCARB 6839-III" (legal memorandum, Hill & Barlow, legal council to NCARB, 1971), 5–6.

6. *Ibid.,* 8; citing *Berman v. Parker,* 348 U.S. 26 (1954).

7. Georgia Bizios, ed., *Architecture Reading Lists and Course Outlines* 3 (Chapel Hill, NC: Eno River Press, 1994), 268.

8. A. Trystan Edwards, *Good and Bad Manners in Architecture: An Essay on the Social Aspects of Civic Design* (London: John Tiranti Ltd., 1945), 1.

9. Alan J. Plattus, letter to Marvin Rosenman, 18 May 1994 (quoted with permission from the author).

GOAL TWO   *Diversity with Dignity*

1. Gregory S. Palermo, "The Faculty Who Teach Architecture," in Special Education Task Force, American Institute of Architects, *Architectural Education Report* (Washington, D.C.: American Institute of Architects, 1991), 31–32.

2. William G. McMinn, "Advocacy for Architectural Education," in Special Education Task Force, American Institute of Architects, *Architectural Education Report* (Washington, D.C.: American Institute of Architects, 1991), 15.

3. Christine Malecki, ed., *Career Options: Opportunities Through Architecture* (Washington, D.C.: American Institute of Architecture Students, 1993), 3–11.

4. Ernest L. Boyer, *Scholarship Reconsidered: Priorities of the Professoriate* (Princeton, NJ: The Carnegie Foundation for the Advancement of Teaching, 1990), 54.

5. *Ibid.*

6. Denise K. Magner, "Engineering and Accounting Professors Earn Most, Survey Finds," *The Chronicle of Higher Education*, 28 April 1995, A46; citing 1994–95 faculty salary survey conducted by the College and University Personnel Association (CUPA).

7. Boyer, 16.

8. *Ibid.*, 17–25.

9. The Carnegie Foundation for the Advancement of Teaching, "Survey on the Reexamination of Faculty Roles and Rewards," item 1. The survey was conducted in the fall of 1994. Instruments were distributed by mail to provosts or chief academic officers of all 1,380 four-year institutions in the U.S. Usable instruments totalling 865 were returned, a response rate of approximately 63 percent.

10. *Ibid.*, item 6e.

11. *Ibid.,* item 10c.

12. Oscar Handlin, "Epilogue—Continuities," in Bernard Bailyn, Donald Fleming, Oscar Handlin, and Stephen Thernstrom, *Glimpses of the Harvard Past* (Cambridge, MA: Harvard University Press, 1986), 131.

13. Boyer, 23.

14. *Ibid.,* 48–50.

15. McMinn, 15.

GOAL THREE   *Standards Without Standardization*

1. The Carnegie Foundation for the Advancement of Teaching, *The Control of the Campus: A Report on the Governance of Higher Education* (Princeton, NJ: The Carnegie Foundation for the Advancement of Teaching, 1982), 21–27.

2. John M. Maudlin-Jeronimo and Peter Hoffmann, "Architectural Education: Accreditation Criteria Revive Some Standards, Tighten Others," *Architectural Record* (Jan. 1987), 33.

3. Robert Geddes, remarks to the AIA/*Architectural Record* Walter Wagner Education Forum, 1995 AIA National Convention, Atlanta, GA, 6 May 1995.

4. National Architectural Accrediting Board, Inc., "1995 Conditions and Procedures," Washington, D.C., 1995, 14–22.

5. The University of Washington Department of Architecture Advisory Council, Executive Summary of Focus Group Meetings, Seattle, WA, Nov. 1995, 4.

6. Richard Chylinski, "Survey of Architectural Professionals in California Department of Architecture," California State Polytechnic University, Dec. 1993, 20.

7. National Research Council, *Education of Architects and Engineers for Careers in Facility Design and Construction* (Washington, D.C.: National Academy Press, 1995), 53.

8. Martin Symes, Joanna Eley, and Andrew Seidel, *Architects and Their Practices: A Changing Profession* (Newton, MA: Butterworth Architecture, 1995); summarized in "The Anatomy of the Architect," RIBA *Journal* (March 1995), 12.

9. National Architectural Accrediting Board, Inc., "The Role and Purpose of the NAAB: Questionnaire Results," Washington, D.C., 19 August 1993, 1–2.

10. National Architectural Accrediting Board, Inc., "1995 Conditions and Procedures," 18–19.

GOAL FOUR   *A Connected Curriculum*

1. John Peter, *The Oral History of Modern Architecture: Interviews with the Greatest Architects of the Twentieth Century* (New York: Harry N. Abrams, Inc., 1994), 30–31.

2. Diane Ghirardo, "Liberal Education: The Basis of a Contestorial Terrain" (paper collected in *The Liberal Education of Architects,* from a symposium sponsored by The Graham Foundation for Advanced Studies in the Fine Arts, the School of Architecture and Urban Design, University of Kansas, 8–9 Nov. 1990), 22.

3. Harold L. Adams, "1993 Gold Medallist" address, Tau Sigma Delta Honor Society in Architecture and the Allied Arts, 15.

4. Michael Bailey, "Department of Architecture: 1994 Survey of Employers of Recent Graduates," Office of Educational Advancement, Kansas State University, Manhattan, KS, Feb. 1995, 1.

5. Timothy Fuller, ed., *The Voice of Liberal Learning: Michael Oakeshott on Education* (New Haven and London: Yale University Press, 1989), 27–28.

GOAL FIVE   *A Climate for Learning*

1. Kathryn H. Anthony, *Design Juries on Trial: The Renaissance of the Design Studio* (New York: Van Nostrand Reinhold, 1991), 129.

2. *Ibid.,* 120.

3. *Ibid.,* 132–33.

4. *Ibid.,* 125.

5. Association of Collegiate Schools of Architecture, "Code of Conduct for Diversity in Architecture Education," Washington D.C., 1990, 2.

6. Commission on Professionals in Science and Technology, *Professional Women and Minorities: A Total Human Resource Data Compendium,* eleventh edition (Washington, D.C.: Commission on Professionals in Science and Technology, 1994), 94–96, 240. According to federal and industry data cited in the report, women accounted for 28.7 percent of

the degrees awarded in architecture between 1970 and 1991, compared with 38.9 percent in accounting, 35.3 percent in business and management, 29.9 percent in law, and 34 percent in medicine.

7. John Morris Dixon, "A White Gentleman's Profession," *Progressive Architecture* (November 1994), 55.

8. D. X. Fenten, *Ms.-Architect* (Philadelphia: The Westminster Press, 1977), 11.

9. Sherry Ahrentzen and Linda N. Groat, "Rethinking Architectural Education: Patriarchal Conventions & Alternative Visions from the Perspectives of Women Faculty," *Journal of Architectural and Planning Research* 9, no. 2 (summer 1992), 98.

10. Linda N. Groat and Sherry Ahrentzen, "Reconceptualizing Architectural Education for a More Diverse Future: Perceptions and Visions of Architectural Students," *Journal of Architectural Education* 49, no. 3 (Feb. 1996), 177.

GOAL SIX   *A Unified Profession*

1. Joseph Bilello, "Issues in Architectural Education: An Introduction," *Professions Education Researcher Quarterly* (July 1992), 6–7.

2. National Council of Architectural Registration Boards, *Intern Development Program Guidelines, 1994–1995* (Washington, D.C.: NCARB, 1994), 5.

3. American Institute of Architecture Students, "AIAS Architectural Internship Experience Questionnaire," Washington D.C., 1991.

4. Marvin Malecha, "Continuing Education: Research Transfer," in American Institute of Architects, *The Roles of the Schools of Architecture in Continuing Education* (Washington, D.C.: American Institute of Architects, 1993), 51–52.

5. American Institute of Architects, *The Roles of the Schools of Architecture in Continuing Education* (Washington, D.C.: American Institute of Architects, 1993), 4–5.

6. Ibid., Appendix One: 1992–1993 Survey of Architecture School Administrators, 89.

7. Donald Watson, "The Notion of Critical Practices: Roles for Educators in Continuing Education" (paper presented at the 82nd Association of Collegiate Schools of Architecture Annual Meeting, Montreal, Quebec, 1994), 151–52.

GOAL SEVEN   *Service to the Nation*

1. Ernest L. Boyer and Fred M. Hechinger, *Higher Learning in the Nation's Service* (Princeton, NJ: The Carnegie Foundation for the Advancement of Teaching, 1981), 3.

2. Michael Tomlan, "Historic Preservation Education: Alongside Architecture in Academia," *Journal of Architectural Education* (May 1994), 188.

3. Robert Geddes, remarks to the AIA/*Architectural Record* Walter Wagner Education Forum, 1995 AIA National Convention, Atlanta, GA, 6 May 1995.

4. Dru Meadows, "Fast Track Vs. Green Track," *Corporate Real Estate Executive* 10, no. 2 (Feb. 1995), 27–29.

5. Chris Santilli, "Kids for Architecture," *The Construction Specifier* (Oct. 1993), 80–82.

6. Lawrence Biemiller, "An Architecture School Develops Environmentally Friendly and Affordable Starter Houses," *The Chronicle of Higher Education*, 26 Jan. 1994, A51.

7. Boyer and Hechinger, 61.

8. Harry G. Robinson III, "Preserving the Historic Trust of Architectural Education in Howard University" (Howard H. Mackey Lecture delivered at Howard University, Washington, D.C., 10 Feb. 1992), 12.

EPILOGUE   *Building Community*

1. John Peter, *The Oral History of Modern Architecture: Interviews with the Greatest Architect of the Twentieth Century* (New York: Harry N. Abrams, Inc., 1994), 180–81.

2. William H. Whyte, *City: Rediscovering the Center* (New York: Doubleday, 1988), 112.

3. David W. Orr, *Earth in Mind: On Education, Environment, and the Human Prospect* (Washington, D.C.: Island Press, 1994), 112–13.

INDEX

INDEX